Praise for *E*

"De Grey is hardly just another fountain-of-youth huckster. His it-might-work ideas are based on existing, published, peer-reviewed research. He thinks more like an engineer than a scientist. If even one of his proposals works, it could mean years of extended healthy living."

—Paul Boutin, *The Wall Street Journal*

"Biomedical gerontologist Aubrey de Grey and science writer Michael Rae take a detailed look at techniques and technologies that, they say, may not only extend life but also could theoretically halt and reverse aging altogether. . . . The authors explore the nitty-gritty of how we age and posit an audacious blueprint for cheating the reaper, suggesting that biomedical technology may one day be harnessed to reverse physical aging by repairing damage that builds up in cells over time."

—Janet Cromley, *Los Angeles Times*

"[De Grey is] a serious thinker who had enough courage to break with the crowd. A lot of people who are not conventional are not serious. But the real breakthroughs in science are made by serious thinkers who are willing to work on research areas that people think are too controversial or too implausible."

—Peter Thiel, cofounder of PayPal,
quoted in *The Washington Post*

"*Ending Aging* guides the reader through a maze of advances in molecular and cellular biology that could lead to anti-aging therapies, which the authors term SENS (for 'strategies for negligible senescence'). The targets are a range of cancer-causing nuclear and mitochondrial mutations, intracellular and extracellular debris, and molecular cross-links that contribute to pathological decay over a lifetime."

—Dr. Judy Iles, *Nature*

ENDING AGING

ENDING AGING

The Rejuvenation Breakthroughs That Could Reverse Human Aging in Our Lifetime

Aubrey de Grey, Ph.D.,
with Michael Rae

ST. MARTIN'S GRIFFIN NEW YORK

www.stmartins.com

LIBRARY OF CONGRESS CATALOGING-IN-PUBLICATION DATA

De Grey, Aubrey D. N. J., 1963–
Ending aging : the rejuvenation breakthroughs that could reverse human aging in our lifetime / Aubrey de Grey ; with Michael Rae.—1st St. Martin's Griffin ed.
p. cm.
ISBN-13: 978-0-312-36707-7
ISBN-10: 0-312-36707-4
1. Longevity. 2. Aging—Molecular aspects. 3. Biotechnology. I. Rae, Michael. II. Title.

QP85.D348 2007
612.6'8—dc22

2007020217

First St. Martin's Griffin Edition: October 2008

10 9 8 7 6 5 4 3 2 1

Aubrey dedicates this book as follows: "To the tens of millions whose indefinite escape from aging depends on our actions today."

Michael dedicates this book as follows: "To the two tenders of the flames that have inspired me throughout this work. To April Smith, for erupting, like Athena, out of the secret depths of my mind, raining Greek fire on my Manichee heart, reigniting smoldering embers I had thought long extinguished, and opening up the promise of a shared indefinite tomorrow; and to Dr. Aubrey de Grey, for tirelessly and courageously bearing Promethean fire to a world yet shivering under the winter of age-related death and decay, kindling the sparks that we must fan into a blaze that will cast out its obscuring darkness and melt its frozen grip."

Contents

Preface

The biomedical revolution described in this book is still some way off—at least a few decades, maybe more. Why, you may then ask, should you concern yourself with it now?

The answer is simple: Once you know what I have to tell you, you'll want to make it happen sooner, and some of you will put that desire into action. The more people are aware of what has now become foreseeable in the fight against our oldest foe, aging, the faster it will become acceptable to "come out" as an ardent opponent of aging, and then unacceptable not to. We aren't close enough to this revolution to put accurate timescales on its arrival, but we *are* close enough that our action (or inaction . . .) today will affect the date at which aging is defeated.

In fact, we've been at that point for a few years now. It could, therefore, be argued that I should have written this book sooner. Well, maybe I should have—but there's a trade-off: with every year that has passed since I developed the key concepts described here, progress has been seen in the laboratory. Every step of this progress has strengthened the case that the overall scheme will succeed, so the book as a whole is more compelling than it could have been a year or three ago. In fact, without the diligent efforts of a large number of scientists within and beyond biogerontology, my plan for defeating aging could not exist.

Another reason this book has only been written now is the usual one: books don't write themselves, and I've been spending every waking hour engaged in other work to further the anti-aging mission. Without doubt, you would not have this book in your hands today if it were not for the diligent work of my research assistant Michael Rae, who dedicated much of 2006 to it: he can take credit for most of the text of Part 2.

Michael is not the only person without whom this book could not have come to pass. Thanks to Peter Ulrich for painstakingly going over the fascinating history of patient work, inspired reasoning, and scientific serendipity behind the development of alagebrium. Any misunderstandings of this story are Michael's. Special thanks go to our graphics team, who prepared the illustrations: Bram Thijssen, Bryan English, Benjamin Martin, Tyler Chesley, Zachary Bos, Hoyt Smith, and their coordinator, Jeff Hall. Additionally, Michael and I received outstanding editorial help from Methuselah Foundation volunteers Reason, Anne Corwin, and David Fisher. Our agent, John Brockman, and his staff were tremendously efficient in shepherding the book through the worldwide publication process, and our editor at St. Martin's, Phil Revzin, also provided invaluable editorial input. And finally, my work on this book has, as with all my contributions to the crusade against aging, depended hugely on the unswerving intellectual and emotional support of my beloved wife, Adelaide Carpenter.

I hope that this book will enjoy a wide readership; if it does, most readers will be nonbiologists and certainly nonbiogerontologists. Some, however, will be people who do possess expertise in these areas. To that group I would like to make clear at the outset that, in presenting SENS, the "Strategies for Engineered Negligible Senescence," to a general audience, I have not been able to delve into *every* nook and cranny of the relevant science, and you will surely identify aspects of SENS that, if what you read here were all there was to it, would seem flawed. I merely remind you now that this book is *not* all there is to SENS, and that, if you see what seems to you to be a slam-dunk objection to what I say, you should consult my published academic work (and, preferably, consult me personally, too) before dismissing it.

However, the above applies only to "errors" of omission, of course. Any errors of commission are, I fully accept, my responsibility and mine alone.

ENDING AGING

Part One

1

The Eureka Moment

Marriott Hotel, Manhattan Beach, California.

June 25, 2000.

Four o'clock.

In the morning.

It was 4 A.M. in California, but my body insisted on reminding me that it was noon in Cambridge. I was exhausted from the intercontinental flight and by a day spent in debate with some of the most influential personalities in biogerontology, at an invitation-only brainstorming workshop on ideas to combat aging. Evolutionary biologist Michael Rose was there. So were calorie restriction researchers Richard Weindruch and George Roth, nanotechnologist Robert Freitas, and several others. But I couldn't sleep: On top of the mismatch between biological and geographical clocks, I was frustrated at what I saw as the day's failure to make any real progress toward a concrete, realistic anti-aging plan. As I dozed and pondered, a question on the nature of metabolism and aging wormed its way into my brain and wouldn't let go.

In my bleary irritation, I sat up, ran my hands over my beard, and began pacing the room, turning over the quandary in my mind. "Normal" metabolism was just so *messy,* and the raging debates in the biogerontology literature showed how difficult it was to determine what paced what: which metabolic disruptions were causes of aging, and which were effects (or secondary causes) that would simply disappear if the underlying primary causes were addressed. How could we make a positive difference in such a complex, poorly understood system? How could any meaningful change made in metabolism not be like a butterfly flapping its wings—apt to cause large, unwanted storms further down the line?

Then a second line of thought began to form in my mind—idly at first, just as a notion. The real issue, surely, was not which metabolic processes cause aging damage in the body, but the damage itself. Forty-year-olds have fewer healthy years to look forward to than twenty-year-olds because of differences in their molecular and cellular composition, not because of the mechanisms that gave rise to those differences. How far could I narrow down the field of candidate causes of aging by focusing on the molecular damage itself?

Well, I thought, *it can't hurt to make a list . . .*

There are mutations in our chromosomes, of course, which cause cancer. There is glycation, the warping of proteins by glucose. There are the various kinds of junk that accumulate outside the cell ("extracellular aggregates"): beta-amyloid, the lesser-known transthyretin, and possibly other substances of the same general sort. There is also the unwholesome goo that builds up *within* the cell ("intracellular aggregates"), such as lipofuscin. There's cellular senescence, the "aging" of individual cells, which puts them into a state of arrested growth and causes them to produce chemical signals dangerous to their neighbors. And there's the depletion of the stem cell pools essential to healing and maintenance of tissue.

And of course, there are mitochondrial mutations, which seem to disrupt cellular biochemistry by increasing oxidative stress. I had for a few years felt optimistic that scientists could solve this problem by copying mitochondrial DNA from its vulnerable spot at "ground zero," within the free-radical generating mitochondria, into the bomb shelter of the cell nucleus, where damage to DNA is vastly rarer.

Now, if only we had solutions like that for all of this other stuff, I mused, *we could forget about the "butterfly effect" of interfering with basic metabolic processes, and just take the damage ITSELF out of the picture.*

Hmm.

Well, I thought, *why the bloody hell not?*

I went back over my list. Protein glycation? A biotech startup was already running clinical trials using a drug that had been shown to break the dysfunctional handcuffing of the proteins that this process caused. The extracellular aggregates? Here again, animal studies had shown that you could just remove the damage, in this case by vaccinating against the amyloid plaque and letting immune cells gobble the stuff up. In theory, at least, there were all kinds of ways to deal with cellular senescence, though I wasn't sure which of them would ultimately pan out. Anyone who'd read a newspaper in the last year knew that scientists were hotly pursuing a way to deal with the loss of cells: stem cells, cultured in the lab and delivered as a rejuvenating cellular therapy. Lipofuscin? It was at this point in my survey that I began to feel I might really be on to something, because just a year previously I'd come up with a way to eliminate lipofuscin that, although extremely novel, had already secured the enthusiastic interest of a few of the top researchers in that area. I didn't have any radical new ideas up my sleeve for cancer; it was going to have to rely (for now, at least) on other people's ideas. But that was okay: after all, there was already a huge effort under way to deal with it. And as for other problems arising from nuclear mutations, I had recently come to the admittedly counterintuitive conclusion that they were not in fact a major cause of age-related cellular dysfunction.

I went over my list again and again, and as I did so I became ever surer that there was no clear-cut exception. The combination of my own idea for eliminating intracellular garbage like lipofuscin; the idea I'd been championing for a few years for making mitochondrial mutations harmless; and the various other therapies being worked on by others around the world for addressing glycation, amyloid accumulation, cell loss, senescent cells and cancer—it seemed that this was really and truly an adequately exhaustive list. Not necessarily *totally* exhaustive—there certainly might be other things going wrong in the body—but very possibly comprehensive *enough* to give a few decades of extra life to people who are already in middle age before we start the treatments. And that was certainly a much more promising first step than anything that had been suggested the previous day, or in the many conferences and articles that I'd devoured over the previous few years.

For decades, my colleagues and I had been earnestly investigating aging in the same way that historians might "investigate" World War I: as an almost hopelessly complex historical tragedy about which everyone could theorize and argue, but about which nothing could fundamentally be done. Perhaps inhibited by the deeply ingrained belief that aging was "natural" and "inevitable," biogerontologists had set themselves apart from the rest of the biomedical community by allowing themselves to be overawed by the complexity of the phenomenon that they were observing.

That night, I swept aside all that complexity, revealing a new simplicity in a complete redefinition of the problem. To intervene in aging, I realized, didn't require a complete understanding of all the myriad interacting processes that *contribute to* aging damage. To design therapies, all you have to understand is aging damage *itself*: the molecular and cellular lesions that impair the structure and function of the body's tissues. Once I realized that simple truth, it became clear that we are far closer to real solutions to treating aging as a biomedical problem, amenable to therapy and healing, than it might otherwise seem.

Grabbing a notepad, I jotted down the molecular and cellular changes that I could confidently list as important targets for the new class of anti-aging therapies that I would soon call SENS, the "Strategies for Engineered Negligible Senescence." Each of them accumulated with age in the body throughout life and contributed to its pathological decay at later ages. As far as I could tell, the list was exhaustive, but I'd present it to my colleagues and see if they could add to it. I rushed downstairs before breakfast to transcribe my scrawled notes onto a flipchart in the meeting room. I was bursting to present my new synthesis to my esteemed colleagues. But truth be told, I already knew full well that at this first hearing they'd greet it with blank stares. The paradigm shift was just too great.

2

Wake Up—Aging Kills!

How many lives do you think you could save, in your life?

This is not a trick question. But in order to make it even more precise, I'm going to modify it a little. When we speak of saving lives, we mean giving the beneficiaries of our action the chance to live longer than they could otherwise have lived. However, when we ask in detail about the importance of saving a life, we may not regard all lives equally. For example, saving an eighty-year-old from drowning may give him only a few extra years of life before he's likely to die of something else, whereas saving a child from drowning gives him a probable seventy years or more of extra life. We may also take into account the quality of life of the persons whose lives we save—predominantly their health. So here's my modified question:

> How many healthy, youthful years in total do you think you could add to people's lives, in your life?

The ultimate purpose of this book is to show you that you could add many more years than you may currently think. So many, in fact, that now is the time to decide whether you want to. The way you can do this is by helping to hasten the defeat of aging. The specifics of how you can help—by donating money or time to the Methuselah Foundation's Mprize fund or its SENS research funding program—will be the topic of Chapter 15; in this chapter I'll restrict myself to communicating the magnitude of what those efforts can achieve in humanitarian terms.

I'll start with some numbers. Around 150,000 people die each day worldwide—that's nearly two per second—and of those, about two-thirds die of aging. That's right: 100,000 people. That's about thirty World Trade Centers, sixty Katrinas, every single day. In the industrialized world, the proportion of deaths that are attributable to aging is around 90 percent—yes, that means that for every person who dies of all causes other than aging added together, be it homicide, road accidents, AIDS, whatever, somewhere around *ten* people die of aging.[1]

And it's worse than that. Look again at my expanded question and you'll notice a couple of adjectives: "healthy" and "youthful." Many people, when thinking about the idea of adding years to life, commit the "Tithonus error"—the presumption that, when we talk about combating aging, we're only talking about stretching out the grim years of debilitation and disease with which most people's lives currently end.[2] In fact, the opposite is true: the defeat of aging will entail the *elimination* of that period, by postponing it to indefinitely greater ages so that people never reach it. There will, quite simply, cease to be a portion of the population that is frail and infirm as a result of their age. So it's not just extending lives that I'll be telling you about in this book: it's the elimination of the almost incalculable amount of suffering—experienced not only by the elderly themselves, of course, but by their loved ones and carers—that aging currently visits upon us. Oh, and there's the minor detail of the financial savings that the elimination of aging would deliver to society: it's well established that the average person in the industrialized world consumes more health-care resources in his or her last year of life than in an entire life up to that point, irrespective of age at death, so we're talking about trillions of dollars per year.

In this book I will explain the scientific and technological basis for my view that we can probably eliminate aging as a cause of death this century—and possibly within just a few decades, soon enough to benefit most people currently alive. But first, I need to get you interested—not just in the sense of entertainment, the sense in which you might read a good story, but in the sense of realizing that as and when this becomes possible it will be rather a good thing. And I've been in this business long enough to know that a description of the level of suffering that would be averted and the number of lives that would be saved does not, on its own, convince most people that it would be a good thing if aging were defeated. So I hope you'll forgive me if I am blunt and to the point in this chapter, before I move on to the science and technology that will get the job done.

Why Did I Write This Book?

I'm a scientist and technologist, and in an ideal world I would spend essentially all my time working on the scientific and technological details of my life's goal, defeating aging. I wouldn't spend much time doing media interviews, or giving public lectures—or even writing books. But there's something about people's attitudes to aging that, for now, changes my priorities. I call it the pro-aging trance. I'm going to start my discussion of the pro-aging trance with a comparison.

Here in the United Kingdom, just as across the whole Western world, there is a campaign of increasing ferocity against smoking. All cigarette packets come with health warnings. Not just tiny inconspicuous health warnings in cautious scientific language, either—warnings of the most hard-hitting and in-your-face nature possible. The simplest and shortest consists of just two words, typically printed in black on a stark white background:

Smoking Kills

And, slowly but surely, smoking is becoming less popular. Just like drunk driving before it, smoking is becoming socially disreputable. It's a long, hard road, though: not just because nicotine is addictive, but because youngsters continue to take up smoking despite the social stigma increasingly attached to it.

It's that latter point, the continued influx of new young addicts, that is my focus here. I've used smoking as my chosen analogy not in order to condemn the smokers among my readers—not at all. No, my focus here is something altogether less controversial, because the battle to protect youngsters from taking up smoking is one that virtually all adults, smokers or not, support. My reason for mentioning it here is *timeliness:* this battle is still being waged, so we can examine at close quarters the contradictions in our attitudes, both as individuals and as a society, that make the battle so hard to win. With specific diseases, there is no argument: the more we can do to defeat them, the better. But with smoking, even though it causes some of those self-same diseases, somehow society is itself subject to an addiction that robs it of its rationality concerning new young addicts. We face every day the brutal disconnect between allowing cigarettes to be advertised and sold widely and seeing how much they blight and shorten the lives of those who fall under their spell. And it's just the same, I claim, with aging.

There are two potential reasons why smoking is declining in popularity and in public acceptability. One is that many people find it unattractive—they don't like the smell (or, in more intimate contexts, the taste). But it's hard to believe that this can be the main trigger for the rather recent change of sentiment against smoking, because today's tobacco is surely no more off-putting than tobacco of a century or three ago. Thus, I think it's clear that the main reason so many people now disapprove of smoking is its other downside, which was not much appreciated even half a century ago: It's really rather bad for you, and also for those around you. Most of all, it massively increases your risk of getting fatal lung cancer, which not only shortens your life but also makes your declining years really miserable.

My goal with this book, as for all my outreach work, is to inject momentum into a similar shift of public opinion concerning aging. I have been aware for many years that most people do not think about aging in the same way that they think about cancer, or diabetes, or heart disease. They are strongly in favor of the absolute elimination of such diseases as soon as possible, but the idea of eliminating aging—maintaining truly youthful physical and mental function indefinitely—evokes an avalanche of fears and reservations. Yet, in the sense that matters most, aging is just like smoking: It's really bad for you. It shortens your life (see Chapter 14 for an assessment of just how much), it typically makes the last several years of your life rather grim, and it also makes those years pretty hard for your loved ones.

So let's look a little more closely at why aging is so passionately defended.

The Motivation for the Pro-Aging Trance

First of all, let me be clear that I realize there's an immense gulf between people's attitude to *modest* postponement of aging and their attitude to the topic of this book, the genuine *elimination* of aging as a cause of infirmity and death. The anti-aging industry is huge, despite the (shall we say) highly variable ability of its products to do what they say they can do, and that can only be because people are not very happy to see themselves falling apart, or to be seen to be falling apart. Yet, the prospect of eventually being able to combat aging as well as we can currently combat most infectious diseases—

essentially to eliminate aging as a cause of death, in other words—strikes terror into most people: Their immediate (and, I must point out, often high-pitched) reaction is to raise the specter of uncontrollable overpopulation, or of dictators living forever, or of only a wealthy elite benefiting, or any of a dozen other concerns.

Now, I'm certainly not saying that these objections are dumb—not at all. We should indeed be considering them as dangers that we should work to preempt by appropriately careful forward planning. No: what shocks me is not that these concerns are raised, but the *way* they're raised. People who are totally rational and open to discourse on any other matter approach the topic of defeating aging with a resistance to debate that virtually defies description. The determination with which people work to change the subject, to relegate the conversation to an exchange of witticisms, or simply to cast the opponent of aging as a deluded nincompoop has to be encountered to be believed.

Perhaps you're wondering whether I've forgotten that I'm talking about you here. But understand that I'm not castigating you at all, because my remarks so far have dealt only with the *logic* of why aging should be fought, and life is not all about logic. There is a very simple reason why so many people defend aging so strongly—a reason that is now invalid, but until quite recently was entirely reasonable. Until recently, no one has had any coherent idea how to defeat aging, so it has been effectively inevitable. And when one is faced with a fate that is as ghastly as aging and about which one can do absolutely nothing, either for oneself or even for others, it makes perfect *psychological* sense to put it out of one's mind—to make one's peace with it, you might say—rather than to spend one's miserably short life preoccupied by it. The fact that, in order to sustain this state of mind, one has to abandon all semblance of rationality on the subject—and, inevitably, to engage in embarrassingly unreasonable conversational tactics to shore up that irrationality—is a small price to pay.

A Word About SENS Skepticism

This book is a description of SENS (Strategies for Engineered Negligible Senescence), my "project plan" for defeating aging. I expect that this will be many readers' first encounter with SENS, but others will have come across it before. In particular, if you have had an interest in life extension for some time, there's a good chance that you've already come across accounts of

SENS in the mainstream media. If so, you'll be well aware that, while many highly credentialed gerontologists have applauded SENS, others have greeted it with strong criticism—even derision. So far in this chapter I have only addressed the flaws in people's reasons for feeling that the defeat of aging might not be *desirable.* But in order to ensure that you read this book with real care, and moreover that you then go out and do something to help the anti-aging effort, I also need to make sure that you understand that the defeat of aging is *feasible.* Therefore, I include here a brief account of where the debate about SENS's chances of success currently stands.

I must first make sure you appreciate that it is the norm for radical new concepts that receive a lot of attention to arouse a sharp division of opinion among expert commentators. In many cases, the establishment detractors are absolutely right and the upstart new idea really is misguided. Very often, however, the detractors have failed to acquire—even avoided acquiring—a detailed understanding of what they are criticizing and have been driven more by vested interests than by scientific argument. If you are not a scientist you may feel that this is an unfair suggestion, but the intellectual and emotional investment that senior scientists have made in their beliefs is a powerful opponent to objectivity: all scientists acknowledge this problem privately, if not publicly. It has been memorably summarized by a number of the world's most eminent scientists over the years; for example, the physicist Max Planck observed over eighty years ago that "science advances funeral by funeral," and the biologist J. B. S. Haldane noted that "there are four stages of acceptance: (i) this is worthless nonsense; (ii) this is an interesting, but perverse point of view; (iii) this is true, but quite unimportant; (iv) I always said so."

Since I work on aging in order to hasten its defeat, and not in order to become rich and famous, I am extremely keen to identify any major holes in SENS so that, if they indeed exist, I can go back to the drawing board without delay. To this end, I talk to my most prominent biogerontologist critics all the time about SENS. I am invariably driven to the view that they are indeed guilty of reacting to my conclusion (that SENS can totally defeat aging) without studying the reasoning behind that conclusion—but I of course appreciate that I, too, may be unable to be objective in this matter. For this reason, and also because the speed of implementation of SENS depends greatly on both public and academic acceptance that it might work, I have worked hard in recent years to generate unbiased evidence as to whether SENS is sense or nonsense. In 2006, I achieved this rather decisively, with the assistance of the prominent magazine *MIT Technology Re-*

view. After publishing a rather negative portrayal of SENS in 2005, *TR* discovered that the mainstream gerontologists on whose opinions it had relied in choosing to do so were unwilling to back up their assessment with any scientific detail. *TR* then admirably put itself at risk of considerable loss of face by organizing a prize challenge to settle the matter.[3]

In order to win the SENS Challenge, one or a group of credentialed biologists had to write a demolition of SENS that I was unable to rebut to the satisfaction of a panel of expert judges. The panel of judges had to be demonstrably impartial, of course, with no connection either to me or to my critics, but yet well versed in biotechnology; *TR* succeeded in appointing a superb five-person panel including the biotech luminary Craig Venter. *TR* put up $10,000 as a prize, and the Methuselah Foundation contributed the same amount. A group of nine highly credentialed biogerontologists obligingly submitted a coauthored entry, as did two other scientists independently. All three entries were unanimously and emphatically judged to fall decisively short of a demonstration that SENS is not worth trying.

Now, I'm certainly not trying to say that this proves that SENS will indeed deliver the defeat of aging: there's only one way to answer that, which is to implement it and see what happens. But my critics made the stronger claim that SENS is so implausible that there's no need to *try* to implement it. That claim has been incontrovertibly refuted by the SENS Challenge process. So, if you find someone still eager to tell you that SENS is fantasy—especially someone who claims to have expert knowledge in this area—you'll know, as *TR* now knows, that asking such people what they *think* about SENS is a great deal less reliable than asking them what they *know* about SENS. And after you've read this book—especially Part 2—you'll be equipped to come to your own conclusions.

Building a Case, Chapter by Chapter

I'm a fighter at heart; I would never have made my peace with aging, however lost the battle seemed. But that's not the life everyone wants, and I respect that. Thus, I would probably not have written this book if I thought we were still too far from defeating aging to have any real chance of success within the lifetimes of anyone alive today. In the next chapter I'll describe why aging is, in principle, just as amenable to modulation and eventual elimination as specific diseases are, and how an inappropriate way of looking at aging has led most gerontologists to favor eventual therapeutic approaches that I consider unlikely to bear fruit. Then, Chapter 4 provides an overview of my scheme for defeating aging within (if all goes well) only a few decades. That concludes Part 1 of the book. In Part 2, Chapters 5 through 12 elaborate on the individual components of that scheme. The book concludes with Part 3, a trio of chapters covering what I predict will be the response of society to initial successes in the laboratory a decade or so from now, how the advances of the next few decades will be progressively refined and aging permanently kept at bay, and how you can already help to accelerate that crusade.

Buried inconspicuously in that last paragraph was something that I want to make sure you don't misinterpret: a tentative time frame. Yes, I consider that if funding is sufficient we have a 50/50 chance of developing technology within about twenty-five to thirty years from now that will, under reasonable assumptions about the rate of subsequent improvements in that technology, allow us to stop people dying of aging at any age—equivalent to the effect of today's antiretrovirals against HIV. There are three big caveats in that statement, though. The first is that it's only a 50 percent chance. Any technological prediction as far in the future as twenty-five to thirty years is necessarily very speculative, and if you ask me how soon I think we have a 90 percent chance of defeating aging I wouldn't even be willing to bet on one hundred years. But I think a 50 percent chance is well worth shooting for—don't you? The second caveat is that aging won't be totally defeated by the initial versions of this technology; we'll have to carry on improving it at a reasonable rate in order to keep aging permanently at bay. I will explain all the details of that in Chapter 14.

But the third caveat is perhaps the most important: the adequacy of research funding. I cofounded the Methuselah Foundation in order to address that problem: at present, the pace of most of the research avenues that we need to pursue in order to combat aging adequately is limited by funding. If you can help to change that—whether by giving money yourself, or by influencing friends, or by writing or broadcasting on the subject—you'll be making as much difference to the speed with which aging is overcome as if you were doing the science yourself.

There's a critical point about funding that I must emphasize here: the pivotal role of relatively small amounts of money at this early stage in the crusade. I've complained at length in this chapter about people's reluctance to treat aging as the curse that it is, and I hope I'm making a difference to that attitude by my outreach activities, but realistically I know that most people are going to sustain their pro-aging trance for a while yet, and that will severely limit the availability of either public or commercial funding for life extension research. The point where that will really change—where the global pro-aging trance will collapse like a house of cards—will, in my view, be when middle-aged mice are rejuvenated thoroughly enough to extend their healthy lives by a large amount. This is a milestone that I've termed "robust mouse rejuvenation," or RMR. The amount of money needed to achieve it is tiny compared to how much we'll then need to spend to get the same result in humans; but when humanity as a whole is behind the effort, willing to pay for it with taxation, there'll be ample funds available. It's now, when private philanthropy is the only major source of funding for such work, that the magnitude of that private philanthropy is so critical. I'll elaborate on this in Chapter 13.

I've discussed in this chapter *why* aging is defended, but I haven't said much about *how*—about the common objections to the prospect of indefinite life extension. In many of my writings and public presentations, and on my Web site,[4] I do address the many questions that arise concerning how society would be different in a post-aging world, and especially how we would handle the transition to that world. This book does not address those issues in detail; I've decided to deal here only with the *practicality* of radical life extension. I hope you'll come away with a pretty good understanding that the genuine defeat of aging is a feasible goal. Whether it's also a desirable goal is a question that you'll then be able to consider more seriously—even, dare I say it, more responsibly and conscientiously—than you could if you still thought it was science fiction.

3

Demystifying Aging

Aging has held us in a psychological stranglehold ever since we realized it existed, and that stranglehold remains intact to this day. I discussed in Chapter 2 the effect that this has on our willingness to think rationally about how terrible a thing aging is, and I explained why this irrationality used to have a valid psychological basis while there was no hope of combating aging, and also why it is now such a formidable obstacle.

There's a complication, though. I've told you that we've recently reached the point where we can engage in the rational design of therapies to defeat aging; most of the rest of this book is an account of my favored approach to that design. But in order to ensure that you can read that account with an open mind, I need to dispose beforehand of a particularly insidious aspect of the pro-aging trance: the fact that most people already know, in their heart of hearts, that there *is* a possibility that aging will eventually be defeated.

Why is this a problem? Indeed, at first sight you might think that it would make my job easier, since surely it means that the pro-aging trance is not particularly deep. Unfortunately, however, self-sustained delusions don't work like that. Just as it's rational to be irrational about the *desirability* of aging in order to make your peace with it, it's also rational to be irrational about the *feasibility* of defeating aging while the chance of defeating it any time soon remains low. If you think there's even a 1 percent chance of defeating aging within your lifetime (or within the lifetime of someone you love), that sliver of hope will prey on your mind and keep your pro-aging trance uncomfortably fragile, however hard you've worked to convince yourself that aging is actually not such a bad thing after all. If you're completely convinced that aging is immutable, by contrast, you can sleep more soundly.

The key qualification in what I've just said, of course, is the phrase "while the chance of defeating it any time soon remains low." Once that chance becomes respectable, you're better off doing your bit to increase it further—not just the actual laboratory work, of course, but also agitating, cajoling, helping others (not least those with influence over research funding) to awaken from their own pro-aging trance. Conversely, if the chance of aging being defeated is really tiny despite whatever you do, the cost-benefit balance of abandoning your comfort zone may tip the other way, in favor of applying the same irrationality to the existence of such a possibility as you may be doing in respect of the pros and cons of aging.

Therefore, in this chapter I'm going to describe what aging is in practical terms, so as to demystify it for you. By doing so I plan to show you that the popular presumption that aging is a phenomenon unlike all other health conditions, somehow beyond even the theoretical reach of medical technology, cannot be reconciled with established fact. Thus, by the end of this chapter I aim to have placed you in the awkward position of still *wanting* to believe (for your own peace of mind) that aging is immutable and thus not worth worrying about, but no longer actually being *able* to believe that. From that point on, my task will be the relatively easy one of explaining why our chances of defeating aging in the foreseeable future are not just non-zero, but high enough to justify my having broken your pro-aging trance in the first place. Justify, because once your pro-aging trance is no more, you—yes, you—can make a difference to how soon aging is defeated, and the fulfilment you will derive from that effort will far outweigh any comfort you may have found in your previous certainty that aging can never be combated.

The Illusory Boundary Between Aging and Disease

It used to be the case that people died of aging, but, if you believe what's written on death certificates, these days they rarely do. The phrase "natural causes" was the accepted term for the cause of death when it occurred at an advanced age and in the absence of clearly defined pathology. These days, however, that's considered inadequately informative, and coroners or their equivalent are encouraged to enter something more specific.[1]

We all know, however, that quite a few people do indeed die in that way—not from a heart attack, not from pneumonia or influenza, not from cancer, not even from a stroke, but just peacefully, often in their sleep, because their heart simply stops. These relatively lucky people indisputably die of aging.

That brings me to the first of several times in this book when I must engage in the unpleasant business of exposing a serious distortion of the facts that has been perpetrated—often unintentionally, I realize—by a large number of senior researchers in the field of biogerontology, the study of how aging works. This distortion has by now been generally seen for the awful error that it was, but the disastrous consequences for the field are still being felt, and probably will be for many years to come. Through the 1950s, '60s, and '70s, while gerontology was making its big push for recognition as a legitimate biological discipline, rhetoric developed to the effect that the infirmities of aging should be viewed as separable into two distinct phenomena: on the one hand, age-related diseases, and on the other hand, "aging itself." This distinction was publicly *defended* mainly on the basis that everyone has aging, whereas no age-related disease is universal. The *motivation* for this distinction, on the other hand, was purely pragmatic: by ring-fencing their area of work intellectually, gerontologists hoped to ring-fence it financially, too.

And ring-fence it they did, most notably with the creation (while President Richard Nixon was paying limited attention, so it is said) of the National Institute on Aging.[2] So far, so good. However, it's not good enough. All gerontologists know full well that it's no accident that age-related diseases are age-related: they appear at advanced ages because they are consequences of aging, or (to put it another way) because aging is no more and no less than the *collective early stages* of the various age-related diseases. Gerontologists knew this back then, too. Thus, they also should have seen back then that, by trumpeting the short-termist rhetoric that "aging is not a disease," they were constructing an immense obstacle for themselves in the longer term: the response from policy makers that, well, if it's not a disease, why should we spend money on combating it? The era of that backlash began decades ago and shows no sign of ending. Gerontologists these days point out over and over again that if we could just postpone aging even a little bit we would derive far more health benefits than would result from even the most decisive breakthroughs against specific diseases, but over and over again their paymasters fail to get the message.[3] I maintain that it is overwhelmingly the inaccurate rhetoric of gerontologists, resulting from their misguided policy of previous decades, that has brought about such entrenched resistance to a simple, obvious and (within the field) universally agreed-upon truth about the potential value of postponing aging.

I told you just now that age-related diseases are merely consequences of aging; now I'll tell you why we know that. In the process, I'll also tell you why aging has the range of speeds that it does—within a single individual, and between individuals, and also between species.

Why Aging Doesn't Need a Timer

The fact that a fair proportion of people die of natural causes, rather than of any specific disease, might at first sight imply that aging is a process independent of diseases: something that increases people's *vulnerability* to disease (thus making diseases more common among the elderly) but that also kills us itself if no disease does so first. This is only partly correct. The elderly are indeed more vulnerable to infectious diseases, because one aspect of aging is the decline of the immune system. However, most diseases of old age have only a minor, if any, infectious component: they are mostly or wholly intrinsic. Take cancer, for example. A few types of cancer affect young people, but most types are never seen in people below the age of forty or so (except for people with very rare congenital DNA repair deficiencies). Some cancers are caused by viral infections—the best known of these is cervical cancer, caused by the human papilloma virus. But the major underlying cause of cancer is the simple accumulation over time of mutations in our chromosomes. Mutations are inevitable: they happen as a purely intrinsic side effect of our biology. The time they most often happen is when the DNA of our chromosomes is replicated during the process of cell division. The accumulation of mutations is, therefore, part of aging, and cancer is predominantly a consequence of aging—or, if you prefer, part of the later stages of aging.

Sounds pretty simple, doesn't it? And yet, there is a pervasive presumption—one shared even by some biologists—that aging is some kind of mysterious phenomenon qualitatively different from any disease: something that has eluded, and thus may forever elude, biological elucidation.

There are a few main reasons for this presumption, so I'll briefly describe those reasons and why they're wrong.

The first is that aging proceeds so much more slowly than specific diseases. So slowly, in fact, that we hardly notice its progression, whereas we are much more keenly aware of the more rapid development of conditions like cancer or diabetes. This is a conspicuous difference, but in fact it's just what one would expect, because aging is a downward spiral. The more we age, the more our self-repair functions decline, so the less able our body is to stop us aging, so we age faster and faster. So it's to be expected that the late stages of aging, the diseases, would go faster than the earlier stages.

Another thing that confuses people about aging is that it proceeds at very different rates in different species but at pretty similar rates in all members of a given species. This might be thought to imply that there is some kind of internal clock driving the process, which is set at different speeds in different species. The inference is that this clock is somehow immune to biomedical intervention, because changing its speed would require us to stop being human. But that's not correct either, for two reasons. First, even if there were such a timer, we could in principle postpone the later stages of aging without changing the speed of the timer itself—I'll be elaborating on this below. And second, if there were such a clock, why shouldn't it be amenable to biomedical intervention anyway? The fact that organisms of the same species tend to age at the same rate is just one consequence of the fact that they're genetically very similar to each other. It says nothing about what can or cannot be altered by biomedical technology.

Perhaps the most common reason for the belief that there is an "aging clock" is the fact that the various outcomes of aging (including age-related diseases) all tend to appear at more or less the same age in different individuals within a given species. Surely this must mean that there is indeed a central aging clock, which has ticked down enough to set these diseases on their way, right? No—and, again, this is for two main reasons.

First, this is exactly what one would expect if the debilities of old age were just the later stages of a multifaceted decay process, just so long as that system has one key feature: a rich degree of *interconnection* of the various chains of cause and effect. If lots of things are going quietly wrong throughout life, and their accumulation is feeding back on themselves *and each other* to accelerate them, they'll necessarily all proceed at more or less the same rate and all "go critical" (explode into clinically identifiable disease) at about the same age. And that interconnectedness is, indisputably, indeed present in aging.

Second, if we think about the evolutionary basis of aging for a moment we can easily see that, even *without* much interconnection between the chains of events that lead to the various diseases of aging, we'd still expect to have them all emerge at roughly the same age. This is because, if we had genes that defended against one particular cause of death so well that everyone was dead from other causes before they died of that one, those genes would not be protected by evolutionary selection and would accumulate random, mild mutations from one generation to the next. Over evolutionary time, therefore, the quality of those genes would thereby sink down to the point where the disease they protected against occurred at the same age as all other age-related diseases.

Another common but incorrect reason for thinking that aging is somehow special is that it is "universal"—it happens to everyone. Well, yes: If you live long enough, you'll exhibit signs of aging. But this is only a corollary of my earlier point about rates—that aging is really slow compared to age-related disease. Because age-related diseases progress from diagnosability to death rather quickly, many people die of one such disease before the others emerge, or at least while they are still too early-stage to have been diagnosed. But if those people hadn't suffered the disease that killed them, they'd have lived long enough to suffer others. In fact, *all* the diseases of aging are universal in the sense in which the question ought to be asked: namely, you'll definitely get them if you don't get something else first.

Thus, in concluding this section I hope to have convinced you that aging is not something inherently mysterious, beyond our power to fathom. There is no ticking time bomb—just the accumulation of damage. Aging of the body, just like aging of a car or a house, is merely a maintenance problem. And of course, we have hundred-year-old cars and (in Europe anyway!) thousand-year-old buildings still functioning as well as when they were built—despite the fact that they were not designed to last even a fraction of that length of time. At the very least, the precedent of cars and houses gives cause for cautious optimism that aging can be postponed indefinitely by sufficiently thorough and frequent maintenance.

The Corollary That Even Most Experts Overlook

Everything I've explained above is well known to biogerontologists, the people who study aging. From the way that most biogerontologists go about exploring how to postpone aging, however, you might think they didn't know this at all. People who work to combat specific diseases explore the way in which the disease progresses and look for ways to disrupt that pathway. In gerontology, however, the predominant modus operandi for designing interventions is to compare organisms that age at different rates—different species, or individuals of the same species in different conditions—and to come up with ways to copy or extrapolate those differences so as to make aging happen more slowly. This is effectively an a priori capitulation, not even trying to dissect and disrupt the process but rather treating it as a black box. It's especially surprising when you bear in mind that biogerontologists certainly do work hard to dissect the aging process in order to understand it—just not in order to combat it. (Unfortunately, these two goals motivate different types of dissection.) Rather, the most promising ways to postpone aging are by disrupting the pathways underlying it, just as we do for specific diseases. Thus, since aging is just the accumulation of damage, we should be looking at ways to alleviate that accumulation. I'll return to this in greater detail in the next chapter and beyond.

Why Fixing Aging Is *Easier* than Fixing Similarly Complex Machines

Now let's move on to another reason that people often give for clinging to the belief that aging is inherently inaccessible to biomedical intervention. If aging is just damage, and the body is just a complex machine, it stands to reason that we can apply the same principles to alleviating the damage of aging as we do to alleviating damage to machines. But people sometimes point out that the body has a host of self-repair and self-maintenance processes, which machines basically don't have, hence we're not really machines at all. Thus, they claim, the maintainability of machines is no basis for confidence that the body is in principle similarly maintainable.

Well, I invite you to think about that logic for a moment. We have built-in repair and maintenance machinery. Why on earth would that make it *harder* to maintain our bodies in good working order? Clearly the opposite is the case: if our bodies are doing most of the job automatically, that leaves *less* for us to do with biomedical technology.

Let me stress that I'm not saying the task is easy. The body is a great deal more complicated than any man-made machine—and what's more, we didn't design it, so we have to reverse-engineer its workings in order to understand it well enough to keep it running. But that doesn't change the above logic: the natural capacity for self-repair that we're born with is our *ally* in the anti-aging crusade, not our enemy.

Postponed Aging in the Lab: No Longer Just Theory

By now I may have satisfied some readers that, indeed, aging is not a mystical phenomenon beyond the reach of mere, um, mortals. I'm well aware, however, that many people find theoretical arguments only modestly persuasive, even if no holes in those arguments seem evident. Such people—you, perhaps—feel altogether more comfortable with a conclusion if it is backed up by hard evidence. You'll be pleased to discover, then, that for several decades scientists have been finding ways to lengthen the lives of various organisms in the laboratory. Best of all, they've done this not by extending those organisms' period of declining vigor at the end of life, nor (by and large) by keeping them immature for longer, but by extending the period of peak health and vigor between maturity and frailty.

One highly robust life-extension technique was discovered over twenty years ago by a young Canadian researcher named Michael Rose, who is now a professor at the University of California, Irvine. Rose is an evolutionary biologist, and at that time he already had a thorough knowledge of the ways in which evolution optimizes a species' longevity for its ecological niche. He realized that it might be possible to *breed* longer lived organisms, rather in the vein of the Howard families in Robert Heinlein's *Lazarus Long* books, by maintaining them over many generations and only allowing those with the longest lives (actually, strictly speaking the longest reproductive lives) to contribute to the next generation. It would take many more generations than Heinlein described, but Rose was working with fruit flies, which reach maturity only a week after their own conception. And it worked, spectacularly: Rose was eventually able to achieve average lifespans twice those in his starting population.[4]

This approach, impressive though it was, has a fundamental and rather relevant limitation—a limitation that has probably not escaped you. Specifically: it can't be applied to you, only to your great-great-great . . . great-grandchildren. Rose knew this, too, of course, and more recently he's been working hard to identify the genetic, and thence molecular, basis for this life extension with a view to eventual therapies that might work on those of us unfortunate enough to be already alive. But thus far, all he has are long-lived *distant descendents* of short-lived flies.

Luckily, other laboratory life-extension successes have not shared this drawback. The first and best-known way to delay aging in the laboratory was discovered way back in the 1930s by a researcher named Clive McCay, working with laboratory mice.[5] It is called calorie restriction—or sometimes dietary restriction, energy restriction or food restriction. It's an extraordinarily simple concept: If you feed rodents (or, in fact, a wide variety of other animals) a bit less than they would like, they tend to live longer than if they have as much food as they want. This is not simply because such animals tend to overeat given the chance and become obese: animals that "eat sensibly" and maintain a constant body weight throughout most of their lives still live less long than those given less food.

The next researcher (not counting Rose) to take the postponement of aging a major leap forward was a geneticist working with a third, almost equally widely studied, model organism: the nematode worm *Caenorhabditis elegans*. His name is Tom Johnson. He was not, strictly speaking, the discoverer of the phenomenon I will describe here—that honor goes to one of his coworkers—but he spearheaded the work on it for some years and that work has become identified with him, so I'll focus on him for the moment. What Johnson and his colleagues discovered and researched was a mutation in a single, identified gene, which *on its own*—without any of the sustained selective pressure employed by Rose—added at least 50 percent to the youthful adult lifespan of his worms.[6] This was an immense breakthrough, because single genes can be modified in the test tube and then introduced into the body by *gene therapy:* either germline gene therapy, which affects only the recipient's descendents, or somatic gene therapy, which affects the organism that receives the treatment. Somatic gene therapy for humans is still taking its baby steps, but there is widespread confidence that it'll work well eventually. And human germline gene therapy raises ethical concerns (though there are technical approaches to avoiding these). But as a proof of principle, the postponement of aging by a single, defined genetic alteration is vastly closer to clinical applicability than something accomplished by selection over many generations and affecting an unknown number of genes.

Perhaps because of this, and also partly because of the experimental methods involved, Johnson's result initiated a massive surge in attempts to identify genetic alterations to lab animals that would delay their aging. This surge actually took a few years to get going, but when a second laboratory (that of Cynthia Kenyon at the University of California San Francisco) identified a mutation in a different gene, also in nematodes, that extended their lives even more than Johnson's mutation did, the topic became one of the hottest in the whole of biology.[7] Kenyon and other top biogerontology researchers have been able to publish nearly all their best work in the very top few journals ever since—journals that scientists in most fields are lucky to publish in even a couple of times in their whole career.

Johnson's and Kenyon's mutations were in different genes, but these genes participate in largely the same range of metabolic processes. In particular, they help to mediate an alternative developmental trajectory that normal, nonmutant nematodes can follow, termed the *dauer* pathway. When a nematode larva follows the dauer pathway, it suspends its development for a period than can last much longer than the entire lifetime of a nematode that follows the normal, non-dauer trajectory. What, you may ask, triggers this developmental choice? And what "restarts" development and the resumption of the path toward normal nematode adulthood? Well, it just so happens that the usual trigger for entry into the dauer pathway is starvation, and that exit from dauer is stimulated by the presence of food. In other words, the dauer pathway is neither more nor less than nematodes' extreme version of rodents' response to calorie restriction.

Since Johnson's and Kenyon's breakthroughs, many other mutants have been discovered—not only in nematodes but also in fruit flies and mice—that have extended lifespans, and nearly all of these mutations have also disrupted genetic machinery that mediates the sensing or metabolism of nutrients. In general, the mutations confer a delay of aging at most equal to that achievable by simply restricting calorie intake.[8] A few publications have appeared in the past few years reporting life extension in mice caused by reducing *oxidative stress,*[9,10,11] but I am currently cautious about the reproducibility of these findings, because a huge number of other attempts to postpone mouse aging in the laboratory in that way has failed.

At this point, therefore, I can point to a pretty compelling, double-whammy argument that aging is worth trying to tackle. On the one hand we should *in principle* be able to postpone aging by a large degree; moreover, we have *actually done so* in the laboratory. This is surely great cause for optimism that we will do so in the clinic in the not-too-distant future.

Isn't it?

Well, I would hardly have written this book if that were not indeed my ultimate conclusion. However, the operative word here is "ultimate." Before closing this chapter, I must explain why calorie restriction and its genetic emulation are not, in fact, pointers to the most promising route to combating human aging.

Calorie Restriction and Its Emulation: A False Dawn

Do you know any perfectionists?

I do—and I always have, because my mother is one. I certainly wouldn't be where I am today without my mother, and that includes her influence on me as well as her sheer hard work and determination to give me the best start in life. But there are certain ways in which her influence on me was to show me a bad example, and her perfectionism is perhaps the most extreme such case. I feel that in many ways it has prevented her from achieving what she might have in her life, so I've never let myself become a perfectionist—and I've certainly never regretted that.

What's wrong with perfectionism? We all know the main problem with it: *Perfectionism takes time.* Most people are interested in getting things done, and there are many circumstances in which a quick and dirty job is the best policy, because the advantages of the "quick" outweigh the disadvantages of the "dirty." There are certainly other circumstances in which the balance is reversed, though—where a more painstaking approach is to be preferred; hence, the ideal is to have good intuition and judgment for how much attention to detail is appropriate in any particular case.

You may think that the above two paragraphs are a remarkably dramatic digression, so let me surprise you by bringing my chain of reasoning straight back to calorie restriction and its limitations in just a single sentence. The life-extending response to nutrient deprivation is neither more nor less than the expression of an organism's genetically programmed intuition regarding the appropriate degree of attention to detail that it should exercise with regard to its day-to-day molecular and cellular functioning—and, because that's all it is, it's not amenable to substantial enhancement by foreseeable biomedical technology.

Some elaboration is in order, so here goes.

I've explained, earlier in this chapter, that there are no genes for aging in most species, simply because genes only survive if they confer enough benefit (and thereby enjoy enough selective pressure for their survival) to outweigh the constant stream of random mutations that all genes experience over evolutionary time, and a gene can't confer any benefit if it only mediates a process that would happen anyway. The only species in which aging is actively driven by genetic machinery are those (such as salmon) in which there is some reason to age and die *rapidly*—something that does not happen by default to a machine that was running well for a long time previously. Slow aging, the sort that we see in nearly all species, is the default scenario, so no genes causing it can survive.

What we most certainly do have genes for, by contrast, is the panoply of interacting processes that turns each of us from a single cell into a fertile adult and that maintain our vigor and fertility until an age at which (in the wild) we're very likely to have succumbed to starvation, predation, and so on. Now, what does that have to do with perfectionism? Well, the reason we have genes to keep us going until we're very likely to have been killed is because the longer our fertile lives continue, the more progeny we'll have time to have, so the greater the chance that our genes will be passed to future generations.

But what about the other end of our fertile life—the beginning? The same applies: the sooner we achieve sexual maturity, the more offspring we'll have time to produce before we die. But here's the problem: the beginning and end of fertile life are not independent of each other. Growth from a single cell to a fertile adult is a process as complex as any known, and mistakes always happen during its execution. You can probably see the light at the end of this logical tunnel now: The organism has a choice between doing a quick and dirty job of its growth, leading to early fertility but sloppy construction, or a more perfectionist job that delays sexual maturity but creates a more smooth-running machine in the end. And a more sloppily constructed animal will on average live less long—partly because it may be less able to defend itself against predators, famine, and such like, but also because the molecular and cellular damage that it laid down during its headlong rush to maturity has effectively given it a head start in the aging process. There's abundant evidence that this is not just a reasonable idea but is also actually borne out in nature: for example, when you compare different species that are same size, the one that matures later tends to be the longer-lived.

So now: What does this have to do with calorie restriction, dauers, and the related genetic manipulations that I surveyed earlier in this chapter? Well, it's actually very simple. In a famine, there are two big problems with passing on your genes. Firstly, gestation consumes a lot of energy, which of course comes from food. And secondly, whatever offspring you do succeed in having during a famine are very likely to die of starvation before they can have their own offspring, which is no better for the survival of your genes than if you hadn't had any offspring in the first place. Thus, the advantage (in terms of your genetic heritage) of maturing quickly is less during a famine than when food is plentiful. But wait: the *dis*advantage of maturing quickly, namely the increased risk of death that results from being sloppily constructed, is unaltered! In fact, that risk may in some cases be amplified: If the duration of a particular famine is a large fraction of the species's lifespan, the period late in life when the well-constructed, late-aging animals are the only ones left to procreate will be the *only* period when successful procreation can occur. In that case, the benefit of being well constructed (i.e., the drawbacks of being sloppily constructed) will be greater in a famine of that duration than when food is plentiful throughout life.

Thus, famine shifts the happy medium toward favoring a more painstaking development process. And since famines are unpredictable events, occurring at irregular intervals, it's not possible for evolution to determine a species' ideal degree of perfectionism in advance: each individual organism must have the ability to respond to its situation. Furthermore, famines have always been like that, ever since organisms started eating other organisms. It's therefore no surprise that, everywhere we look in nature, we find the genetic machinery to respond to a famine early in life by slowing or suspending growth.

You may know that nutrient deprivation in adulthood often has the same effect to a milder extent, a phenomenon that doesn't seem to be explained by what I've just told you. Indeed, there may not be such clear-cut evolutionary reasons why adult-onset calorie restriction postpones aging at all. But there don't need to be, because genetic programs that exist for one time or circumstance are often activated unnecessarily in situations that are similar. Think, for example, of the fact that startling someone causes a mild adrenaline rush, something that exists to facilitate escape from life-threatening situations.

Finally I must explain why the logic I've outlined here implies that manipulating these nutrient-sensing pathways isn't the most promising way to postpone human aging. I actually have three reasons.

First, the degree of life extension that has been obtained thus far in various species exhibits a disheartening pattern: it works much better in shorter-lived species than in longer-lived ones. Nematodes, as I mentioned above, can live several times as long as normal if starved at the right point in their development; so can fruit flies. Mice and rats, however, can only be pushed to live about 40 percent longer than normal. This pattern led me, a few years ago, to wonder whether humans might even be less responsive than that, and I quickly realized that there is indeed a simple evolutionary reason to expect just such a thing.[12] It's a consequence of the fact that the duration of a famine is determined by the environment and is independent of the natural rate of aging of the species experiencing it.

Second, the adjustment of metabolism that organisms undergo when food is scarce causes only a slowdown in the accumulation of molecular and cellular damage, not a repair of damage that already happened. I've already told you that the key "Eureka moment" in my development of SENS was when I realized that repairing the damage of aging (before it progresses into disease) might be simpler than preventing it—but even setting that realization aside, repair is bound to be preferable, simply because any feasible therapy (whether to repair damage or to prevent it) will be only *partial.* That's to say, repair therapies will repair some but not all damage, and prevention therapies will slow but not halt the accumulation of damage. Why does this mean that repair is preferable? The logic is quite simple. In broad terms, if you take a middle-aged person and halve the rate of their subsequent aging, you'll double their remaining lifespan, but that might mean adding only 20 percent to their *total* lifespan. By contrast, if you take that same person at the same age and apply a therapy that halves their *accumulated* damage, and apply that same therapy periodically for the rest of their life, you'll roughly double their *total* lifespan (because their accumulating damage will only consist of the types of damage that your therapy can't repair), which means increasing their *remaining* lifespan (from the point when you first applied the therapy) by a factor of maybe four or five! So prevention-oriented approaches simply don't aim high enough.

But there's a third reason why I don't think nutrient sensing is the most promising target for biomedical intervention in aging, and I would say it's the most decisive. The reason it's been so incredibly easy to extend the lifespans of so many organisms by this one trick is because it's an evolved response to environmental conditions. The machinery that mediates that response is fantastically complex and poorly understood, just like the rest of our biology, but we can manipulate it easily despite that complexity, because its initial step—the sensing of nutrient availability—is simple. Just as you don't need to understand how your computer works to turn it on and off, we also don't need to understand the process of how nutrient deprivation is translated into the adjustment of masses of interacting metabolic pathways in order to turn that process on and off. But therein lies the showstopping problem. You may not need to understand how your computer works in order to turn it on and off, but in order to make it do things that it does not already contain the hardware and software to do, you have to understand a lot more. And if the new functionality requires software that hasn't yet been written or can't be installed, you have to understand a huge amount more, enough to write that software yourself. The human body is, in that sense, like a computer into which new software can't be installed—it's very versatile, but that versatility cannot be extended by the same methods that merely elicit the existing versatility. Therefore, we can be sure that there is a fixed degree of life extension that can be achieved by manipulating the nutrient sensing pathway—whether by calorie restriction (CR) itself, or by drugs that trick the body into thinking it's being starved, or by genetic changes that flip the same switch. As I explained a couple of paragraphs ago, I think that ceiling is very modest, maybe only a two-to-three-year extension; some of my colleagues think it may be as much as twenty to thirty years—but it's still a ceiling. We will never be able to exceed that fixed degree of life extension by such means, however hard we try.

Not Good Enough—But Better than Nothing

I want to end this chapter on a positive note, though. Even though nutrient sensing can only extend life by a fixed maximum amount, and even though it may be a rather small amount, that's still better than nothing! Also, there's a very general finding in laboratory life-extension experiments that animals with some kind of mildly life-shortening genetic problem benefit more from the therapy or regime than congenitally longer-lived individuals. That's quite likely to apply to calorie restriction (CR) in humans, too—which means that doing CR (or taking safe CR-mimicking drugs, as and when they appear) may be a good insurance policy against unknown congenital vulnerabilities. For these reasons, I strongly support the work that many of my colleagues in biogerontology are doing to squeeze the most we can out of that route to life extension.

But in closing, I want to bring you back firmly to the theme of this chapter. Once upon a time, aging was a truly mysterious phenomenon, but that time has passed. We can now reason about the aging of the human body in just the same way, and with just the same confidence, as we can reason about the aging and decay of simple machines. We know why different organisms age at different rates, whether that be because of different genes or different environments. We know that our genes are our allies, not our foes, in our war against aging—that they exist to postpone aging, not to cause it, and we only age because those life-preserving genetic pathways are not comprehensive.

Now—can you still tell yourself, with a straight face, that aging is too mysterious to try to tackle? You may have just one straw to clutch at in your effort to perpetuate your pro-aging trance: you may be telling yourself that the devil is in the detail, detail that I have not yet provided. I'll be snipping that straw in Chapter 4.

4

Engineering Rejuvenation

Let's briefly review what I've told you so far about aging. In a nutshell, it's as follows:

- Aging is really bad for us, however much we like to forget the fact.
- Aging is not a mystery, and we can already postpone it a lot in the lab.
- However, the techniques that have been so successful in the lab do not seem promising for humans.

In this chapter I'm going to expand upon the "Eureka moment" that I related in Chapter 1. I will cover—in still broad, but somewhat more detailed, terms—what each type of damage really is at the molecular and/or cellular level, and also the broad strokes of how I think we can address that damage.

A Caveat: Why Prevention Is Usually Better than Cure

In Chapter 3, I told you two heartening things about combating aging: firstly that in principle it's no different than combating the aging of man-made machines such as cars, and secondly that we've already discovered how to postpone aging by a large factor in the laboratory. However, I then explained that the second of these tidings of good cheer is actually going to be of only very limited biomedical utility. Well, brace yourself, because I'm about to explain that the first piece of good news is not so simple as it seemed, either.

I'll start with a rather more sobering thought about cars. Why are so *few* of them maintained to an age far beyond that for which they were designed, even though we all know they can be?

There are two answers, one reassuringly inapplicable to the analogy with human aging but the other very applicable indeed. The inapplicable answer is: because their owners have the option of getting a new car. All this says is that the chance that you will put in the effort and money to maintain an old and declining machine depends on how much you are in love with it. You may generally choose to junk your car when it starts to malfunction because you're not very attached to it anyway, but if your mother starts to malfunction and the wherewithal exists (even at a hefty price) to repair her, it'll be a different matter.

The other answer is the problem: most people leave the serious maintenance of their car until it's too late. It's obvious that the more damage a machine sustains, the more work is needed to rectify that damage; but more than that, the technology needed to rectify it becomes more and more sophisticated. When a car is really on its last legs, restoring it to full working order requires major attention—replacement of a lot of parts, for example. And unlike the how-much-we-care argument above, in this case the situation is absolutely the same for the human body. The people who know this best are those who work not on the *biology* of aging but on the *medicine* of aging: geriatricians. Geriatricians try to help people whose aging has reached the point where physical or mental function is appreciably impaired. They do their best to apply existing medical technology to postpone the patient's further decline and eventual death. But, as they know and as you also know, it's a losing battle. The damage has already spun out of control: it's feeding back on itself to accelerate the occurrence of additional damage, and the types of damage that are occurring are becoming ever more numerous and varied. All the geriatrician can hope to deliver is a modest improvement of the quality of life of the patient's last years, and perhaps a few months' to a year's postponement of death. It's the age-old rule: Prevention is better than cure.

But Only Usually . . .

But let's not leave it there. There's one thing about geriatrics that it has over gerontology, and I hinted at it above: geriatricians use *existing* medical technology. Why can they do that, when gerontologists can't?

The answer, when you think about it, is simple: To fix a problem that already exists, you don't need to know how it arose. A car mechanic replacing a car component doesn't need to know what type of corrosion wore through a fuel line, or what size rock hit a windscreen; similarly, the geriatrician doesn't need to know anything about free radical chemistry or cholesterol metabolism in order to treat cardiovascular disease or diabetes. But by contrast, *preventing* corrosion or shattered windscreens involves careful analysis of the downstream side effects of salting roads and not clearing debris from the highway; in the same way, the gerontologist needs to know a great deal about extremely subtle and possibly hard-to-discover causal chains of events in order to put "prevention is better than cure" into practice.

So, here we have two alternative approaches to postponing aging, one preventative and one curative; I've explained a problem with each of those approaches that makes them unpromising ways forward; and finally I've pointed out that the problem that each approach has *is not shared by the other approach*—preventing aging is soon enough but too complex, curing the diseases of aging is simple enough but too late. Now, what does that say by way of a possible way forward?

Worst of Both Worlds, or Best?

Well, I'll tell you what it said to me, that early morning in California.

Discussion during the day's roundtable-style sessions had focused on the various theories of aging, and ways to prove or disprove them. This mostly meant running through the multiple metabolic pathways that might contribute to the development of aging damage. I had presented the case that the production of free radicals by *mitochondria*—the tiny "power plants" that extract energy from food and convert it into ATP, a form of energy usable directly by the cell—is at the root of much of the aging process. This was something that most of my colleagues suspected, but I had recently framed it in a novel way that reconciled some unexplained findings in the field. I had confidence in my model, as it was my main specialist area within gerontology at that time: a book-length treatment of it[1] had earned me my Ph.D. But more important to me, it suggested a biomedical solution to what I was convinced was a major cause of aging damage: With some complex but foreseeable gene therapy, the connection between mitochondrial free radicals and pathology could be severed, *without* the need to interfere with the mitochondria's normal energy-producing activity. (I'll say a little more about this below, and lots more in Chapter 6.)

I had come to the conclusion that, in the best case, my mitochondrial gene therapy proposal might (and I emphasize *might*) also slow the rate of aging in humans attributable to most *other* causes by about 50 percent. This would be a massive breakthrough, as it would lead to as good an extension of healthy life as the most severe calorie restriction (even under the CR optimists' scenario) but without CR's side effects. But I was far from sure about that estimate, and in the wee hours of that morning, alone in a hotel room, I was even less confident than usual, because I had spent all day being reminded of just how many things go wrong in an aging body. Many of these problems could be at least partly chalked up to the downstream effects of the insidious, age-related increase in *oxidative stress*—the imbalance between those substances in the body that tend to chemically "need" electrons and substances that chemically "want" to donate them. I believed my mitochondrial gene therapy proposal would nearly eliminate this rise with age, but I couldn't be sure of just how much the rest of the aging process would still go on without additional, targeted therapies—nor what those therapies might be.

The candidates were numerous:

- Inflammatory enzymes essential to the immune system could also oxidize cholesterol, particularly when there's a lot of it around, contributing to atherosclerotic plaques.
- Our bodies' reliance on carbohydrates as a source of fuel exposes us to the reactive chemistry of glucose, causing the "gumming up" (glycation) of cellular proteins.
- Beta-amyloid, an aggregating protein, forms the basis of the "senile plaques" in the brains of Alzheimer's patients. It is the result of abnormal chopping-up of a normal precursor protein in the brain.
- The process of cell division gradually shortens each successive cellular generation's *telomeres*—the protective caps on the DNA double helix that serve the same function as the plastic bits on the end of your shoelaces, preventing the chromosome from "fraying." (See Chapters 10 and 12 for more.)

- Mutations in the cell's genetic database occur when, in the process of creating needed copies of the DNA "instruction book" for the new cell, the body's DNA-replicating machinery makes "typos."
- Tapping into the pools of *stem cells* (the primordial, unspecialized cells that the body holds in reserve and causes to develop into particular cell types as needed to replace cells lost to injury or disease) gradually depletes what is, over a lifetime, a limited source of youthful cellular reinforcements.

The problem had me in its grip—and it wasn't just curiosity that was keeping sleep at bay. While many of my colleagues viewed biogerontology as a phenomenon to study for the sake of understanding it, I saw aging for the humanitarian crisis that it is, the toll of tens of thousands of dead every day ringing in my ears. Abandoning my first career in artificial intelligence research, I had committed my life not just to alleviating the worst of the morbidity and mortality of age-related disease, but to putting an end to the entire horror show. I'd dedicated myself to the "engineering of negligible senescence," as I had first termed the goal in my Ph.D. thesis—to the end of aging.

But my inner dialogue that morning was leading to frustration, and even some despair. Clearly, if real medical control of aging required correcting *all* of these potentially damaging metabolic processes individually, real progress in anti-aging medicine might be like fighting the Hydra: no matter how many heads you put down, more would spring up to take their place. Normal metabolism is such an intricate, finely balanced web of reactions that tweaking one sends perturbations throughout the entire network, usually creating new problems or negating the effect of the intervention by eliciting a counterbalancing metabolic adjustment. For example, chronic inflammation is a source of cellular damage. But if you interfere with inflammation, you might impair immune defenses against pathogens. Equally, free radicals—a by-product of your metabolism—cause oxidative stress and damage over time. But crank up antioxidants, to defend against free radicals, and you might help cancer cells to protect themselves against chemotherapy drugs.

This process of dynamic metabolic adjustment is seen in the aging process, in fact. There are a number of aging changes that, while they might have some pathological consequences, are not themselves forms of damage. Putting it another way: they do not actually accumulate in the body's cells and tissues; rather, they reprsent a shift in the equilibrium between creation and destruction of the molecules involved. It seemed likely to me that such changes, however harmful to the body's youthful functioning, were secondary to something else. This meant that identifying and correcting that "something else" would correct the maladaptive secondary change, rendering moot the question of its contribution to the aging process. For instance, the cell's ability to respond to many hormones and other signaling molecules tends to decline with age. But as we saw in Chapter 3, the logic of evolution seems to dictate that this decline isn't programmed into the body. It must therefore be secondary to some form of damage. Maybe the membranes of the cell lose their fluidity, impairing the ability of receptor molecules to change their shapes to pass on a signal. Maybe the machinery that creates those receptor molecules becomes impaired. Whatever it is, identifying the damage itself would narrow the field of things that directly caused such damage and that were thus at the root of aging.

And, come to think of it, there seemed to me to be far fewer kinds of damage than processes that cause damage—hosts of different mutagens and "pre-mutagenic" changes to DNA, for example, but only two types of mutations: chromosomal and mitochondrial.

Well, I mused, *that's a thought—just how many kinds of aging damage ARE there? And are there similarly promising fixes for the rest of them?*

There are mutations in our chromosomes, as I just mentioned; this sort of damage causes cancer. I didn't (at that time—but see Chapter 12) have any new proposals up my sleeve for that one; it was going to have to rely (for now, at least) on other people's ideas. But there was no shortage of such ideas: Cancer research is among the biggest fields in biomedicine.

What *other* problems could arise from nuclear mutations? It was widely assumed that they were a major cause of age-related cellular dysfunction, but I had been batting around a counterargument in my head for some time—an argument that made me pretty sure that mutations not relevant to cancer would be irrelevant to aging within a currently normal lifetime. Certainly, a noncancerous mutation in a single cell might make that lone cell dysfunctional, but could it really substantially impact the tissue as a whole? Clearly, if every cell in a tissue were misbehaving, a person would be in trouble—but that can't be the case. Why not? Well, if it were that easy for an average cell to pick up a mutation, then everyone would be riddled with cancer by the time they were adults, because it takes just one cancerous cell to be allowed to grow to make a life-threatening tumor. What that suggested was that nearly all cells are kept genetically intact into and beyond a person's forties, and that the great majority of cells continue to be so throughout the "normal" life span. In other words, in order to prevent us from dying of cancer before puberty, our DNA maintenance machinery has to be so good that mutations not relevant to cancer just don't happen often enough to matter. Better yet, the exact same logic seemed to work for what biogerontologist Robin Holliday had memorably termed "epimutations"—changes not to the DNA sequence itself, but to the structure of the individual bases or the proteins around which the double helix is normally wrapped. Epimutations can do great harm, because they change the rate at which genes are decoded into proteins, but epimutations can cause either cancer or other problems, just as bona fide mutations can, so the "cancer is a bigger problem than anything else" conclusion applies to them too. I'll tell you more about this line of reasoning in Chapter 12.

In addition to chromosomal mutations, there are mitochondrial mutations, which may be a major part of the problem caused by free radicals. (Mitochondria are the only cell components that contain their own DNA independently of our chromosomes.) *Luckily,* I thought, *I believe I already know a feasible solution to mitochondrial mutations.* My solution was totally unlike the problematic approaches that were being proposed by other researchers, and I felt that it was much more powerful. It didn't rely on souped-up antioxidant defenses, an idea which was still being pursued not only by vitamin salespeople but even by some biotech companies. (This despite the fact that biogerontology specialists had long ago concluded that antioxidants are a dead end after they had failed, again and again, to affect aging one whit.[2] A better demonstration that our ambivalence about aging is only skin deep is hard to find.) Free radicals are just too reactive to be effectively mopped up with vitamins, nor even with the novel free radical scavengers that were coming out of pharmaceutical labs around this time (with names like MnTBAP and EUK-134, synthetic versions of the antioxidant enzyme superoxide dismutase). Or if not too reactive, they might be too *necessary*—it had recently become clear that cleaning up too many free radicals would cause new headaches for the body. After millennia of exposure to their reactive chemistry, evolution has learned to harness free radicals as signaling molecules,[3] so a heavy-handed repression of the cell's exposure to them would actually harm cellular metabolism, not aid it. The body might even react to antioxidant supplements by reining in its natural antioxidant defenses to compensate.

Trying to reduce free radical *production* was a job that many of my colleagues considered to be the best way to slow down aging damage, but (for the reason just given) actually pulling it off without seriously impairing the organism's ability to carry on with life's many duties would be extremely tricky. Not only that, most free radicals are produced in the mitochondria in the process of making ATP from food energy, and trying to mess around with that central feature of metabolism is bound to create side effects.

As I alluded to a few paragraphs ago, I had already proposed avoiding these problematic approaches by a strategy I'll cover in detail in Chapter 6. Briefly, the idea is to let metabolism proceed as normal—accepting that some free radicals will be generated and some biomolecules damaged—but to sever the link between free radicals and oxidative stress at its nexus. In my Ph.D. thesis, I had argued that (contrary to the prevailing view at that time) mitochondrial free radicals do not drive a systemwide rise in oxidative stress with age by damaging the rest of the cell *directly.* Instead, the damage that they cause to the mitochondrial DNA causes the mitochondria to enter a maladaptive state that spreads oxidative stress out *beyond* the cell. This, I had reasoned, meant that scientists could solve the problem of mitochondrial mutations by copying mitochondrial DNA from its vulnerable spot at "ground zero," within the free-radical generating mitochondria, into the bomb shelter of the cell nucleus, where damage to DNA occurs far less frequently. The proteins they encoded would have to be constructed in a manner that induced the cell to transport them to mitochondria, but the procedure to achieve that had been mostly understood for some time. In this scenario, the nuclear copies would act as a "backup" for the mitochondrial DNA: the mitochondria could operate as normal even if their DNA became damaged, so they would not cause long-term harm to the organism as a whole. Mitochondria would still suffer damage, but would not enter the maladaptive state I just mentioned, so they would not cause the creeping, destructive slide into oxidative stress in the rest of the body.

Okay, two down (chromosomal and mitochondrial mutations); how many to go? There is *glycation,* the warping of proteins by glucose. Well, that seemed relatively easy, because it was well known in the field that a biotech startup called Alteon was already running clinical trials using a compound called ALT-711, which appeared to reverse the protein cross-linking that this process caused. While the effect was weak, it was significant: the compound had a limited ability to restore some of the flexibility that's lost to glycation with age in the heart and blood vessels, and also showed promise for kidney damage in diabetics. It was proof-of-principle that without interfering with glucose metabolism, you could allow the *formation* of protein cross-links but prevent the pathological results by *undoing* the damage after the fact. (This is an important and common theme, as you will see—don't interfere with the process, but rather repair or clean up the damage that has accumulated.) See Chapter 9 for lots more detail about the glycation problem.

What else? There are the various kinds of junk that accumulate outside the cell: beta-amyloid, the lesser-known transthyretin, and possibly other substances of the same general sort. Here again, recent studies in the private sector—this time by a Californian company named Elan—had shown that you could actively *remove* the problem, in this case by vaccinating mice against the amyloid plaque and letting their immune cells gobble the stuff up. The concept had shown such rapid success in the lab that it was already close to clinical trials. Will I be telling you more about this, later in the book? You bet—see Chapter 8.

We must also address the unwholesome goo that builds up *within* the cell, such as lipofuscin. I started to get quite excited at this point, because just a year previously, in Dresden in June 1999, I'd come up with a new proposal to eliminate such material, involving the identification and engineering of enzymes from soil bacteria. (This was a classic case of someone not immersed in their own experimental work being able to bring together ideas from very distant disciplines to form a new approach to an existing problem—a critical element of modern scientific progress, which has been sadly neglected in many areas of medicine and biology.) The concept of using soil bacteria to degrade long-lived organic material had been around for decades, but not in gerontology, or even in any biomedical field. Rather, it was a mainstay of environmental decontamination, where it is known as "bioremediation." No one in gerontology had really even heard of it, let alone seen its biomedical potential. If you're intrigued, well, you only have to wait until Chapter 7.

Another item that must be added to the list is cellular *senescence,* the "aging" of individual cells. Senescence, in this meaning of the word, is a state of arrested growth in which the cell produces chemical signals dangerous to their neighbors. In theory, at least, there are all kinds of ways to deal with cellular senescence, though I wasn't sure which of them would ultimately pan out. Senescent cells express distinctive marker proteins, which should allow them to be targeted for selective destruction. Alternatively, once researchers tease out the damage or gene expression shifts that keep cells locked in this abnormal, arrested state, it might be possible to restore senescent cells to normal functionality. This was all still speculative, of course, but Judy Campisi at Berkeley and others were already hot on the trail. Chapter 11 will reveal all.

There's also the *depletion* of cells—of nondividing cells like neurons or heart cells, which are not naturally replaced when they die, and also the more paradoxical depletion of stem cell pools essential to healing and maintenance of tissue. Anyone who'd read a newspaper in the previous few years knew that scientists were hotly pursuing a way to deal with the age-related loss of cells, including stem cells: more stem cells, cultured in the lab and delivered as a rejuvenating cellular therapy. There were several viable approaches to this, different ones probably suited to different conditions. One was extracting the adult stem cells already in the patient, growing more of them, and then reinfusing them into the patient. Another was harvesting some of the more versatile embryonic stem cells that were already sitting, waiting to be thrown out as medical waste, in fertility clinics all across the world. The most complex was "nuclear transfer," in which a person's old, specialized cells could be transformed into young, versatile stem cells again via the environment of a woman's egg and a quick jolt of electricity. Researchers were already showing in animal models that these cells could be used to cure age-related diseases and trauma, and there was every reason to expect that the same techniques, once perfected, could be used to replace cells lost to age-related decay.

What else? Er . . .

I couldn't think of any more categories of damage! Try as I might, I really couldn't. There were a couple of other examples of molecular changes that accumulated throughout life, but I had reasons to believe that they were in the same boat as non-cancer-causing chromosomal mutations: they might be harmful if we lived hundreds of years, but they very probably weren't harmful in a normal lifetime. Other than that, everything I had learnt about during my five years of study and conference-hopping seemed to be covered.[4]

I stepped back for a moment and articulated the logic I had been developing those past few hours. At root, I was addressing a simple question: if geriatrics fails because prevention is better than cure, and gerontology fails because our understanding of metabolism is so limited, then might an intermediate target be the best of both worlds? Might it be possible to repair damage after it's been laid down (hence avoiding the need to understand the details of how it's laid down) but before it spirals out of control (hence also avoiding the losing battle that is geriatrics)? See **Figure 1**.

I could only answer this question in the affirmative if I could make a specific, extremely bold claim: that these intermediates, these proximate side effects of metabolism that accumulate in the body throughout life,

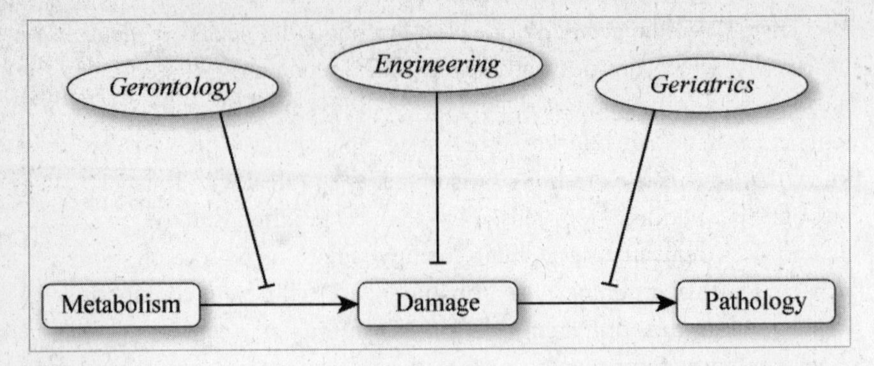

Figure 1. The "engineering approach" that I conceived in June 2000, as an intermediate, best-of-both-worlds alternative to gerontology and geriatrics as a strategy to combat aging.

could *all* be either (a) ruled out of relevance to late-age pathology (as I felt I could do for mutations that don't cause cancer) or (b) repaired or made harmless by foreseeable therapies. If some could be repaired, and some were definitely harmless or could be made so, but some fell into neither such category, the idea would fail. Like any machine, the body is only as robust as its weakest link, so partial maintenance will have little or no effect on longevity.

But I went over my list again and again, and as I did so I became ever surer that there was no clear-cut exception. The combination of my own idea for eliminating intracellular garbage, the idea I'd been championing for a few years for making mitochondrial mutations harmless, and the various other therapies being worked on around the world to address glycation, amyloid accumulation, cell loss, senescent cells, and cancer . . . that was really and truly an exhaustive list. **Figure 2** shows my enumeration of the problems and solutions that constitute the SENS (Strategies for Engineered Negligible Senescence) plan as it stands today.

As I mentioned above, there may well be other problems that will emerge if we succeed in solving all of these and thereby live a great deal longer. I felt, however, that my list might very well be comprehensive enough to give a few decades of extra life to people who are already in middle age before we start the treatments. And that was certainly a much more promising first step than anything that my colleagues had reviewed the previous day or in the many conferences and articles that I'd devoured over the previous few years.

The California sun was rising, and with it my spirits. It was clear that

Damage	Could be fixed or made harmless by	For details see chapter
Cell loss, cell atrophy	Cell therapy, mainly	11
Junk outside cells	Phagocytosis by immune stimulation	8
Crosslinks outside cells	AGE-breaking molecules/enzymes	9
Death-resistant cells	Suicide genes, immune stimulation	10
Mitochondrial mutations	Allotopic expression of 13 proteins	5.6
Junk inside cells	Transgenic microbial enzymes	7
Nuclear [epi] mutations (only cancer matters)	Telomerase/ALT gene deletion plus periodic stem cell reseeding	12

Figure 2. The seven parts of SENS.

daunting technical hurdles would need to be overcome if the therapies I envisaged were to start saving lives in the real world. But even so, I recognized that the line of thought I'd followed had the potential to paint the broad strokes of a revolution in biogerontology—and hopefully, in due course, in the future of human life. Repairing (or, in the case of mitochondrial mutations, obviating) accumulating damage was a genuine best-of-both worlds middle ground between the traditional gerontology and geriatrics approaches. It focused on a weak link in the chain of events leading from metabolism to pathology: it was early enough in that chain to avoid the downward spiral that doomed geriatrics to be forever a losing battle, but yet it was late enough in the chain to avoid the perturbation of metabolism that doomed the "over-preemptive" gerontology approach.

This idea would be easily grasped by my former colleagues in the computing field, or indeed by most engineers. In engineering, it's *routine* to design technologies before a full theoretical understanding of the underlying physics is achieved. Engineers were making workable use of electricity, superconducting magnets, and even nuclear energy (in the form of weapons) long before they had a coherent theoretical explanation for the forces they were manipulating. Even in medicine, the effective use of treatments had historically often long preceded our mechanistic understanding of them. Salicylates from willow bark have been used as anti-inflammatory treatments for centuries, and Bayer chemist Felix Hoffmann was even able to modify these natural compounds to make them more palatable and less prone to upset the stomach, yet the molecular basis for the action of the new wonder drug (aspirin) would not be understood for seven decades. Of course, even more effective drugs can often be designed from the ground up once the key enzymes and genes upon which they might act are sequenced—but that level of detail was *not* needed to *get started* in developing effective medicines.

Making this reorientation was dizzying—but once you accepted it, I realized, the whole project suddenly became tractable, and the way forward clear. You could stop thinking of aging as a hopelessly complex *theoretical* problem to solve, and get on with attacking it head-on, as an *engineering challenge* that needed to be overcome. "Engineered negligible senescence," a phrase that I'd previously used offhandedly, suddenly presented itself as the most precise description possible of the task ahead.[5]

In fact, I realized, the problem might even be thought of in terms of the way we prevent "aging" in other physical structures, such as houses or cars. As I discussed in Chapter 2, evolution's priorities for nearly all organisms stop them from living indefinitely without aging: mutations in the genes involved would not be removed by selection when the ageless organism was eaten by predators or otherwise succumbed in just a tiny fraction of "forever." This is much like cars, which are designed to meet the opposing priorities of durability and low cost at some midway point that is acceptable to the consumer. Thus, our bodies—like our other vehicles—are designed to survive for a biological "warranty period": they are given enough robustness and self-repair capability to function at peak performance for as long as they can reasonably be expected to stay alive in the wild, but no longer.

But of course, individual users of cars or of bodies may have very different priorities from those of Detroit or of our "selfish genes." If you want a car to last much longer than manufacturers of cheap cars typically intend, you have two options. One is to pick up a better model in the first place: buy yourself a Volvo instead of a Chevy Cavalier. This is all well and good for cars, but it isn't an option for those of us who have only the genes we were born with. And of course, even Volvos will still eventually break down, only a few years after a more economical product would do.

That's why, when we want to keep a car on the road for an exceptionally long time, we actually choose the other option: we *fix damage as it hap-*

pens. Whether it's a poor laborer keeping his ancient VW bug running because it's the only car that he'll ever be able to afford, or a wealthy collector maintaining an old MG for the sheer love of it, we all know that a car can be kept going more or less indefinitely with sufficient maintenance. We don't have to keep the cars off the road in climate-controlled garages, and we don't rely on the latest gasoline additive: we simply repair worn-out parts when they begin to fail. As I saw then, and as I will describe in the chapters ahead, the analogy to humans (at the cell, tissue, and organ level) is strikingly exact.

The Devil Is in the Detail

At the end of Chapter 3, I explained that the purpose of this chapter would be to remove people's last hope for maintaining their pro-aging trance: the belief that my breathlessness about the recent progression of aging from mysterious to manipulable might be all talk and no substance. I hope I've done that—but I've come up against the pro-aging trance often enough to know it's sometimes very hard to break.

That's why, for most of the rest of this book, I'll be wading deep into the fine scientific detail of the seven SENS categories and the remedies for them. I know that most readers of this book will not be scientists, so this may be intimidating. But Michael Rae and I have worked hard in Part 2 to present cutting-edge science in a manner comprehensible to any educated layman who's willing to put in the time to read it carefully. I therefore urge you to dive in and learn in detail about the types of damage that comprise aging and the foreseeable technologies that will, I am confident, enable us to repair or obviate that damage comprehensively enough to avoid age-related physical and mental decline indefinitely.

Part Two

5

Meltdown of the Cellular Power Plants

The cellular components—"organelles"—known as mitochondria play a large role in aging, hand in hand with reactive chemicals known as free radicals. When I came into the field in the mid-1990s, however, this role was not clearly defined; evidence and interpretations were contradictory and awaiting synthesis. In response, I developed what is now a widely accepted mitochondrial free radical theory of aging. Read on: In order to understand aging, you must understand a little of the way in which our cells work.

In Chapter 4, I explained that there are seven major classes of lifelong, accumulating "damage" that we must address if are to uncouple their causes—the processes of life—from their eventual consequences, the pathology of aging, and thereby prevent those consequences. Six of those seven are the subject of one chapter each in this part of the book—Chapters 7 through 12. But the first one I'm going to address, mitochondrial mutations, is going to take two chapters. That's because the question of whether mitochondrial mutations matter *at all* in aging is actually a really complicated one, and we must do our best to answer it in order to know whether we need to worry about them. Remember that in Chapter 4, I briefly noted that SENS does not incorporate a plan to address mutations in our cell nucleus at all unless they cause cancer, because non-cancer-causing mutations accumulate too slowly to matter in a normal lifetime. (I will explain this logic much more thoroughly in Chapter 12.) A lot of gerontologists feel that way about mitochondrial mutations. I disagree with them, so I need to tell you why. For each of the other six SENS damage categories, by contrast, there's no argument: at least one of the major pathologies of aging is *clearly* caused or accelerated by that type of damage. So those six categories will only require one chapter each, focusing mainly on the solution and with a relatively brief description of why there's a problem to solve.

Free Radicals: A Brief Primer

Almost everyone has heard of *free radicals* by now. Their involvement in aging is asserted so often and so confidently in popular press articles—especially articles trying to promote the latest "antioxidant" nutritional supplement—that you'd think the matter was done and dusted. As we'll see, however, the exact roles played by free radicals in the aging process—and the best ways to deal with the problems they cause—are a lot more complicated, and more controversial, than these articles let on.

Free radicals in biology are, for the most part,[1] oxygen-based molecules that are missing one of the *electrons* in their normal complement. Electrons are charged particles that surround the central nucleus of the atom, and they occupy well-defined locations (you can think of these as distances from the nucleus) termed *orbitals*. By their nature, molecules can only remain chemically stable when each of the electrons in the orbitals of their constituent atoms has a paired twin to complement it; an orbital with only one electron is unstable. So when a molecule *loses* one half of an electron twosome, it becomes chemically reactive until it gets that electron back. Usually, the free radical's stability is restored when it tears an electron out of the nearest available normal, balanced molecule—but with this electron stolen from it, that second molecule generally loses *its* chemical stability, and will seek in turn to restore its balance by a similar theft. It's a chain reaction.

Some unusual molecules—antioxidants—finesse this logic and are relatively stable even when they contain an unpaired electron. These molecules can "quench" free radical chain reactions. Until they do, however, free radicals will tear their way through your body like biochemical vandals, trashing whatever essential biomolecule they bump into: the structural proteins that make up your tissues, the fatty membranes that compartmentalize and facilitate your cells' various specialized functions, the DNA code that holds the blueprints of the enzymes and proteins required by the cell, and so on. In biology, function follows structure, so the ability of these molecules to support metabolism and hold you together is impaired when they are chemically deformed by this process.

This is obviously not good for you—and unfortunately, it's unavoidable. Free radicals are part of being alive.

While popular press articles on aging often give the impression that free radicals come mostly from environmental pollutants or toxins from an impure diet, the fact is that the overwhelming majority of the free radicals to which your body is exposed are generated in your very own cells—in the *mitochondria,* our cellular "power plants." Mitochondria are one of several types of "organelle," or self-contained cellular component that exists outside the nucleus. Each cell has hundreds to thousands of mitochondria. Man-made power plants take energy that is locked away in an inconvenient form of fuel—such as coal, natural gas, the strong nuclear force which holds atoms together, or wind—and convert it into a more convenient medium, electricity, which you can use to run your blender or computer. In just the same way, mitochondria convert a difficult-to-use energy source (the chemical energy locked up in the glucose and other molecules in your food) into a more convenient one: *adenosine triphosphate,* or ATP, the "universal energy currency" that your cells use to drive the essential biochemical reactions that keep you alive.

Mitochondria generate most of their cellular power using almost identical principles to the ones used by hydroelectric dams—right down to the turbines (see **Figure 1**). Using a series of preliminary biochemical reactions (each of which generates a small amount of energy), energy from food in the form of electrons is transferred to a carrier molecule called *NAD^+* (and a similar one called *FAD*). These electrons are used to run a series of "pumps" called the *electron transport chain* (ETC) that fill up a "reservoir" of *protons* held back by a mitochondrial "dam" (the *mitochondrial inner membrane*).

The buildup of protons behind the "dam" creates an electrochemical force that sends them "downhill" to the other side of the mitochondrial inner membrane, just as water behind a dam is drawn downward by gravity. And just as a hydroelectric dam exploits the flow of water to run a turbine, the inner membrane contains a quite literal turbine of its own called "Complex V" (or the "F_o/F_1 ATP synthase") that is driven by the flow of protons. The rushing of protons through the Complex V turbine causes it to spin, and this motion is harnessed to the addition of phosphate ions

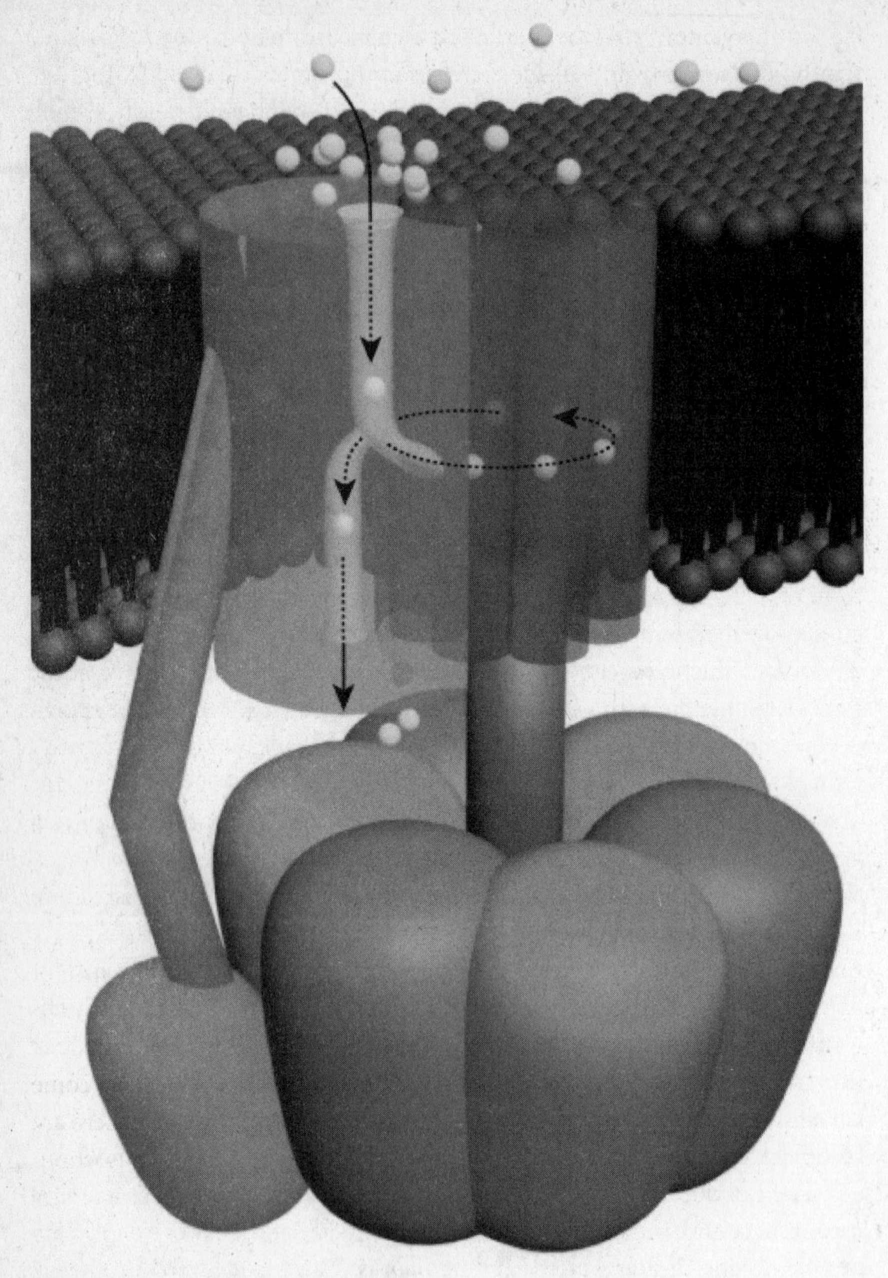

Figure 1. The F_o/F_1 ATP synthase.

("phosphorylation") to a carrier molecule (*adenosine diphosphate*, or ADP), transforming it to ATP.

Unlike hydroelectric dams, however, the use of chemical energy in food to generate ATP via this system is a *chemical* reaction. As with the burning of coal or wood to release energy, the powering-up of ADP into ATP consumes *oxygen,* which is why we have to breathe to keep the whole system running: oxygen is the final resting place of all those electrons that are released from food and channelled through the proton-pumping electron transport chain. Thus, the whole cycle is called *oxidative phosphorylation* (OXPHOS).

But while hydroelectric dams are (for the most part) environmentally benign, mitochondria are in one key aspect more like conventional power sources. Just like coal or nuclear power plants, mitochondria create toxic wastes during the conversion of energy from one form into another. As the proton-pumping complexes of the electron transport chain pass electrons from one to the next, they occasionally "fumble" an electron here or there. When this happens, the electron usually gets taken up by an oxygen molecule, which suddenly finds itself with an extra, unbalanced electron. (I just mentioned that oxygen is also the sink for the electrons that are *not* fumbled—that are properly processed by the mitochondria—but that process loads *four* electrons onto each oxygen molecule, not just one, so there's no problem of electron imbalance.) Adding *one* electron, by contrast, transforms benevolent oxygen into a particularly important free radical, *superoxide.* With your mitochondria generating ATP day and night continually, the ongoing formation of superoxide is like having a constant stream of low-grade nuclear waste leaking out of your local reactor.

Once scientists established that mitochondria were the main *source* of free radicals in the body, it was quite quickly realized that these organelles would also be their main *target*. Free radicals are so rabidly reactive that they never travel far, attacking instead the first thing that they come across—and the mitochondria themselves are at ground zero. And there are plenty of potentially sensitive targets for these radicals in the mitochondrion. Free radicals produced in the mitochondria are right next to the very membranes and proteins on which ATP production depends, and also within spitting distance of the *mitochondrial DNA.* What's that, you say? Well, whereas other components of the cell have all their proteins coded for them by the cell's centralized genetic repository in the nucleus, mitochondria have their *own* DNA for thirteen of the proton-pumping, ATP-generating proteins in their membranes.[2] If that DNA is significantly damaged, the mitochondrial machinery will go awry. Unfortunately, it's clear that mitochondrial DNA does suffer a lot of self-inflicted damage, taking as much as a hundredfold more initial oxidative "hits" than the cell's central, nuclear DNA, and suffering many times more actual, enduring *mutations* with age.

Starting with a classic paper put out in 1972 by chemist Denham Harman[3] (who already had the distinction of being the father of the original "free radical theory of aging"), researchers put these facts together with a range of experimental findings and came up with several variations of a "*mitochondrial* free radical theory of aging." Let's briefly survey that experimental evidence.[4]

First of all, there was evidence from comparative biology. Slower-aging organisms, relative to faster-aging ones of similar size and body temperature, are always found to have slower mitochondrial free radical damage accumulation. They produce fewer free radicals in their mitochondria; they have mitochondrial membranes that are less susceptible to free radical damage; and sure enough, they accumulate less damage to their mitochondrial DNA. Calorie restriction (CR)—the only nongenetic intervention known to slow down aging in mammals—improves all of these parameters: it lowers the generation of mitochondrial free radicals, toughens their membranes against the free radical assault, and above all it reduces the age-related accumulation of mitochondrial DNA mutations—the irreparable removal or overwriting of "letters" in the genetic instruction book.

Leading toward the same conclusion from the opposite direction, CR *does* slow down aging, yet it has no consistent effect on the levels of most self-produced antioxidant enzymes. The enzymes that people examined in this regard in the 1980s were ones that are found predominantly in the rest of the cell, not in the mitochondria. This again suggests that free radical damage outside of the mitochondria is not a directly important cause of aging, since aging can in fact be slowed down (via CR) without doing the one thing that would most directly interdict that damage.

Fast-forward, for a moment, to 2005. In that year, the most direct evidence so far on this point came to light. It involved mice that had been given genes allowing them to produce extra amounts of an antioxidant enzyme (catalase), specifically targeted to different parts of their bodies.[5,6] There was little to no benefit provided by giving these organisms catalase to protect their nuclear DNA—the genetic instructions that build the entire cell and determine its metabolic activity, except for those parts that the mitochondria code for themselves. And there was also no benefit observed from targeting catalase to organelles called *peroxisomes,* which are involved in processes that produce hydrogen peroxide (the molecule that catalase detoxifies) and accordingly are already stoked up with the enzyme. Yet, delivering catalase to the animals' mitochondria, which significantly reduced the development of mitochondrial DNA deletions, extended their *maximum* lifespan by about 20 percent—the first unambiguous case of a genetic intervention with an effect on this key sign of aging in mammals.

These mice were no more than a twinkle in their creators' eyes a decade ago, when I first addressed the question of mitochondrial oxidative damage. But even back then, it seemed unassailable that free radical damage to the mitochondria was a key driver of aging. The question was: what linked the one to the other?

That might sound like a stupid question, given that free radicals are obviously toxic, but it turned out to be decidedly tricky to come up with a coherent, detailed, mechanistic explanation for the connection. Scientists convinced that mitochondrial free radicals play a role in aging all begin with the undeniable observation that free radicals spewing out from the mitochondrion damage the membranes and proteins that it needs to generate ATP, and also cause mutations in the mitochondrial DNA that codes for some of those same proteins. But any such theory must explain how this self-inflicted damage contributes to the *progressive, systemic* decay that constitutes biological aging. Until recently, nearly all such theories postulated the existence of some form of mitochondrial "vicious cycle" of self-accelerating free radical production and bioenergetic decay.[7,8]

In these highly intuitive schemes, the mitochondrion is pictured to be like a hydroelectric dam whose turbines become rusted, worn, or broken by the forces to which they are subjected every day. Thanks to free radical damage to their membranes, proteins, and DNA, mitochondria in cells throughout the body would become progressively less able to pump protons and to keep a lid on the "hot potato" chain of ATP synthesis with age, leading to inefficient energy generation and increased production of free radicals as more and more electrons escape from increasingly banged-up transport complexes. Thus would begin the "vicious cycle," as more and more renegade electrons tore into more and more mitochondrial constituents, leading to further damage to those same constituents, causing yet more inefficient, dirty energy generation, and so on and so forth, spiralling downward and ultimately starving the cell of energy and rendering it a hazardous waste site. See **Figure 2.**

Some version of this scenario is repeated in nearly all popular books and articles about the role of mitochondria in aging, as well as most scientific

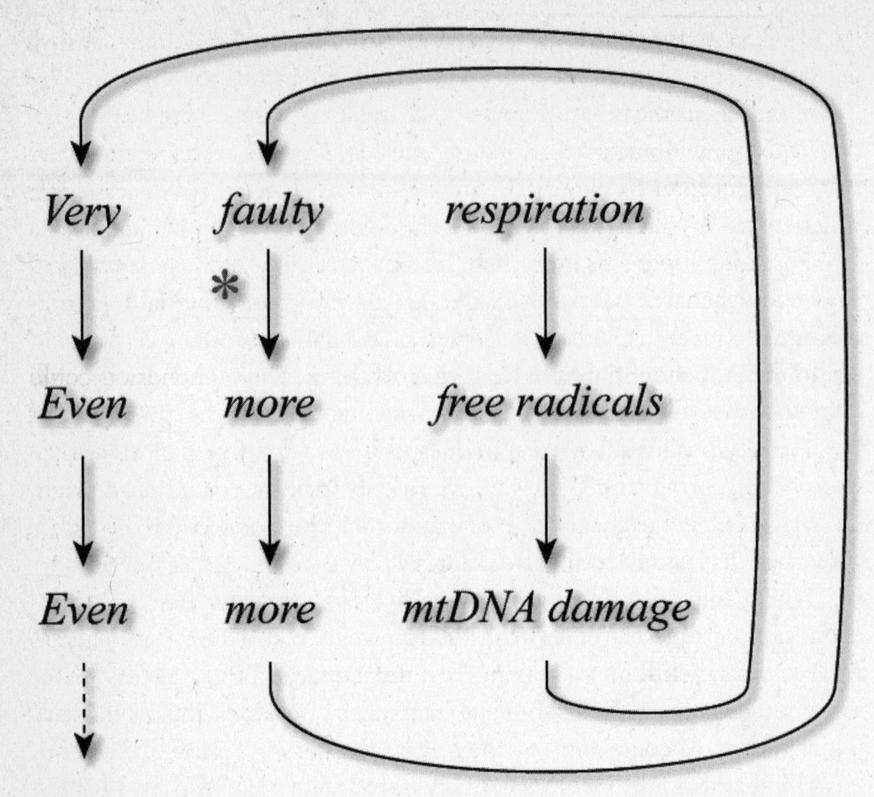

Figure 2. The "vicious cycle" theory of mitochondrial mutations accumulation. The key tenet of the theory is denoted by the asterisk: that typical mitochondrial mutations raise the rate of release of free radicals.

journal publications. Yet we've had evidence for nearly twenty years showing conclusively that it can't be right.

Everything You Know Is Wrong

The "vicious cycle" theory of mitochondrial decay paints a picture that sounds seductively plausible, but it just isn't compatible with the data. While many scientists remain oblivious to the glaring inconsistencies in these theories even today, a few mitochondrial specialists and biogerontologists have been pointing these problems out since the mid-1990s. So different were the predictions of "vicious cycle" theories from the experimental findings that many of these researchers went so far as to suggest that the findings simply ruled out a role for mitochondria free radicals in aging—forgetting that such a role might exist but *not* via a vicious-cycle mechanism.

The first problem with at least some versions of the vicious cycle theory had actually been pointed out decades earlier—just after Harman had originally put forward the first version of the mitochondrial free radical theory, in fact—by none other than Alex Comfort. (Yes, that's the same Alex Comfort who wrote *The Joy of Sex*. He was a true polymath: he was also a controversial anarchist agitator, a poet, and a highly distinguished biogerontologist.)

In 1974, Comfort pointed out that while each mitochondrion could *temporarily* suffer progressively increasing damage to the proteins and membranes that make up its ATP synthesis machinery from the ongoing free radical barrage, no theory based on the idea that this damage would get worse and worse with age could fly, for the simple reason that the cell is constantly replacing and renewing those very components.

Periodically, old mitochondria are tagged for destruction in yet another type of organelle, the *lysosome* (the cellular "toxic waste incinerator," about which I'll be talking a great deal in Chapter 7). Then, to make up for the loss of energy factories, the cell puts out a signal for the remaining mitochondria to replicate themselves. During replication, each mitochondrion duplicates its DNA, and then that essential core "grows" itself a new "body," including pristine new proton-pumping proteins and membranes. Whether you're five years old or fifty, any given mitochondrion in your cells contains membranes and proteins that are on average only a few weeks old. Thus, the newest and the oldest of these mitochondrial components are present in the same proportions in the very old as in the very young. It just can't be that aging is driven by a *progressive* process of degeneration in components that undergo a continuous process of renewal.

However, this objection wasn't necessarily a problem for the more popular versions of the vicious cycle theory: those that assert that free radical damage to the mitochondrial *DNA* drives aging. Although mitochondrial membranes and proteins are periodically replaced, all mitochondria inherit their DNA directly from their "parent" power plants, which faithfully copy out their DNA and pass it on to their "children"—and just as whole organisms pass on any mutations that they may harbor in their DNA to their offspring, so errors in "parent" mitochondria appear in the next mitochondrial generation. If the mitochondria that have damaged DNA are preferentially destroyed by lysosomes, the effect will be as before—damage will be removed as fast as it spreads—and that's what Comfort reasoned would be occurring. But if there's no such bias in which mitochondria are and are not destroyed, DNA damage will accumulate.

These theories received a *superficial* plausibility boost from studies conducted in the 1990s, which showed that aging bodies do, indeed, accumulate cells populated with mutant mitochondria. However, the same studies shot entirely new—and even more deadly—holes into mitochondrial DNA-based "vicious cycle" theories.

For one thing, it was found that all of the mutant mitochondria in a given cell contain *the same mutation.* This is exactly the opposite of what the "vicious cycle" would predict. If each mitochondrion individually decayed as a result of a self-accelerating cycle of oxidative "hits" to its DNA, then each one would display an unique, random profile of mutations. In the same way, if one day—independently—two disgruntled librarians were each to snap, going on automatic rifle rampages through the collections in their care, you would naturally expect to see that the bullets would have hit *different books,* even if the *collections themselves* were the same: one would have put a bullet straight through the spine of a copy of *Finnegans Wake,* while the other would have punched an off-center dot atop the "i" in *Bridget Jones' Diary,* and so on, at random, until each mad bibliophile ran out of bullets.

Instead, wherever cells that contain defective mitochondria are found, the mutants all harbor an *identical* DNA mutation. It's as if librarians across the state had marched into work and each fired his or her guns exactly once, in every case into copies of *The Catcher in the Rye.* Random mutations happening continuously in each mitochondrion cannot reasonably result in each of the damaged mitochondria in a cell containing the *same* error in their DNA; a random mutation-creating process can't explain the complete takeover of cells by such mitochondria, or the presence of other cells that contain nothing but healthy ones.

In fact, it's even weirder than that, because the presence of mutant mitochondria turns out to be an all-or-nothing affair. That is: near enough, not only do all of the mutant mitochondria in a given cell turn out to contain the exact same mutation, but those cells that harbor *any* damaged mitochondria are found to contain nothing *but* mutants—while the other cells have nothing but pristine, youthful "power plants."

But wait: the findings are even more bizarre yet. While each cell is full of mitochondria that all bear the same mutation, the mitochondria in different cells contain *different* mutations. It's as if all of the librarians in Delaware had pumped their local branch's copies of *War and Peace* full of lead, while their colleagues in California had simultaneously displayed an equally single-minded determination to purge their collections of *Lady Chatterly's Lover*. It's a case in which even the most skeptical conspiracy-theory debunker would be forced to admit that a "random act of violence" was not a credible explanation for the crime scene.

The *nature* of these mitochondrial mutations also proved to be inconsistent with vicious cycle theories of mitochondrial priority in aging. The assumption had been that damage to the DNA blueprints for mitochondrial proteins would most often result in minor defects in the instructions by which those proteins are coded. The ensuing proteins would be close enough to their proper structure to remain more-or-less capable of carrying on, but would be dysfunctional, "fumbling" more electrons into free radicals and generating less ATP. Instead, the mutations that accumulate in cells' mitochondria were found to be overwhelmingly *deletions* of large blocks of DNA, which *completely shut down* the creation of all the mitochondrially-encoded proteins.

The vicious cycle theory proposed that mitochondria would make progressively less ATP and more free radicals as they accumulated more and more of these defects in their DNA. Instead, it was found that mutant mitochondria produce essentially *no* free radicals, and that the change in each mitochondrion could be attributed to a single, catastrophic event, rather than to a "death from a thousand cuts."

Forget the Quality, Feel the Quantity

An additional finding seemed to leave no room for *any* way, vicious or otherwise, in which mitochondrial mutations could be involved in aging: very, very few cells actually contain mutant mitochondria at all. The vast majority of cells remain in perfect mitochondrial health well into old age. Yes, a tiny proportion of older people's cells—about 1 percent—could be shown to be completely taken over by mitochondria that all suffer from an identical defect in their DNA; but 99 percent of cells were fine. How could 1 percent of cells matter?

Many biogerontologists concluded that these findings put the kibosh on any theory that asserted that mitochondrial decay was important in aging. If nearly every cell in the body still enjoys the same level of ATP output as it had in its youth, and suffers no more free radical damage than it did in its prime, how could a tiny proportion of cells that are low on power, but whose mitochondria produce no more free radicals than their neighbors—in fact, produce *no free radicals at all*—possibly have much negative effect on the function of the tissue in which they reside or the organism as a whole? To these scientists, the mitochondrial free radical theory of aging seemed dead in the water.

This was where the field sat in the mid-1990s, when I first became aware of the unsatisfactory state of aging science and decided to try to do something to change the situation. When I saw the confused state of the mitochondrial free radical theory of aging, I felt that the field looked ripe for a new synthesis. On the one hand the evidence supporting the existence of a central role for mitochondrial free radicals in aging seemed strong; on the other hand, the standard vicious cycle theories simply could not be reconciled with the emerging results. It was into this fray that I stepped with my first formal scientific papers in 1997[9] and 1998,[10] and where I have made my most widely acknowledged contributions to biogerontology. My key insights were, firstly, an explanation of how aging cells accumulate mitochondria that share one mutation in common instead of the random set predicted by the vicious circle theory, and secondly, an explanation of how only a small number of cells taken over by such mutant mitochondria could drive aging in the body as a whole. Let's walk through these ideas one at a time.

SOS: Survival of the Slowest

The vicious cycle theory had assumed that each mitochondrion would slowly accumulate minor, random mutations over the course of its lifetime. The fact that, in those cells where there *were* mutant mitochondria, all the mutants shared the *same* mutation—and that the mutants had *completely replaced* all healthy mitochondria in the cell—proved that assumption wrong.

The only reasonable alternative seemed to be "clonal expansion": the idea that a single mitochondrion had originally gone bad, and that its progeny had slowly taken over the entire cell. Remember, mitochondria reproduce themselves by splitting themselves in half, much like amoebae: the original mitochondrion makes a copy of its DNA, and then forms two identical genetic "clones" of itself. This means that each clone will contain exact copies of any mutations present in the original organelle. So it seemed inescapable that the strange mitochondrial monocultures found in the all-mutant cells were the result of *one* mitochondrion initially acquiring a mutation, passing it on to its offspring, and then having its lineage somehow outcompete all of its neighbors until it eventually becomes the only game in town.

However, the idea that mitochondria with mutated DNA could somehow win a battle for dominance within the cell was itself a bit of a paradox. After all, these mitochondria are *defective,* with one or more enormous chunks blasted out of their DNA by free radicals or replication errors. While it's true that once in a blue moon a mutation turns out to be beneficial—this is, after all, what allows evolution to happen—it's supremely unlikely that it would happen over and over again, such that random mutations occurring in mitochondria in widely separated cells would turn out to be so beneficial to the mitochondrion as to give it a Darwinian "fitness advantage" over its fellows. And indeed, the mutations in question were *known* to be deleterious: they completely knock out the mitochondrion's ability to perform oxidative phosphorylation, and thus turn off the great majority of their contribution to the cellular ATP supply.

The "clonal expansion" explanation was also hard to reconcile with the fact that *many different* mutations can cause a specific mitochondrial lineage to replace all other alternative lineages in the cell. That is: while the mutant mitochondria in a given cell all contain the same, specific mutation, a second such cell often contains mitochondria that all harbor a *completely distinct* mutation from the one that was found in the first. So it wasn't that there is a single, specific mutation that gives the mutants their selective advantage over their neighbors: numerous mutations, independently arising in single mitochondria within widely separated cells, confer the same competitive edge. Was it really likely, I asked myself, that there were this many advantageous, yet unrelated, mutations to be had?

Yet, these various mutations do have one thing in common. They aren't mild mutations, damaging just one protein: all of them are of a type that prevent the synthesis of *all* thirteen of the proteins that the mitochondrial DNA encodes. This shared property, I felt, might be the key to how they managed to take over the cell.

I set myself to thinking what would distinguish such mitochondria from their healthy counterparts. They would not generate nearly so much ATP, of course: only the small amount of cellular energy that gets produced in the initial stages of extracting chemical energy from food, which was a fraction of the total that a functioning oxidative phosphorylation system could churn out. This would be unhealthy for the *cell,* certainly, but I realized that it would have little negative impact *on the mitochondrion itself,* which normally exported nearly all of the ATP that it produced anyway. So, while I couldn't see how this reduced energy output could explain the selective advantage enjoyed by the mutants, I saw that—contrary to what one might initially think—it really wasn't a direct disadvantage *relative to other mitochondria in the cell* that might hinder them from rising to dominance in the host.

The other thing that would set mitochondria with no oxidative phosphorylation capacity apart from other mitochondria in a cell seemed more likely to be advantageous: such mitochondria would no longer be producing free radicals. Remember that mitochondrial produce free radicals when electrons leak out of the regulated channels through which they pump protons into the "reservoir" that drives the "turbines" of the mitochondrial inner membrane. If you aren't feeding electrons into the pump system because the system itself is missing, then clearly there will be no leakage—and no free radicals.

Not having to deal with constant free radical vandalism sounded like it might be good for the mitochondrion—but it was not clear exactly how it might lead to an actual competitive advantage vis-à-vis the healthy mitochondria with which it was surrounded. True, its DNA would stop being bombarded—but of course, by this time it would *already* be suffering from a gaping hole in its DNA.

It was also obvious that the mitochondrion's inner membrane would no longer suffer free radical damage—but again, it didn't seem that this would matter in aging, since mitochondria are constantly having their membranes torn down and replaced anyway, either during replication or at the end of their brief individual lives, when mitochondria with defective membranes are sent off to the cellular "incinerator" in any case.

Now hang on a minute, I thought.

Like other researchers who had puzzled over this question, I'd been trying to imagine some improvement to the mitochondria's *function* conferred by the mutation—the equivalent, in the microscopic evolutionary struggle, to sharper teeth, faster running, or greater fecundity. But what if, instead, the important thing about the mutation was not that it made its carriers "better," but that it *prevented them from being destroyed?*

Recycling Cellular Trash

There's still a lot of work to be done to explain what exactly causes mitochondria to be sent to the cellular garbage disposal system. Still, even as early as Alex Comfort's criticism of the original mitochondrial free radical theory of aging, it was widely believed that there was some *selective* process that specifically targeted old, damaged organelles for destruction. This could not be taken for granted, however. It was long believed that some components of the cell are turned over by an ongoing process of *random* recycling, in which the lysosome (strictly, a special sort of pre-lysosome called an *autophagosome* or *autophagic vacuole*) simply lumbers about the cell, swallowing a given number of various cellular constituents at random each day, so that everything is ultimately turned over sooner or later.

It is now widely accepted that this isn't the way the lysosome works: the engulfment of proteins and other cellular components is known to be a highly directed process. In part, this is simply a matter of good use of scarce resources. Imagine if, in order to ensure that old, decaying vehicles were taken off the road (to improve the nation's overall air quality and greenhouse gas emissions, eliminate the eyesore of old cars rusting on blocks, and bring down the price of recycled steel), the government were to send its agents wandering aimlessly through economically depressed neighborhoods to randomly select cars to be sent off to the scrap heap. Such a program *would* achieve some of its goals, but it would hit too many fully functional vehicles to make it viable, even setting aside the question of individual property rights.

But in some cases there is an even more powerful reason than mere efficiency to be sure that *specific* organelles get sent to the scrap heap. Some cellular components can become actively toxic to the cell if they aren't quickly degraded once they've outlived their usefulness. Like the enchanted broom in *Fantasia* that continues to fill the vat in the sorcerer's hall with water until it overflows and floods the room, many proteins and organelles are only useful for a limited period; when their job is done, they must be "put away," and in the cell, this means torn apart for recycling. For instance, producing a pro-inflammatory enzyme can be critical to mobilizing an immune reaction against an invading pathogen, but leaving that enzyme to keep generating inflammation after the invader has been defeated would lead to a destructive, chronic inflammatory state with effects similar to those of autoimmune diseases like rheumatoid arthritis or lupus.

It then occurred to me that there were very good reasons for the cell to take care that its mitochondria were destroyed when their membranes had suffered free radical damage. Recall that the inner mitochondrial membrane acts as a "dam" to hold back the reservoir of protons that powers the energy-generating turbine of Complex V. Holes in that membrane would be "leaks" in the dam, depleting the reservoir as ions simply seeped through the holes without generating ATP. Evidence to confirm this basic scenario, in the form of leaks created from damaged membrane molecules, had actually been uncovered as early as the 1970s.[11]

This would make the "leaky" mitochondrion a serious drain on scarce resources, as the electron chain would continue to consume energy from food in a furious, futile attempt to refill the reservoir. Food-derived electrons would continue to be fed into the chain, which would use them to keep pumping protons across the membrane, but these ions would leak back across as fast as they were pumped "uphill," without building up the electrochemical reservoir needed to create usable energy for the cell. This would drain the cell of energy, turning nutrients not into ATP but instead into nothing more useful than heat.

Moreover, the damage to the inner membrane might also allow many of the smaller proteins of the mitochondrial inner space to be released out of the mitochondrion and into the main body of the cell. If they continued to be active, these components could well be toxic to the cell when released from the controlled environment of the mitochondrion.

It would make sense, then, for the cell to have a system in place that would ensure that mitochondria are hauled off to the lysosome for destruction when their membranes become damaged by their own wastes. This prediction seems to have been fulfilled with the recent discovery of a specific targeting protein that "tags" yeast mitochondria for lysosomal pickup.[12] We still don't know for sure what makes the cell decide *which* mitochondria to "tag," but it has now been shown that the formation of holes in the mitochondrial membrane does send a signal that increases the rate at which these organelles get sent to the scrapyard.[13]

I had no doubt that this was all to the good: I'm all for the removal of defective and potentially toxic components from the cell, and as usual nature has evolved an ingenious way to make sure that it happens. But I saw that, ironically, large deletions in the mitochondrial DNA would actually allow them to escape from the very mechanism that cells use to ensure that damaged mitochondria get slated for destruction. When mitochondria suffer the mutations that have been shown to accumulate with aging, they immediately cease performing oxidative phosphorylation (OXPHOS)—and with it, generating the resultant free radical waste. But reduced free radical *production,* in turn, should lead to less free radical *damage* to their membranes. Don't forget that the prevailing vicious cycle theory proposed that mitochondrial mutations proliferate by causing their host mitochondria to make *more* free radicals than nonmutant mitochondria do. That, I saw, was where the proponents of the vicious cycle theory had gone wrong.

Hiding Behind Clean Membranes

Having advanced this far, I immediately saw how the mutants gained their advantage over their healthy counterparts. Even perfectly functional mitochondria constantly produce a steady, low-level stream of free radicals, leading to membrane damage. Every few weeks, this damage builds up to a level at which the mitochondria are consigned to the rubbish tip, and then the cell sends out a signal for a new round of mitochondrial reproduction to replace the decommissioned "power plants."

But this process only weeds out mitochondria with damaged membranes—which will overwhelmingly be power plants whose DNA is still healthy enough to allow for the very electron transport that leads to the free radical damage to their membranes in the first place. Mitochondria with *intact* membranes, but *damaged* DNA, would not show outward signs of their internal injuries, and so would be passed over by the Angel of Death.

After a certain number of damaged mitochondria have been hauled off to the lysosome, the cell will send out the signal for mitochondria to replicate. Some or all of the remaining mitochondria—the genetically healthy and the mutants alike—will reproduce themselves, and because those mitochondria that bear large DNA deletions will almost always have survived the purge of outwardly damaged power plants, they will enjoy the opportunity to reproduce. But many of the membrane-damaged—but genetically intact!—mitochondria will already have been removed before replicating themselves. This will give the mutants a selective advantage over the nonmutants: every time replication happens, more and more of them will have survived a cull that has sent many of their genetically healthy competitors to the garbage disposal unit.

This is exactly how evolution works in organisms, of course. Animals that are slower runners, or less able to find food, or have poorer eyesight are more vulnerable to death from predators, exposure, or disease, which prevents them from successfully reproducing and passing on their genes. Meanwhile, a disproportionately high number of better-adapted organisms get the chance to breed, leaving behind progeny that carry their genetic legacy into the future. Over time, the genes that are best adapted to the specific threats in their environment come to dominate in the population.

In the cell, the threat to mitochondrial survival is the lysosome—a "predator," which is supposed to ensure that only mitochondria fit to safely support cellular energy production survive. What the mutant mitochondria evolve (yes, evolve) is, in effect, *camouflage* that masks them from the eagle eye of this predator. Thanks to their undamaged membranes, these highly dysfunctional mitochondria appear healthy to the cellular surveillance system. Like the proverbial pharisees, their outsides are clean—but inwardly, they are full of ravening and wickedness.

I concluded that this "camouflaged mutant" concept provided the first consistent, detailed explanation for the takeover of cells by defective mitochondria. I named it "Survival of the Slowest" (SOS), because it postulates that the *quiescent* ("slow") mitochondria enjoy a fitness advantage in the Darwinian fight for survival in the cellular jungle. See **Figure 3**.

But now, having explained *how* a small number of cells become a monoculture of defective mitochondria, there remained a question which might be thought to be more important—namely, *how does that tiny fraction of the body's cells drive aging across the entire body?* It wasn't long before I had a good explanation for that, too.

The "Reductive Hotspot Hypothesis"

The old vicious cycle models didn't need to invoke any additional mechanism to explain how mitochondrial mutation could contribute to aging, because they assumed that there would be an accumulation of increasingly defective mitochondria in *lots* of cells with age. As more and more of their mitochondria randomly went on the blitz, the cells would suffer more and more oxidative hits, and be more and more starved for ATP, as the power plants' efficiency sank with every new defect. It was a nice, common-sense explanation for the role of mutant mitochondria in the aging of the body.

But, as we've seen, it was also clearly wrong. Most cells in the body simply *do not* accumulate mutant mitochondria as we age: at most, about 1 percent of all of the trillions of cells in the body do so. Most cells and tissues suffer no decline in production or availability of ATP, and far from increasing free radical production, most of the mutant mitochondria that accumulate in the body produce *no free radicals at all,* because their radical-generating electron transport chains are simply absent.

It was hard to see how so few cells, containing mitochondrial mutants that were not harming nearby cells in any obvious way, could possibly be driving aging in the body. In fact, these findings were enough to make plenty of biogerontologists talk about the death of the mitochondrial free radical theory of aging. Yet, as we briefly discussed earlier in this chapter, the circumstantial evidence that mitochondrial mutations somehow contribute to

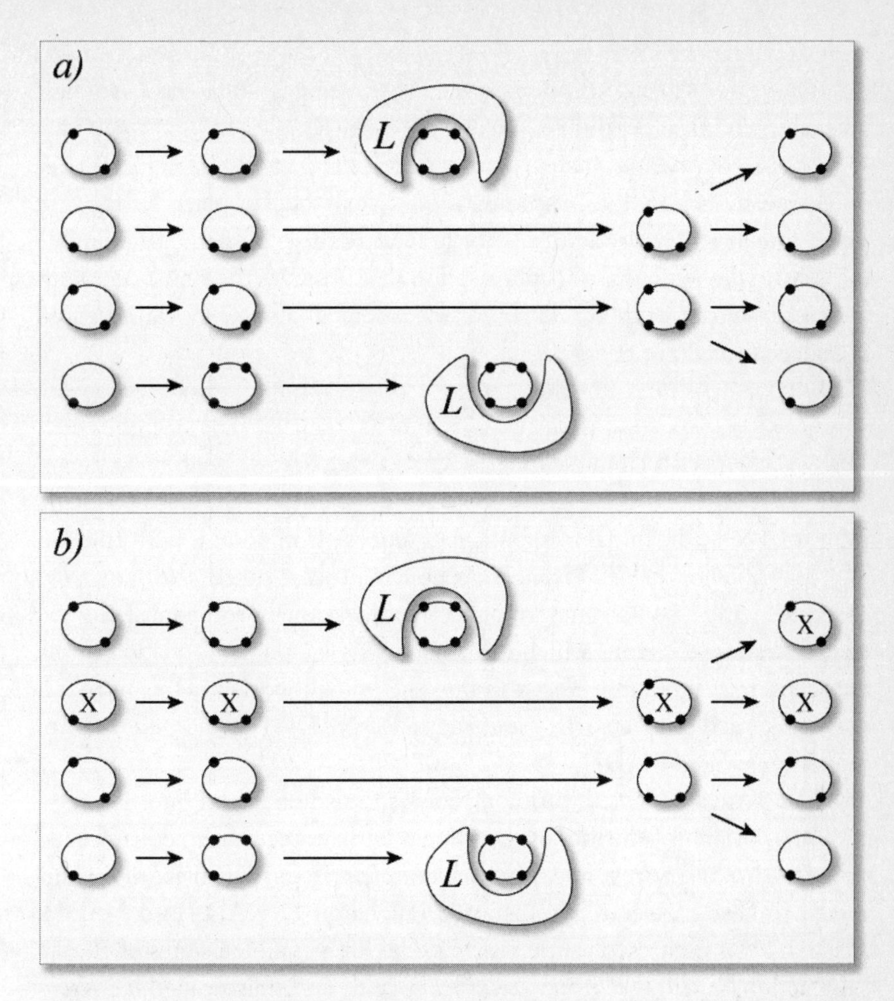

Figure 3. The "Survival of the Slowest" model for mitochondrial mutation accumulation. (a) The proposed normal mode of turnover and renewal of non-mutant mitochondria; spots denote membrane damage. (b) The clonal expansion of mutations (denoted by X) resulting from low free radical damage to membranes and slow lysosomal destruction.

aging is too strong to dismiss. To reconcile the two sets of data would require a truly novel solution to the puzzle.

I saw that any refined version of the mitochondrial free radical theory of aging would have to do two closely related things. First, since so few cells are taken over by these burned-out power plants, it would have to show that cells harboring mutant mitochondria somehow spread toxicity *beyond* their own borders. And second, it would have to explain the *nature* of that toxicity, since the usual suspect—free radicals—appeared to have been ruled out by the fact that the mitochondria in these cells would have their normal free radical production turned off at the source.

I began by trying to work out just what cells that had been taken over by mutant mitochondria were doing to survive in the first place. What were they using as an energy source? Not only were these mitochondria unable to perform the oxidative phosphorylation that provides their host cells with the great majority of their ATP, but it was not at all obvious how they could produce any cellular energy at all.

Upstream of a Blocked Dam

In normal cells, the initial metabolism of glucose from food is performed in the main body of the cell through a chemical process called *glycolysis*. Glycolysis generates a small amount of ATP, a breakdown product called *pyruvate,* and some electrons which can drive oxidative phosphorylation in the mitochondria. To shuttle electrons into the mitochondria for this purpose, they are loaded onto a carrier molecule called NAD^+. The charged-up form of NAD^+ is called NADH.

The pyruvate formed during glycolysis is also delivered into the mitochondria, where it is further broken down into another intermediate called *acetyl CoA.* This process releases some more electrons, which are again harvested for use in electron transport by "charging" NAD^+ into NADH. Acetyl CoA is then used as the raw material for a complex series of chemical reactions called the *tricarboxylic acid (TCA) cycle* (also called the *Krebs cycle* or *citric acid cycle*), which liberates many times more electrons (again leading to creation of NADH) than has been created in previous steps.

Finally, all of the NADH charged up via all these processes—glycolysis, the breakdown of pyruvate into acetyl CoA, and the TCA cycle—is delivered to the electron transport chain, which uses this electron payload to generate the proton "reservoir" that drives the generation of nearly all the cell's energy.

This was well-understood biochemistry, taught in its simple form to students in middle-school science classes. But it's all predicated on being able to feed these electrons into the electron transport chain machinery. So, I asked myself, what would happen when that machinery was shut down, as it is in mitochondrially mutant cells?

It seemed to me that the whole process might grind to a halt. Every step along the way—from glycolysis to the TCA cycle—loads electrons onto waiting NAD^+ "fuel tankers" for delivery to the electron transport chain. There is, of course, only a limited supply of NAD^+ "carriers" available to be charged up into NADH, but normally that isn't a big deal: there are always plenty of these carriers available, because NADH is recycled back into NAD^+ when it releases its electron cargo to the electron transport machinery in the mitochondria.

But with that natural destination shut off, there is no obvious way for NADH to relieve itself of its burden of electrons. (Similarly, you can imagine how, if all the refineries on Earth were suddenly decommissioned, the taps on the world's oil wells would quickly need to be turned off. With nowhere to deliver their oil for processing, fuel tankers could only be filled once before their capacity would be taken out of circulation, and continuing to pump oil would lead to a logistical nightmare.) And since every step of the process—glycolysis, intermediate metabolism of pyruvate into acetyl CoA, and the TCA cycle—needs NAD^+ to proceed, a lack of NAD^+ would be expected, at first glance, to lead to the deactivation of the entire process, leaving *no* mechanism for the cell to produce even the small quantities of ATP energy that result from these early processing steps.

In fact, I could see how mitochondrially mutant cells might be even worse off than this. NAD^+ is required for a wide range of cellular functions unrelated to energy production—and each time these functions utilize NAD^+, they not only reduce the pool of available NAD^+, they also convert it to yet more NADH, further upsetting the cell's metabolic balance. In fact, some researchers believe that many of the complications of diabetes are caused by an excess of NADH and a lack of NAD^+, leading to disruption of these various metabolic processes (although the imbalance of NAD^+ and NADH in diabetics has different causes than the loss of OXPHOS capacity).

Despite all this, however, cells that have been taken over by mutant mitochondria *do* survive, as is shown by their gradual accumulation with aging. So they have to be getting ATP from *somewhere*. At the time I was exploring this matter, the general presumption in the field was that these cells could survive by shutting down the TCA cycle and relying entirely on glycolysis for energy production. This is what happens in muscle cells as a brief stopgap during intense anaerobic exercise, when the cell is working so hard that it uses up all of the available oxygen and can't keep oxidative phosphorylation going. Glycolysis would provide the cell with a small but just adequate amount of ATP, and this school of thought suggested that the cell could deal with the resulting small excess of NADH through a biochemical process that converts pyruvate into lactic acid—the biochemical origin of the famous "burn" that weightlifters suffer during the very last possible rep in their set.

But I realized that this theory didn't match the evidence. For one thing, the expected rise in lactic acid didn't seem to happen. And even more bizarrely, rather than a *shutdown* of TCA activity (as one would expect because of the lack of the required free NAD^+), enzyme studies had strongly suggested that mitochondrially mutant cells have *hyperactive* TCA cycles. So, I wondered, how do they keep this seemingly unsustainable process going?

Learning from the Great Mr. Nobody

A big step toward understanding this phenomenon was achieved with the creation of so-called ρ^0 ("rho-zero") cells, whose mitochondria are *completely lacking* in DNA. This condition renders cells functionally very similar to cells that have been overtaken by mutant mitochondria, because the deletion mutations found in these mitochondria actually shut down the ability to turn *any* DNA instructions into functioning proteins. If your DNA can't be decoded into usable blueprints, it might as well not be there, like instructions on how to build a bridge that are written in a language you don't understand. So having these DNA deletions puts mitochondria in just the same condition as having no mitochondrial DNA at all.

One of the first things that scientists working with ρ^0 cells discovered was that they did, in fact, quickly die—unless their surrounding bath of culture medium contained one of a few compounds which are not normally present in the fluid that surrounds cells in the body. Intriguingly, however, *some of these compounds are unable to enter cells,* which meant that whatever it was that these compounds do to keep cultured ρ^0 cells alive, it must be something that can be accomplished from *outside* the cell. This fact made little sparks go off in my brain, because I was looking for a way to explain how cells that had been overtaken by a clonal brigade of mutant mitochondria could export some kind of toxicity outside themselves to the body at large. Might these compounds rescue ρ^0 cells by unburdening them of this same toxic material?

And might this toxic material be none other than . . . *electrons*?

I immediately drew a connection between the predicted excess of

NADH in cells that could not perform oxidative phosphorylation, and the dependence of ρ^0 cells on the presence of the "detoxifying" compounds in their medium. What the mitochondrially mutant cell needed to do was to dump electrons, so as to recover some NAD^+—and the "rescue" compounds for ρ^0 cells were all electron acceptors, and they worked even if they were kept outside the cell's boundaries. My hypothesis: Mitochondrially mutant cells prevent a crippling backlog of unused electrons by exporting them out of the cell, via a mechanism similar to that which is essential to the survival of ρ^0 cells in culture—and this export somehow spreads toxicity to the rest of the organism.

The Safety Valve

To turn this idea into a reformulation of the mitochondrial free radical aging theory, I needed clear answers to three questions. First, how were these cells delivering electrons to acceptors that were located outside their own membranes? Second, since the electron acceptors used in the ρ^0 culture studies are normally not found in bodily fluids (or not at adequate concentrations), what electron acceptors are available to do the same job for mitochondrially mutant cells in the body? And third, could these processes provide an explanation for these cells' systematic spread of toxicity throughout the body, as seemed required in order to accept that they might play a significant role in aging?

The first question turned out to have been answered already. For decades, scientists had known of the existence of an electron-exporting feature located at the cell membrane that we today call the Plasma Membrane Redox System (PMRS). While little was understood about its actual purpose in the body, its basic function was well established: it was known to accept electrons from NADH inside cells and to transport them out of the cell, thereby recycling the NADH to NAD^+. This export allows even normal, healthy cells to have better control over the balance of chemically oxidizing and reducing factors within their boundaries, and to keep tighter control over the availability of NAD^+ and NADH for essential cellular biochemistry. In other words, it does exactly what mitochondrially mutant cells would need to be able to do in order survive.

And the PMRS turned out to be an almost impeccable candidate for the job. PMRS researcher Dr. Alfons Lawen, of Monash University in Australia, had by this time already shown (with no thought of its application to aging, mind you) not only that the PMRS is able to deliver electrons to the same membrane-impermeant electron acceptors that allow ρ^0 cells to survive, but that PMRS activity is *required* for the survival of these cells. This proved both that the export of electrons is a requirement for cell survival, and that the PMRS is the dock that sets them loose into the ocean of surrounding bodily fluids.

The ability of mitochondrially mutant cells to recycle NADH back to NAD^+ would allow them to carry out their normal cellular processes, and without becoming so burdened with extra electrons as to create an internal environment in which other critical cellular chemistry becomes impossible. Moreover, I realized, this is a plausible explanation for the fact that these cells have an unusually active TCA cycle. With oxidative phosphorylation shut down, putting the TCA cycle into overdrive would allow the cell to double its production of precious ATP from sugars (and to do many other metabolic jobs in which the TCA cycle participates). The PMRS would make this possible, by recycling the greatly increased amounts of NADH that would be created and thereby providing the cell with the extra NAD^+ required to keep the process going.

But the drastic increase in PMRS activity required to make the increased TCA activity sustainable would make the surfaces of mitochondrially mutant cells positively bristle with electrons undergoing export, forming a hotspot of electrically "reducing" pressure (i.e., an unstable excess of electrons). The next question, therefore, was onto what molecules the PMRS was unloading the electron surfeit. None of the electron acceptors being used to keep ρ^0 cells alive in culture existed in adequate concentrations in the body to accomplish the task, so something else had to be performing this essential role. For example, some of the burden might be taken up by *dehydroascorbate,* the waste product of vitamin C that is created after it is used in quenching free radicals, but there wasn't enough of that to deal with the powerful "reductive hotspot" created by these cells.

At this point, an attractive candidate that could tie the story together flashed into my mind: our old two-faced friend, oxygen.

Don't Throw Your Junk in My Backyard, My Backyard . . .

Oxygen, of course, can and does absorb surplus electrons in its environment. As we discussed above, this is exactly how mitochondria generate free radicals during oxidative phosphorylation, as "fumbled" electrons slipping out of the electron transport chain are taken up by the oxygen dissolved in the surrounding fluid. And oxygen is the only such molecule in the body's bathing fluids that is present in sufficient quantity to be able to sponge up the huge electron leak that mitochondrially mutant cells would be predicted to generate.

This uptake of electrons by oxygen could, in an ideal world, be safe: the PMRS might load *four* electrons onto each oxygen molecule, turning it into water, just as the electron transport chain does with electrons that it doesn't "fumble." But it was eminently possible that the PMRS would fumble some electrons—maybe quite a lot of them. If so, it would generate large amounts of superoxide radicals at the surface of mitochondrially mutant cells. The consequences of this would clearly be bad. But actually, maybe not *so* bad: one might think that the negative effects would mostly be confined to the immediate locality. Like nearly all free radicals, superoxide is highly reactive, and therefore short-lived. Either it would be dealt with by local antioxidants, or else it would attack the first thing with which it came into contact (a neighboring cell's membrane, for instance)—but its aggressiveness would be quenched in the process. Superoxide certainly couldn't remain a free radical for long enough to reach the far corners of the body, as the implications of the mitochondrial free radical theory required.

But what if, instead of attacking the components of a cell's immediate neighbors, superoxide generated by the PMRS were to damage some *other* molecule that *was* then stable enough to be carried throughout the body? There was an obvious suspect: serum cholesterol, especially the LDL ("bad") particles—low-density lipoproteins, to give them their full name—that deliver their cholesterol payload to cells all over the body.

Oxidized (and otherwise modified) cholesterol was already known to exist in the body, and everyone now accepts that it's the main culprit behind atherosclerosis (a subject to which we'll return in Chapter 7). It was quite plausible, I realized, that superoxide originating at the surface of mitochondrially mutant cells could be oxidizing LDL as it passed by, not only because LDL is ubiquitous and therefore an easy target, but also because the presence of loosely bound reactive metals such as iron ions would multiply superoxide's potential virulence. One might think that the presence of antioxidants—such as the vitamin E that's dissolved in LDL—would prevent this from happening, but researchers had already discovered that it didn't. Not only is oxidized LDL found routinely in the body, but studies using the most accurate available tests of lipid peroxidation had shown that vitamin E supplements were unable to reduce the oxidation of fats in healthy people's bodies.[14] In fact, the lack of antioxidant partners in the inaccessible core of the LDL particle means that when it is subjected to anything more intense than the most trivial free radical challenge, the particle's vitamin E can actually *accelerate* free radicals' spread to its center through a phenomenon called "tocopherol-mediated peroxidation."[15] (Tocopherol is the technical name for vitamin E.)

I saw the light at the end of the logical tunnel now. The oxidation of LDL would provide a very plausible mechanism to explain the ability of mitochondrially mutant cells to spread oxidative stress throughout the aging organism. Despite its ability to promote atherosclerosis when present in excessive amounts in the blood, LDL cholesterol serves an essential function in the body. Cells need cholesterol for the manufacture of their membranes, and LDL is the body's cholesterol delivery service, taking it from the liver and gut (where it is either manufactured or absorbed from the diet) out to the cells that need it.

But if its cholesterol consignment became oxidized by mitochondrially mutant cells along the way, LDL would become a deadly Trojan horse, delivering a toxic payload to whichever cells absorbs its cargo of damaged cholesterol. This would spread free radical damage into the incorporating cell, as the radicalized fats propagated their toxicity through the well-established chemical reactions that underlie the rancidity of fats. As more and more cells were taken over by mutant mitochondria with age, more and more cells would accidentally swallow oxidized LDL, and oxidative stress would gradually rise systemically across the entire body. See **Figure 4.**

The New Mitochondrial Free Radical Theory of Aging

I walked myself through the entire scenario again and again, and gradually satisfied myself that I had indeed developed a complete, detailed, and consistent scenario to explain the link between mitochondrial free radicals and the increase in oxidative stress throughout the body with aging. This model answered all of my key questions and resolved all of the apparent paradoxes that had led so many of my colleagues to abandon the mitochondrial free radical theory of aging entirely. Mitochondrial free radicals cause deletions in the DNA. These mutant mitochondria are unable to perform oxidative phosphorylation, drastically reducing their production of both ATP and free radicals. Because they are not constantly bombarding their own membranes with free radicals, the cell's lysosomal apparatus will not recog-

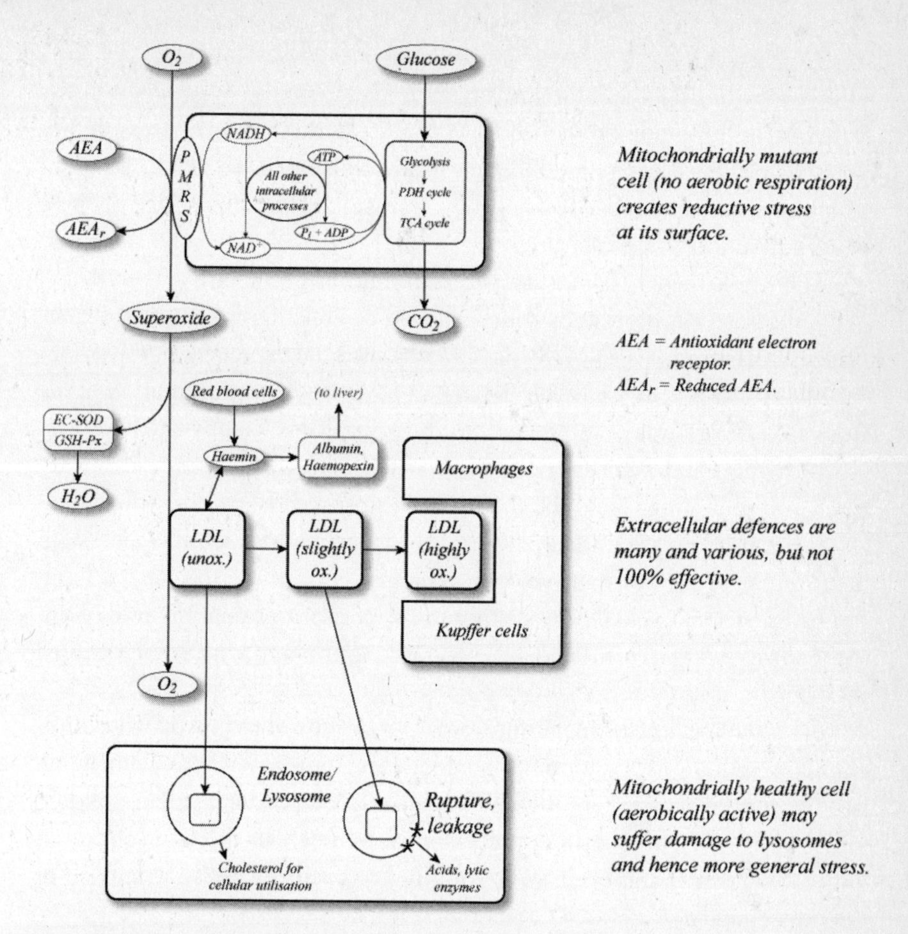

Figure 4. The "reductive hotspot hypothesis" for amplification of the toxicity of rare mitochondrially mutant cells.

nize them as defective and they will gradually drive out their healthy neighbors. (This was the maladaptive evolutionary process that I had already, a year previously, termed "Survival of the Slowest.")

To continue producing the ATP and other metabolites their host cells need for survival, these mitochondria must maintain the activity of their TCA cycle—but this, combined with other cellular processes in the absence of oxidative phosphorylation, will quickly deplete the cell of needed NAD^+ carriers unless a way is found to relieve them of their electron burden. This is accomplished via the "safety valve" for excess electrons located at the cell membrane—the Plasma Membrane Redox System (PMRS).

The snarling buzz of electrons congregating at these cells' outer surfaces turns them into "reductive hotspots," generating a steady stream of superoxide free radicals. These radicals contaminate passing LDL particles, which then spread to cells far away in the body, driving a systemic rise in oxidative stress throughout the body with age. With oxidative stress comes damaged proteins, lipids, and DNA, as well as inflammation, disrupted cellular metabolism, and maladaptive gene expression. This could certainly be a central driver of biological aging.

The whole theory is unattractively elaborate, as you'll have gathered, and as I immediately appreciated—but as far as I can tell, it's the only hypothesis that can accommodate all of the data. I published the twin arms of the theory in rapid succession in the late 1990s, and a more detailed presentation of the integrated theory was accepted as my Ph.D. thesis for Cambridge and published in the Landes Bioscience "Molecular Biology Intelligence Unit" series.[16] In the ensuing years the theory has enjoyed widespread appreciation and citation in the scientific literature. Unfortunately, despite this fact, both the popular press and many biogerontologists[17] continue to cite the long-disproven vicious cycle theories either to support or to refute the role of mitochondria in aging, instead of seriously grappling with this detailed mechanistic account.

So now you know, in minute detail, my interpretation of why mitochondrial mutations are probably a major contributor to mammalian aging, and therefore why they are included as a category of "damage" in the definition of SENS. The question, then, is what to *do* about the toxic effects of mitochondrially mutant cells. The solution to this problem is the subject of the next chapter.

6

Getting Off the Grid

There are good reasons to believe that most present attempts to modify metabolism to produce less damage to mitochondria (and thereby to the body at large) are a poor use of resources. Fortunately, a better path forward exists, promising far greater results for the same application of time and money. It seems possible and plausible to prevent damage to mitochondria from harming us as we age—and scientists are already working on many options for the first steps of this process.

In the previous chapter, I explained in excruciating detail my views about the complex mechanisms whereby mitochondrial DNA deletions may act as a major engine of aging. I must now tell you that in a very real sense, it simply *does not matter* if that hypothesis is correct or not.

This point applies equally strongly to the other SENS interventions, and it is so central to the engineer's approach to anti-aging medicine—and so enormously counterintuitive—that I must beg your indulgence if you find me to be repeating it too often. We must all keep it in the forefront of our minds when thinking about these problems. If our purpose were simply to *understand* aging, then teasing apart the specific pathways that lead to the accumulation of age damage would indeed be absolutely imperative. But that is *not* our purpose. Our purpose is to put an end to aging's consequences: the daily descent into decrepitude, and subsequent deaths, of tens of thousands of people.

Aging is a deadly pandemic disease, and I believe that our understanding of its mechanisms, while still highly imperfect, is now good enough that we are in a position to intervene in it. We need only to identify *the nature of the damage itself*—the accumulating lesions that are the source of age-related loss of functionality in the organism—and then either to reverse that damage, or to eliminate its threat to our health and life expectancy. This goal should become the central focus of biogerontological work, and the major target of biomedical funding generally.

The problem of mutant mitochondria is a case in point. Mitochondrial DNA mutations are a form of molecular disorder that distinguishes the biologically young from the biologically old, and there is powerful evidence that they are deleterious.[1] So, whether mutant mitochondria take over their host cells via "Survival of the Slowest" or through some other mechanism, and whether these cells exert their toxic effects on the rest of the body by way of the export of electrons through the PMRS or via a completely unrelated process, the nature of the task at hand is ultimately the same. Our therapeutic goal is clear: either to fix the mutations themselves, or to make them functionally irrelevant. How to achieve that goal is the subject of this chapter.

Before laying out my proposals to accomplish this goal, however, I must first spell out why the appealing-sounding solutions that many biogerontologists would propose are probably wastes of time and scarce resources.

You Can't Stop a Moving Train (Safely!)

I termed the "over-preventative" approach to combating aging the "gerontology" approach because biogerontologists predominently favor it. By and large, when my colleagues think seriously about actually *doing something* about aging rather than just refining their understanding of it, their first instinct is to find some way to make metabolism run more cleanly. After all, aging is the result of the accumulation of the deleterious by-products of our metabolic processes; surely, the thinking goes, if we could just tweak or dampen down those processes a little bit, we could reduce the exposure of the organism to metabolism's reactive by-products, reduce the rate at which our cells and tissues accumulate microscopic damage over time, and thus slow down the gradual decay of our bodies into age-related frailty and accelerating vulnerability to death.

This approach has a strong intuitive appeal, reinforced by the continuous stream of encouragement from supplement vendors and public health authorities alike. We are constantly urged to clean up our lifestyles and practice preventive medicine: surely, we think, it makes more sense to put one's energies into attempting to interfere with the *causes* of aging and age-related diseases than to try to undo an established molecular mess. But, as we saw in Chapter 3, the causes of aging lie in the fundamental chemistry of life, and our capacity to interfere beneficially with that chemistry is limited by what the organism's underlying biology will accommodate.

I gave examples of this general principle in Chapter 3, but let's now look at the more concrete case of intervention into the problem of mitochondrial mutations. The obvious, old-school approach would be to try to reduce the formation of mutant mitochondrial DNA by cutting back on the bombardment of the mitochondrial DNA by free radicals. Just such a trick has been pulled off with some success in mice[2] by giving them a copy of the gene for the antioxidant enzyme *catalase,* specially targeted to their mitochondria. Catalase breaks down the free radical-like molecule *hydrogen peroxide,* turning it into harmless water before it can become more vicious and do serious molecular damage. Animals that received the targeted catalase gene enjoyed a fifty-fold jump in the activity of the enzyme within their mitochondria, preventing a great deal of mitochondrial DNA damage—including some of the mutations that initiate the whole destructive process of decay that I described in the last chapter.

Compared to the repeated, abject failures of dietary antioxidants to extend lifespan, or the ambiguous[3,4,5] or negative[6] results of previous attempts to hold back the free radical onslaught in mice using genetic manipulations, these results are quite impressive. Mice given mitochondrially-targeted catalase gained a 20 percent extension not only of average, but of *maximum* lifespan—the gold standard for data relevant to aging. And while no detailed analysis of the level of age-related pathology in these animals has been published, we do know that they suffer less age-related heart degeneration and fewer cataracts.

This being so, why should we not vigorously pursue the development of a targeted boost in mitochondrial catalase for *humans*? Well, in part this goes back to the testability (and, thereafter, the possibility of full clinical development) of this intervention. It suffers the same sorts of weaknesses on these fronts that characterize all "preventive" anti-aging medicines, as outlined in Chapter 3. Firstly, because there is no short-term disease against which extra catalase could be tested as a cure, regulatory bodies won't let clinical trials be performed using it, and will never approve it for human use; and secondly, the timescales required to prove its effectiveness against aging make it too expensive and risky for venture capital to touch it. That probably means that no amount of agitation by scientists or the public will actually put mitochondrially targeted catalase into the hands of clinicians to save people's youth, health, and lives: the interests of those with the power to fund or halt development are aligned against moving forward.

But even if these structural hurdles did not exist, I would still conclude that there are more fruitful soils into which to plow our limited resources for dealing with mitochondrial mutations. Boosting these mice's catalase supply *reduces* the incidence of mitochondrial mutations—but it doesn't *eliminate* or *obviate* them. Thus, it *slows,* but cannot *treat,* the progressive accumulation of this form of aging damage. I think we can do a lot better than this. My evaluation of the evidence indicates that it is not worth diverting our intellectual and financial capital into an intervention that might yield a 20 percent increase in lifespan (making the average person in a developed country live into his or her late nineties), because the same bricks, boards, and brains could be put to work in the development of an intervention that would not prevent this damage from happening, but would instead render it harmless. My analysis suggests that this intervention might, in turn, not only slash the rate of aging damage primarily caused by *other* mechanisms in half, but—when combined with a panel of similar therapies—would ultimately give us an *indefinite* youthful lifespan. More on that in Chapter 14.

In fact, if we were to eliminate, one by one, the *other* forms of molecular damage that cause us to wither and to die as we age, but were to "only" slow down the incidence of mitochondrial mutations by the degree seen in the mitochondrially targeted catalase mice, we might find ourselves stranded one step away from a future that stretches out further than our eyes can see. When advanced glycation endproduct (AGE)–breaking drugs reverse the gumming up of our structural proteins; when careful exploitation of the immune system clears out the extracellular junk that impedes cellular function; and when dormant cells that cause the decay of our immune system itself have been removed; at that time, with all the other key SENS platform interventions in our hands, we would still have mitochondria that are playing a game of Russian roulette with their DNA. However much more slowly they may be spinning the chamber, they could still end up being the weak link in a chain that could otherwise give us indefinite youthful lifespan.

We could apply these same objections to any approach to anti-aging biomedicine based on prevention of aging damage instead of genuine remediation. But there is a more specific objection to the "catalase solution." Relying on an extra dose of catalase to deal with mitochondrial mutations could actually become *harmful* in important ways in a person whose body had already been cleared of—or rendered immune to—all of the other identified aging damage.

Catalase cleans up hydrogen peroxide, which can be damaging when it is the result of imperfections in oxidative phosphorylation—as most of it is. But evolution is, over the long term, an extremely clever engineer, and has learned ways of making the best of a bad job, harnessing hydrogen peroxide for its own purposes. While randomly spewing the stuff out of the cellular power plants does us no particular good, our cells also generate some hydrogen peroxide *on purpose,* for use as a chemical signal that regulates everything from glucose metabolism to cellular growth and proliferation.[7] In this way, damping down the level of hydrogen peroxide in a cell—even using a technique designed to focus on the mitochondria—might well be expected to interfere with the functioning of our complex intracellular machinery.

One of the most dramatic examples of this potential problem is the need for hydrogen peroxide to reach the mitochondria themselves as part of the signaling cascade that triggers *apoptosis,* or "programmed cell death." Apoptosis is important during embryonic development as part of a "remodelling" process that rids the nascent organism of excess cells that are only required during specific phases of its growth. But its main role in the body is similar to the self-destruct mechanisms built into James Bond's Aston Martin or the Starship *Enterprise:* to give your cells a way to destroy themselves if they have been hijacked by "enemy forces" (viruses or cancer, for example) before they can threaten the integrity of the organism as a whole. When the cells of our immune system detect a hijacked cell, they bind to its surface, flipping a switch that signals its mitochondria to "blow their tops" and destroy it. Hydrogen peroxide is a player in that signaling system, and studies have shown that antioxidants—including catalase—can block the proper activation of this apoptotic program.

Thus, the massive boosting of mitochondrial catalase activity that is required to give these animals their partial protection from mitochondrial DNA damage has a dark side. And while it's clear that the net effect of this boost is good for *them*—as evidenced by their extended lives and reduced age-related pathology—the overall balance of risks and benefits would in all probability be totally thrown off in an organism in which all other types of aging damage had been eliminated. In this otherwise rejuvenated body, a chronic dysregulation of cell signaling pathways would be a high price to pay for lower oxidative stress.

Finally, there is reason to think that the catalase boost given to these mice—which was performed in the mice while they were still early embryos, rather than in adult organisms—might *not even work* if done in adult organisms. Catalase genes only expressed the enzyme in some cell types, not in others, and the scientists who performed the study suggested that this might have been the result of a form of evolutionary selection during their development in the womb. The idea is that the extra catalase might be beneficial in some kinds of cells but harmful in others, and therefore those cells that were expressing a lot of the enzyme would tend to die off if they were of a type in which the extra catalase was deleterious. Meanwhile, other cells of the same type that did *not* express the new gene would survive, replicate, and come to dominate. But performing the same trick in mature organisms would short-circuit this internal evolutionary process, supplying the new gene to all cell types indiscriminately, so that the benefits of the catalase gene therapy in some cell types might be outweighed by negative effects elsewhere. This would mean that people unfortunate enough to have already been born would not be able to reap any benefit from such therapy—and no proposal to insert the gene into healthy developing infants, let alone embryos, is going to get past a medical ethics board.

None of this should dishearten us; it should simply remind us of the need to focus our efforts elsewhere. As I outlined in Chapter 4, I advocate a fundamentally different approach to dealing with age-related molecular damage. Instead of trying to "mess with metabolism" in ways that might *prevent* aging damage such as mitochondrial DNA deletions, it is my contention that we need to focus on developing anti-aging biomedicine that can *repair or render harmless* any mutations that may occur in mitochondrial DNA. While most people—whether laypersons or professional scientists—tend to assume that this must be far more difficult to achieve than a preventive strategy, there are in fact several promising techniques sitting on the drawing board that require no biotechnology more advanced than that which would *already* be required to put catalase into the mitochondria—namely, gene therapy. This suggests that we could actually have either type of intervention in the clinic at about the same time. In fact, it really tells us that the *remedial* technologies should be able to reach people suffering aging damage sooner than preventive ones, for the regulatory and pragmatic reasons I outlined earlier.

The Best Bet: "Head for the Bomb Shelter!"

As I discussed in the last chapter, the main reason for mitochondrial DNA's unusual vulnerability to oxidative damage is its location: being next door to a leaking nuclear reactor (the mitochondrial electron transport chain) is a sure way to increase your risk of mutations, whether you're a growing child or a tiny biological machine. Metabolism clean-up approaches try to make this reactor run more cleanly (so that it produces less damaging waste), or to install "pollution control" equipment that would catch more of its by-products before they do any harm (the cellular equivalent of the "smoke-stack scrubbers" on coal-fired power plants). My preferred approach is completely different: that we should put these mutations "beyond use" of harm. This could be accomplished by putting *backup copies* of the genes that are currently housed in the mitochondria in the safe haven of the cell's nucleus, far from the constant bombardment of free radicals from the mitochondrion itself. This solution is called "allotopic expression" of the proteins these genes encode—i.e., expression from a different (Greek *allo-*) place (*topos*).

Let's be clear about this. Allotopic expression would do absolutely nothing to prevent the native mitochondrial genes from suffering mutations: free radicals would hit the vulnerable mitochondria just as often, and mutations would occur at exactly the same rate, as they did before. But with a nuclear backup copy of these genes, any such mutations would be rendered functionally irrelevant, because the cell would be able to keep producing the proteins that the knocked-out genes in the mitochondrion had previously encoded. These mitochondria would thus enjoy functional proton-pumping, electron-transporting proteins, and would therefore behave *exactly like* mitochondria with intact DNA, just as if they had not suffered mutations in their "local" DNA. Electrons would continue to flow into the electron transport chain from NADH; protons would continue to be pumped; free radicals would continue to leak out of the system at random. The concept is illustrated in **Figure 1.**

Not only that: because such mitochondria would continue to damage their mitochondrial membranes, the cellular "incinerator system" (the *lysosome*) would be able to tell when they got old and haul them off for destruction. Therefore, the Survival of the Slowest mechanism that leads mutant mitochondria to take over their host cells would never take place, nor would cells be forced to shift into the abnormal metabolic state that cells with mutant mitochondrial DNA require in order to deal with an

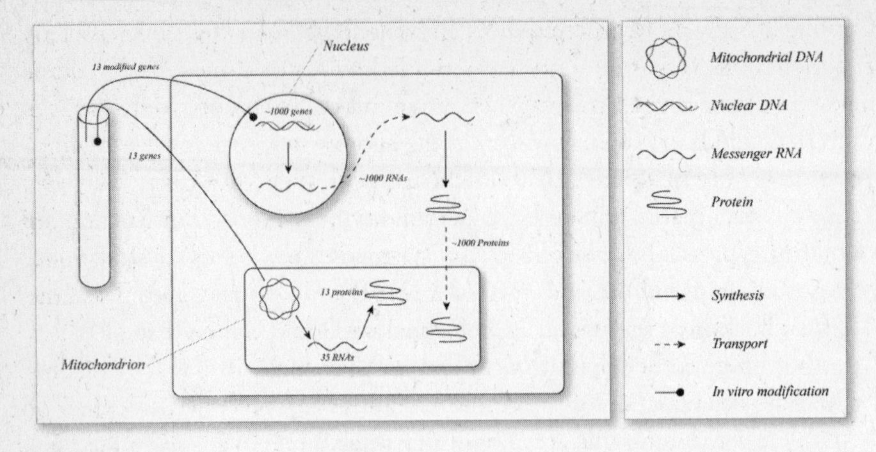

Figure 1. The concept of allotopic expression to obviate mitochondrial mutations.

imbalance in their NADH-to-NAD+ ratio. Thus, despite the fact that these cells contain mutations in their mitochondrial DNA, they would not be unloading their excess electrons into LDL, would not spread oxidative stress to the rest of the body, and would make no more contribution to the aging of the organism as a whole than cells with perfectly intact mitochondrial DNA.

"But," I hear you ask, "what if one of these backup genes is *itself* mutated? Won't we be facing the same catastrophe?" Fortunately, no! There is, in fact, no real risk of a functionally meaningful failure of this backup system occurring, even over the course of a lifespan that has been dramatically extended by a full panel of SENS anti-aging interventions.

To understand why this is so, let's look at what would be required for such a failure to occur. First, in order for a cell with an allotopic copy of a mitochondrial gene to slide into Survival of the Slowest, it would have to have suffered mutations in *both* copies of the gene: the mitochondrial original *and* the duplicate copy that we would have placed in the nucleus.

This is less likely to happen than it may initially sound. It is already unusual for DNA located in the mitochondria to suffer permanent damage (remember that as things stand, less than 1 percent of cells are overtaken by mutant mitochondria), and the odds of a backup copy located in the nucleus being mutated are much lower. Aside from being better shielded from free radicals because of its location (DNA housed in the nucleus is many times less susceptible to mutations than its mitochondrial counterpart), there are many more proteins encoded by genes located in the nucleus: tens of thousands, versus only thirteen encoded by genes that are in the mitochondria themselves. So even when a free radical does get into the nuclear DNA, the odds of it damaging one of the allotopically expressed mitochondrial genes are many times lower than the odds that it will hit some other gene. Indeed, many such free radicals will not hit a gene (an instruction for building a protein) at all, but one of the many stretches of nonfunctional "junk" DNA. Therefore, the chances of *both* the mitochondrial copy *and* the nuclear backup of the *same gene* being mutated are vanishingly small.

Moreover, while the unusual design of the mitochondrial DNA ensures that large deletions in its structure wipe out its ability to synthesize *any* of its proteins, the same will not occur in the nuclear case: only the protein for the *specific* mutated gene will be affected. Of course, your mitochondria can't function without all thirteen proteins, but we could help reduce the risk of any actual shutdown of oxidative phosphorylation—and the resulting clonal expansion of a mutant mitochondrion—by providing a double or even triple set of functioning backup copies.[8]

The "Mitochondriopathies"

One category of hurdle facing the clinical development of anti-aging biomedicines is structural: aging is not a recognized disease, so the FDA will not allow trials to be performed on interventions claimed to cure it. This is obviously a bucket of water thrown directly onto the heads of venture capitalists who might otherwise be interested in investing in startups working on a treatment for age-related mitochondrial DNA mutations. From the perspective of getting effective anti-aging interventions into the clinic as quickly as possible, allotopic expression has the advantage that it is *already* being pursued as a treatment for a recognized group of diseases: the *mitochondriopathies*.

These diseases are caused by defects in the mitochondrial DNA that are *inherited* (or, more rarely, acquired through causes independent of the aging process). These mutations lead to a failure of energy production that causes a spectrum of dysfunctions in various organs, depending on the exact disorder: the brain, heart, and muscles tend to be the most vulnerable, but damage can also extend to the liver, kidneys, lungs, and certain glands. Because allotopic expression is a promising therapy for mitochondriopathies, government funding (albeit not nearly enough) is already available for work on its development; and once it is ready to move into the clinic, there will be an incentive for venture capital to invest in its development, giving a clear route forward for near-term testing in FDA-approved clinical trials.

In turn, once allotopic expression has been proven to be safe and effective as a treatment for *inherited* mutations of the mitochondrial DNA, we will be in an excellent position to make the small tweaks needed to adapt it for use as a treatment for mutations acquired during the aging process. This parallel applicability is a feature of most of the anti-aging interventions included in the SENS platform—and indeed, prototypical versions of several of the proposed interventions are already in clinical trials today.

To Boldly Go Where Evolution Has Gone Before

The other hurdles facing allotopic expression are the more purely scientific ones. Fortunately, as we will see, progress on these problems has been rapid in the last decade. But it's better than that: evolution has been working on a similar job for untold millennia.

In ages long past, the ancestors of the mitochondria occupying our own cells were not just *components* of cells as they are today, but organisms in their own right, which formed an I'll-scratch-your-back-you-scratch-mine relationship with our one-celled ancestors. Because they were independent organisms, these proto-mitochondria naturally had a full complement of their own DNA—at least one thousand genes.

But precisely because the hazardous environment of the mitochondrion put the genes housed there at extremely high risk of mutation, evolution has been performing allotopic expression on mitochondrial genes since long before humans appeared on the scene. Over the glacially slow timescales of evolution, organisms have copied mitochondrial genes that code for mitochondrial proteins into their cell nuclei, after which the original mitochondrial genes became redundant components and mutated into oblivion.

And to give Mother Nature her due, evolution has gone a long way in this direction. Out of more than one thousand original mitochondrial proteins, all but thirteen have had their genetic instructions moved into the nucleus.

Starting in the mid-1980s, scientists started showing that they, too, could perform allotopic expression of some mitochondrially coded proteins using biotechnology, albeit initially only in yeast—a crucial series of proofs-of-concept.

Obstacles, Evolutionary and Otherwise

But things get a lot trickier when we start trying to do the same thing with the thirteen protein-encoding genes that are still located within the mitochondria in *human* cells. The reason why evolution hasn't finished the job for us already is a matter of some debate, but everyone agrees that there must be some kind of "forces" holding the process back. Whatever those forces are, this job is probably not going to be easy. What we have to do is figure out *what* forces are keeping those genes in the mitochondrion, and then devise ways of overcoming them. I have advanced the case that there are only two such forces that need concern us.[9]

One, which doesn't apply to all organisms but does apply to us, is that the DNA "languages" of mitochondria and of the cell nucleus have evolved slightly different "dialects," so that an exact copy of a given mitochondrial DNA sequence becomes indecipherable when it is dropped into the nucleus. This problem is called *code disparity*.

The case is rather like the changes that have occurred over time in the writing of the letter "s" in English. Up until the nineteenth century it was common for an "s" occurring in the middle of a word to be written in an elongated fashion that looks much more like a modern "f" than an "s." Gradually, as writing became more widespread and irregularities in the written language more standardized, the elongated "s" came to be replaced by the shorter, more curved version of the letter that we use today. Thus a modern reader of an Enlightenment-era order to launch a naval attack ("*Sail* for the enemy") might mistake it for a command to "throw" the battle ("*Fail* for the enemy"), and in other cases an instruction might be reduced to pure gibberish.

This disparity in the genetic codes of mitochondria and nucleus makes moving mitochondrial genes that contain such discrepant lettering into the nucleus a near impossibility for evolution. Indeed, all the genes still housed in the mitochondria contain such quirks, and this fact alone can explain why they haven't made the jump to the nucleus. But code disparity doesn't pose any serious problem for biotechnology: with our outsider's understanding of the discrepancies in the two codes, we can simply create the new, allotopic gene using the nuclear version of the code (substitute "s" for "f") and rest assured that it will be translated and turned into a protein just like any of the other genes for mitochondrial proteins that are already housed there.[10]

The second problem appears to have been a somewhat *less* imposing barrier to the *evolutionary* transmigration of mitochondrial genes to the nucleus, but is a much *greater* challenge for the anti-aging biotech engineer. It is the repulsion by water ("hydrophobicity") of many of the mitochondrial proteins whose genes are still located within the mitochondria themselves. These proteins have sites in their chemical structures that have such a "fear" (phobia) of water (hydro) that, like a phobic human, they will literally curl themselves up into a ball in response to it.

Hydrophobicity is no problem for proteins when they are built from DNA that is already located within the mitochondrion, because the final three-dimensional shape of the protein is *supposed* to be twisted up, and there are special enzymes that guide such proteins into the proper conformation. But it becomes a showstopper when the same protein is constructed from DNA located in the nucleus. The genes for such proteins are translated into their products in the main body of the cell, and the proteins must then be moved from the fluid environment outside of the mitochondrion, through the outer and inner mitochondrial membranes, and into their final location in the mitochondrion's core.

Of course, the membranes won't just let proteins pass *freely* through them, or the integrity of the mitochondrion—and its ability to preserve its proton reservoir—would be compromised by the constant leak of material into and out of its chambers. But mitochondria do need to be able to import hundreds of proteins: for example, the dozens of the subunits of the electron transport chain whose genes have already been moved successfully into the bomb shelter of the nucleus by evolution. Mitochondria have therefore evolved elaborate machinery that specifically moves (translocates) proteins through these membranes. These are sensibly called the "Translocase of the Inner Mitochondrial" membrane (TIM) and the "Translocase of the Outer Mitochondrial" membrane (TOM), giving the whole system the glorious name *TIM/TOM complex*.

The problem is that once a protein has been gnarled into a balled-up Twister configuration by its repulsion to the water in the fluid compartment of the cell, the cell becomes unable to jam it through the pores of the TIM/TOM machinery—rather like trying to force a wildly bent-out-of-shape coat hanger down a drainpipe. This would be not just a failure, but actually *counterproductive:* not only would it prevent the *newly* allotopically expressed proteins from getting where they have to go, it would also "clog up" the TIM/TOM complex, disrupting the import of the many proteins that were previously being *naturally,* successfully imported.

In fact, *all* of the thirteen proteins that are still coded directly in our mitochondria are very hydrophobic. The instructions for several of those proteins have never been moved into the nucleus in *any* species, and those proteins are the most hydrophobic of all. This suggests that hydrophobicity is indeed the biggest hurdle to the allotopic expression project. There are several cases that *seem* to violate this rule in other species, but I have published detailed analyses[9] that explain why all of these apparent counterexamples are misleading—why they would have happened in the course of evolution despite the fact that hydrophobicity really is the most important barrier to making the move.

So, if we are to obviate mitochondrial mutations via allotopic expression (my preferred solution), then in addition to the relatively simple task of editing the code in cases where the DNA "language" of mitochondria and nucleus are mismatched, we will have to find ways of tweaking the proteins that, in their present form, can't be imported into the mitochondria.

When I first began contemplating this problem, I came up with a workable-sounding but technically challenging way of dealing with it, which I published in the journal *Trends in Biotechnology* in 2000.[11] I'll describe this approach later on. The reason I'm putting off discussion of this solution is that recent experiments suggest that we may not need to go to the lengths that I then proposed in order to overcome the hydrophobicity problem. There are at least two alternative ways to engineer these proteins to make them importable—ways that appear to be much easier.

Pirating Mother Nature's Intellectual Property

The first solution to the hydrophobicity problem is to look around for cases in which evolution has already done the yeoman's work for us—in other species. We humans (and our evolutionary ancestors) have not yet enjoyed the simultaneous good fortune of the right random mutations, the right environment and the right selective pressures to pass nuclear versions of any of these genes along to us as their descendents. But that doesn't mean that the same happy confluence of circumstances has never occurred in other species' evolutionary history. Natural selection has been working on the hydrophobicity problem in many species *independently* of our own, and has in several cases come up with workable solutions—solutions that we did not inherit, simply because they occurred in a separate evolutionary lineage. By looking beyond our own mitochondrial DNA into the evolutionary inheritance of other species, we might find viable solutions that we only need to tweak slightly for use in our own cells.

Other species' mitochondrially encoded proteins are not identical to our own, of course, but their structure is close enough to make it reasonable to believe that they could stand in for the originals if inserted into our mitochondria, or at least show us how to modify the sequences of the human counterparts to render them importable. If species could be found whose genes for their versions of some of the thirteen proteins had spontaneously moved into the nucleus, we would expect to be able to put those same genes into our own cells' nuclei with only minimal modification. Those proteins would be constructed in our cell bodies, imported into our mitochondria, and take the place of the native version if and when mutations shut down the mitochondria's ability to do the job themselves.

This idea is not just a fancy of mine: it's been done in isolated human cells already. Work on such a project began in 1998, shortly after I started shouting about the importance of a discovery that had been made eight years earlier, when the mitochondrial genetic library of the green alga *Chlamydomonas reinhardtii* was sequenced. When the proteins for which these creatures' mitochondrial genes code were identified by comparison with the equivalent genes in other species, it was found that they are missing versions of *six* of our thirteen hydrophobic electron transport protein genes.

In fact, of course, these genes are rather like your mysteriously "missing" car keys: you haven't actually *lost* them, they just aren't where you thought you'd left them. In these organisms, the changes needed to make these proteins less hydrophobic *have* taken place, because the greatest barrier to the change—code disparity—was never erected. These algae are so close to the "root" of the evolutionary "tree" that the disparity between the DNA coding systems of the nucleus and the mitochondrion never took place in them. Without that hurdle to leap, evolution has only had to address the much less challenging hydrophobicity problem—and it has done so with some success. The genetic instructions for these proteins are now housed in the algal cells' nuclei. The cell's machinery reads those instructions, manufactures the proteins in the main body of the cell, and then the mitochondria import them—the same thing that happens with most of our own mitochondrial proteins.

At this point Dr. Mike King at Thomas Jefferson University, Philadelphia, comes into the picture. King was not originally interested in aging, but in the *inherited* mitochondrial diseases (mitochondriopathies). Researchers had long dreamed of gene therapy for these disorders, but there are immense technical challenges to putting genes directly into the mitochondria. King thought that allotopic expression in the nucleus could provide a faster route to a cure.

But the hydrophobicity problem loomed over this potential solution to inherited mutations in the genes for the thirteen mitochondrially coded proteins, just as it does for the plan for mitochondrial genes mutated during the aging process via free radical damage. When he heard about the existence of nuclear-coded versions of some of these proteins in the algae in the late 1990s, Mike saw that they offered a potential blueprint for replicating the algae's tricks in human patients with mitochondrial diseases.

What followed was an astonishing surge of progress. In 1998, King began a fruitful collaboration with Dr. Diego Gonzalez-Halphen of the Department of Molecular Genetics at the Autonomous National University of Mexico, to identify and clone the algae's genes for the six analogous proteins. Within three years, these scientists had identified three of them. One of these (ATP6, a component of the mitochondrion's Complex V "turbine") was of particular interest because inherited mutations in it cause two rare but extremely serious disorders of the brain and muscular system in humans: NARP (Neuropathy, Ataxia, and Retinitis Pigmentosa) and maternally inherited subacute necrotising encephalomyopathy. In these diseases, ATP synthesis is reduced by 50 to70 percent, leading to severe dysfunction of the neuromuscular system. Thus, the identification of an importable version of the mutated protein held forth therapeutic potential for people suffering from these diseases, as well as for all of us as we age.

Picking up the ball, Eric Schon and his coworkers from the Department of Neurology at Columbia took the next step, inserting a cloned copy of the algae's ATP6 gene into the nucleus of human cells whose mitochondrial DNA harbored the same mutations that cause these neuromuscular diseases in humans. The cells decoded the genetic instructions, turned out the protein in the main chamber of the cell, imported it into the mitochondria, trimmed off its *targeting sequence* (a special string of amino acids that, when appended to the "nose" of a protein, directs it into the mitochondria), inserted it into the electron transport chain, and apparently took the place of the mutated protein, rescuing the cells from the destructive effects of the mutation.[12]

In other words, these researchers did exactly what I had been calling for someone to do. They found, in an alien species, a nuclear gene for a mitochondrial protein whose human counterpart is located in the mitochondrion itself; they inserted it into the nuclear DNA of human cells; and they showed that those cells could use it in just the way that the algae do, restoring near-normal functionality to cells with otherwise disabling mutations.

There is still a lot of work ahead, of course. To turn this into a viable therapy for people with inherited or age-related mitochondrial mutations, we will have to do two things. One is to identify in other organisms, or engineer ourselves, genes that can be put in the nucleus for the *rest* of the mitochondrially encoded proteins, and show that they restore function to mutant cells. And the other will be to do the trick in *whole organisms* burdened with these mutations: first in mice (a relatively simple task: genetic manipulation of mice is now relatively routine), and then in grown humans (a technology that we have not yet mastered, but upon which work is proceeding with an intensity that may well yield safe, viable therapies in the foreseeable future).

Borrowed Ideas, Novel Solutions

But of course, there is no guarantee that we will find *all* of the necessary genes in other species. It would be smashing luck if we did, of course, but it's entirely possible that many of the genes for electron transport chain proteins have never been transferred from the mitochondria to the nucleus in *any* species, or that the protein will have been so changed in the intricate branching of life's evolutionary family tree that it will not work in human cells. In that case, we'll just have to figure out for ourselves how to tweak our existing genes to make the proteins they encode importable.

Even here, however, we'll be able to borrow tricks that we've learned from other species. A remarkable example was reported in 2002,[13] when a group at the University of Western Australia discovered that several legume species—including the soybean and the common mung bean (*Vigna radiata*)—are actually in an intermediate evolutionary state, having evolved a nuclear copy of the mitochondrial gene for subunit two of the electron transport chain component *cytochrome c oxidase,* but without having yet discarded their original, mitochondrial copy.

The fact that these organisms survive while expressing the protein from both sites at once is itself good news, as it relieves (to some extent, anyway) a concern that some people have expressed about using allotopic expression as a solution to mutations. The worry is that in cells with healthy mitochondrial DNA *and* allotopic versions of the electron transport chain proteins, the existence of two independent, working copies of the same gene might lead to too many copies of the duplicate-coded proteins being produced, and that this might somehow imbalance or overload the capacity of the mitochondria to fit the various components together into working electron transport complexes. This would be like having two departments in the management of a factory, each using an independent system to order components for their product—a serious glitch in any "just-in-time" inventory system. The observation that no such problem occurs in these organisms suggests that we may not have much to worry about here—and that's good news.

But when other scientists compared the two versions of the legume electron transport gene, they discovered something that makes me even more bullish about our ability to move the full complement of mitochondrial electron transport chain genes into the nucleus. The two versions of the protein differ in twenty-five amino acids (the building blocks of protein) out of hundreds—but only *two* of these differences are necessary to allow the nuclear version to be imported into the mitochondria! This suggests that we may only need to do some relatively minor fiddling with our thirteen proteins of interest in order to make feasible their import into the mitochondria.

Again, I'm no longer just extrapolating from what evolution has achieved in other organisms: progress in adapting these solutions to new problems is definitely under way. Around the same time Schon's group expressed ATP6 in human cells allotopically using the algal version of the gene, they and another group also reported having developed different solutions to the challenge of engineering new, nuclear-coded versions of that protein. The difference was that these new proteins were not taken wholesale from another species, but modified from the mammalian original. As with the success using the algal gene, these human-generated versions were reported to rescue cells bearing inactive versions of ATP6 in their mitochondria.[14,15]

Not long after this, a third group engineered a nuclear-coded version of another mitochondrial gene named ND4, mutations in which cause one of the mitochondriopathies, Leber's Hereditary Optic Neuropathy (LHON).[16] To do this, they first had to find solutions to the problems that had previously been dealt with in the allotopic expression of ATP6. First, the DNA code for the protein had to be altered to make the "spelling" compatible with the nucleus, since ND4 suffers from the "code disparity" problem I discussed earlier. Then, the researchers had to tag on a "targeting sequence" copied from a completely different gene (*aldehyde dehydrogenase*) to guide it into the mitochondria. Next, they had to figure out a way to get the gene into the nucleus in the first place; this was accomplished using a trick borrowed from viruses that sneak their DNA into their infectees' nuclei. And finally, they attached an additional sequence to the gene to allow it to be picked up by the gene-decoding machinery of the nucleus, so that it would be "read" and turned into a protein.

Given the need for so many alterations to the original gene, borrowed from so many different inspirations, you might reasonably be concerned that *something* would fail somewhere along the way. But no. The heavily modified protein, custom-built out of the original by reasoned analysis of what would be required to make it importable (rather than being copied in whole cloth from one of our distant relatives), was successfully incorporated into human cells bearing mutant ND4, which promptly began churning out functional, mitochondrially targeted ND4. The allotopically expressed protein then found its way into the mitochondria, took its place in the electron transport chain . . . and promptly *tripled* the cells' ATP output, bringing it back to levels similar to those seen in normal, nonmutant cells. Not only that: when the cells were subjected to conditions under which they were forced to lean heavily on OXPHOS for energy, these cells with the new allotopically expressed mitochondrial electron chain subunits enjoyed three-fold better odds of survival than mutant cells lacking the engineered gene.

The researchers boldly concluded that "Restoration of respiration by allotopic expression opens the door for gene therapy of Leber Hereditary Optic Neuropathy." I would add that it also puts a large foot into the same door for the solution of the reductive hotspot problem. Doubts have recently arisen concerning the methods used to demonstrate rescue of the treated cells in two of the above studies, but at least two others are still considered clear-cut.

Inteins: Splitting the Difference

I am hopeful that these two approaches may be sufficient to allow us to deal with the hydrophobicity issue; however, it is possible that we may have to go further with some of the very hydrophobic proteins. One potential solution—which I originally put forward before the above successes suggested that it might not be necessary—is to further modify the proteins in question using *inteins.*

Inteins are sequences that are inserted temporarily into some proteins when they are first synthesized, possibly to help the protein mature into its final form properly, and are then snipped out once they've served their purpose. In some cases, the two halves of a final, functional protein are expressed separately, each with a complementary "semi-intein" at the end where the two protein segments must finally be joined. When the two halves of the final protein come together, the two semi-inteins are first bound together, a little like the male-to-female electrical connectors on strings of Christmas-tree lights. But then the united intein sequence is snipped out and the two precursors of the final protein are permanently, directly fused together into the final, completed structure.

We might be able to use inteins of this sort to help us deal with a hydrophobic protein. One approach is to split the protein up close to a site in its structure where it would otherwise curl up when exposed to water, and put semi-intein "caps" on either side of the break. As an analogy, imagine that you wanted to move a long piece of steel with a right angle in its center (the protein) down a straight drainpipe (the TIM/TOM complex). As it stands, the job is impossible. But if you could cut the piece of steel at or near the place where it bends, and then attach a short segment to each of the severed ends of the bar to show your coworker at the other end of the pipe where to weld the two halves together again, you could easily drop the two halves down the tube individually for reassembly at the other end.

Alternatively, whole inteins can actually be built right into the middle of the two halves of the protein, creating one long structure with the intein in its center. I'm afraid the way that this can be exploited is a bit harder to analogize, but I'll try. Imagine if, instead of cutting your bent bar in half where the bend occurs and sending each half down the drainpipe separately, you were to cut the bar, rotate it through a half-twist, and put in a central joining segment, so that the final structure looks more like a stylized lightning bolt than a right angle. You could then drop the "straightened" bar (protein complete with intein) down the drainpipe (TIM/TOM complex) and then excise the joining segment (intein), allowing you to reassemble the bar into its proper configuration after intein removal.

This procedure is a good deal more complicated than switching a couple of amino acids, so I'd rather not have to resort to it if one of the former options will work. When I first came up with the idea of using inteins as a solution to the hydrophobicity problem, I foresaw a whole series of potentially crippling technical hurdles that might have to be overcome to get it to work, and I'm still not sure how easy that would be. Inteins would have to be placed in just the right place, and be of an appropriate length, and this would require a lot of fiddling. Also, natural inteins are designed to be snipped out as soon as the protein that contains them has been constructed, so mitochondrial inteins would need to be designed in ways that prevent them from being removed until the complete protein has been imported into the mitochondria.

Another problem we might face is that of ensuring that, in the "semi-intein" version of this approach, the protein segments are joined to their proper partners. If we have used inteins to break up several proteins that are imported simultaneously, or if one protein has been broken down into more than two pieces (as seems likely to be necessary in at least a couple of cases), the "exposed" ends of protein segments might be mismatched once the inteins are removed. Fortunately, it appears that some such multiple inteins do exist naturally and somehow "know" their proper partners, so this may not really be a problem—and if it is, the alternative solution of putting inteins into the hydrophobic stretches of these proteins is still available.

Yet another potential headache: segments of proteins could begin to fold too early, either before joining to their "other halves" in a way that obstructs access to the fusion site (so that the intended fusion with the mate is no longer possible), or even right in the midst of the TIM/TOM machinery, causing the exact problem that the use of inteins is supposed to prevent. And finally, a protein's separate, newly inteinless segments might be altered by enzymes while they wait to be joined to their "mates"; any such alterations could render them nonfunctional, or prevent them from "hooking up."

The good news is that, despite all of these potential obstacles, at least one split intein job has already been pulled off in cell culture:[17] Yoshio Umezawa and coworkers took an unimportable green fluorescent protein and introduced it into a cell's mitochondria by first adding on a borrowed mitochondrial targeting sequence, and then adding inteins inserted into its structure using a protein splicing system taken from algae, before finally "infecting" the cell's DNA with the composite structure's code. The result was that the protein was indeed expressed and imported into the cell's mitochondria, where it allowed the scientists to detect the presence of *other* proteins that they were trying to direct into the organelle.

Not TINA but TATA

This string of successes indicates that we can expect rapid progress in expressing the remaining mitochondrially coded proteins from nuclear genes in reasonably short order, through some combination of the different techniques I've discussed here. Once we have figured out how to put all thirteen of these proteins' genes into the nuclear "bomb shelter," we will be very close to being able to negate the insidious effects of mitochondrial mutations in aging, turning existing "reductive hotspots" into normal, healthy cells and preventing the formation of new ones. When that is accomplished, we will be able not only to stop but to *reverse* the slow upward creep in oxidative stress—and the accelerating spiral of molecular damage and metabolic disruption—that we think is driven by these mutations today.

But, of course, the precedent of medical history tells us that some pitfalls may still lie in wait for this solution. One can make one's most educated predictions, based on a sober consideration of the published science, and still fail to anticipate an intractable roadblock. My biggest worry is that, even after a bit of tweaking, the volume of TIM/TOM traffic required to import the new allotopically expressed proteins will be so high, and its rate of flow through the system so slow, that we will not be able to transfer the full complement of proteins via this route even if we succeed with each of them individually. I am especially concerned about this since some of the proteins that have been allotopically expressed move quite poorly through the TIM/TOM machinery; one of the examples we discussed above involves a protein that is only incorporated into the mitochondria at 40 percent of the efficiency of the native, mitochondrially-coded version.

Former prime minister Margaret Thatcher, the "Iron Lady," was famous for intoning "There Is No Alternative" (TINA) to the neoliberal economic agenda. I have no intention of painting myself into an ideological corner over my preference for allotopic expression: lives are at stake, not just my ability to admit that I'm wrong. I will instead join the ranks of her anti-globalization critics, who spilled into Trafalgar Square in their thousands chanting "Not TINA but TATA!"—that is, "There Are *Thousands* of Alternatives."

Having made that bold declaration, I hasten to add that I think we can safely trim down "thousands" to "a few, and even fewer that are viable." In fact, quite a number of alternative fixes for mitochondriopathies *have* been advanced, and at first glance these solutions might also be expected to treat age-related mitochondrial DNA deletion accumulation. Unfortunately, I predict that most of those solutions will be of no help in preventing the development of "reductive hotspots," because the mechanism whereby the underlying mutations accumulate (mitochondrial free radical generation and clonal expansion through the Survival of the Slowest) is quite distinct from the mostly inherited problems in mitochondriopathy victims. Research toward cures for mitochondriopathies using these alternatives will still doubtless yield techniques and insights that will in *some* way help us to develop a cure (via allotopic expression or some other therapy), but I do not believe that most of these will be adaptable as anti-aging biotechnology. I shot down the major flawed proposals down in the 2000 article in *Trends in Biotechnology* that I mentioned earlier. Two approaches exist that I feel have much more promise, however.

One, advanced by Drs. Rafal Smigrodzki and Shaharyar M. Khan of the Center for the Study of Neurodegenerative Diseases at the University of Virginia, starts with the ability to introduce into the mitochondria modified versions, not of *individual genes* (as in the classical gene therapy approach), but of complete copies of the *entire* mitochondrial DNA. They accomplished this maneuver—which they have termed *protofection*[18]—using boldly simple techniques that most scientists would have dismissed as unworkable. It's too early to say how versatile and repliable their method is, though—it's too new to have been explored by other scientists.

The other very promising alternative to allotopic expression is to introduce genes for alternative versions of the mitochondrially encoded enzymes, versions that don't quite work the way that the native ones do. The enzymes in question already exist in—and could thus, in theory, be borrowed from—some lower organisms (yeasts and plants). They perform the exact same electron transport activities of the enzymes that we partly encode in our mitochondrial DNA, even though they themselves are not particularly hydrophobic and are encoded in their species' nuclear DNA. The problem—if it is one—is that they *only* do the electron transport, not the proton pumping. But introducing them into our cells might be a fair tradeoff, because while they would not restore the mitochondrially mutant cell to its full capacity, they *would* prevent these cells from impairing the ability of the *rest* of the body to get on with the business of life. This has actually been documented in isolated human cells: when Complex I is chemically inhibited, normal cells quickly die off, but cells given the relevant yeast gene remain viable.[19]

One problem with this proposal is that, if these proteins were to be expressed in cells that had *not* suffered deletions, they would deny the cell of much of its energy source. We would use one gene to bypass the first enzyme in the electron transport chain (Complex I) and another to bypass the rest. Neither of them pumps protons, so having *both* of them substituting for much of the native, proton-pumping complexes in a mitochondrion would seriously impair the buildup of a proton "reservoir" and, thus, ATP production. This could lead to anything from a mild functional impairment to a severe energy deficit—and it would extend through *every cell in the body,* not just the "evil 1 percent" of cells with a population of mutant mitochondria. This would leave us in a very bad way indeed, so we would need to find a way to ensure that the genes for these "bypass enzymes" are expressed only in cells whose mitochondria are no longer producing the native proteins. There is no obvious way to do this at the moment, unfortunately. But luckily, initial research indicates that these enzymes *naturally* possess a system that activates them only when their proton-pumping counterparts are failing. Even expressing one of the two "bypass enzymes" indiscriminately would make cells less efficient at making ATP, so this is actually no surprise.

The Way Forward

Overall, the picture is a rosy one. We have not only a good idea of how mutations in the mitochondrial DNA contribute to the age-related decay of our bodies, but a clear path forward to obviating the problem—even if our understanding of the exact mechanistic link between mutations and pathology turns out to be mistaken. Allotopic expression would allow our mitochondria to keep functioning normally even when their DNA acquired mutations; protofection, alternatively, could simply clear out the old mutant DNA periodically, replacing it with a fully functional new set of genetic blueprints; and the use of easier-to-handle enzymes that pump no protons but keep electron metabolism harmless would at least keep mutant cells from causing trouble outside of their own membranes.

Again, we will need to develop safe, effective, stable gene therapy that works in grown humans in order to turn any of these interventions into a real biomedical intervention against this aspect of the aging process, and that will certainly be a challenge. But, again again, it's a challenge that scientists all over the world are already vigorously pursuing in order to treat genetic disorders—ailments ranging from Huntington's disease, to inherited cancer risk, to familial Alzheimer's disease, to sickle-cell anemia. And we can piggyback even more closely on research that is specific to the mitochondriopathies—a much smaller field, but one that, for the moment, still receives more serious funding and attention than work designed to tackle the slow-motion global plague that is aging.

With the resources already being thrown at advancing gene therapy, we can confidently predict that the clinical readiness of this enabling biotechnology is foreseeable. I am therefore convinced that the major hurdle facing speedy implementation of allotopic expression (or its alternatives) will not be our ability to develop safe gene therapy for patients, but the dearth of investment into the basic science of moving mitochondrial genes into the nucleus.

Remember the positive result using inteins to import proteins into mitochondria in culture? This achievement came about because scientists were seeking a way to get rapid feedback on the results of a completely unrelated project of interest to them. Imagine what could be accomplished with resources *specifically dedicated* to developing allotopic expression for the purposes of reversing aging damage!

How to bring about that change in research priorities is the subject of Chapter 15; but now, let me take you on a tour of the next of the "Seven Deadlies" of aging, and show you what we can do about it.

7

Upgrading the Biological Incinerators

Just like our own households, cells generate garbage as an inevitable result of their normal functioning. Again like households, they are able to dispose of most of this waste—though they recycle a proportion of it that would put the most eco-friendly household to shame. But cells cannot recycle quite all the junk they create, and the portion that escapes destruction accumulates, to the cell's eventual detriment. Several years ago I devised a new approach to this problem that exemplifies, perhaps better than any of my other contributions to this field, the value of the widely cross-disciplinary expertise that is so rare in biology today.

Mary Shelley couldn't ask for a more perfect scene, I thought, as I sank my trowel into the scruffy graveyard sod.

A quick scan of the horizon at Coldham's Common would initially make you think it was a nondescript, even rather dull little field in the heart of England. But knowing its history transforms your view of the spot, opening the mind's eye to a bleak, windswept stretch of near wilderness, dropped as if out of a Gothic horror novella into the midst of a plain bounded by football grounds and parking lots, bisected by a railway line. Though it is sometimes used for public events or cattle grazing, it spends much of the year lonely and abandoned, its sole claim to fame arising from its association with mass death.

In the late seventeenth century, the Great Plague swept its scythe across England. When its icy fingers crept into Cambridge, the plague claimed a third to a half of the residents—including sixteen of the forty professors at the University—and sent the young Isaac Newton fleeing for his life. In its wake, the survivors hastily plowed most of the plague's victims anonymously under the unhallowed ground. Even before it became a mass grave, the area bore the taint of association with infection and death: its most enduring landmark is the remains of Cambridge's twelfth-century Leper Hospital. As if to complete the cliché, on most days of the year Coldham's Common is documentably several degrees colder than the cobbled streets that surround it.

The scene, then, was complete: I, the "mad scientist" (complete with long beard and pale, sunless skin), surrounded by Cambridge's Enlightenment-era faux-Gothic castles and cathedrals, had hopped my way with an irrational furtiveness over multiple fences into the last resting place of the mortal remains of untold scores of lives, and was now digging into the soil of a mass grave, in hot pursuit of the secrets of Life and Death.

Victor Frankenstein, eat your heart out.

I must confess that the above account incorporates a *small* amount of poetic license: the person who performed the above task was not I but a graduate student in my University of Cambridge department, and actually she retrieved the soil sample from Midsummer Common, not the nearby Coldham's Common. But that's by-the-by. To understand what she was doing there, let's take a detour out of the graveyard and into the junkyard.

The Waste of Your Life

Whether we toss things into the garbage bin without a thought, or painstakingly wash and sort our recyclables, we in the developed world generate an astounding amount of waste material every day. When we decide that we don't need something anymore, or that it's too damaged or rotten to be worth our efforts to salvage, we simply stuff it into a bag or bin and put it out for pickup, confident that the unsung heroes of the sanitation department will take it away and out of our concern. Thanks to an efficient waste-disposal infrastructure, a truly remarkable volume of waste material can pass through our homes, workplaces, and streets—yet these places remain clean, pleasant-smelling, and sanitary.

It wasn't always this way, of course. For most of the history of civilization, the streets of our cities were literal cesspools, into which the citizens hurled their trash and human waste directly out of their windows without care for what—or even who!—was below them. Most of us truly cannot imagine what foul, malodorous, and *dangerous* places cities were until quite recently. The toll of living in such a toxic environment can be seen in the disparity of life expectancy for people living in different environments in seventeenth-century England. An Englishman would typically live to be thirty to forty years old if he lived in the countryside, but if he lived in London, he could expect just twenty-one to thirty-four years of life.

Anyone who's lived through the kind of big-city garbage strike that nearly paralyzed London in 1976 has an idea of just how vital a functioning waste-collection system is for health and the carrying-on of the business of daily life. In shockingly short order, trash can literally be piled ten feet into the air, in precarious piles that fall apart in the wind or as new bags are added to the crude structures. And the mountains of garbage are not merely unsightly: Aside from the smell, the garbage attracts vermin, and with it, disease—particularly when the contents of the trash bags begin to spill onto the street because of attacks on the bags by animals, the elements, or the putrefaction and liquefaction of its contents. Sidewalks become increasingly impassable, and even street traffic may be impeded. People become less willing to leave the house or go into shops. A strike that lasted just nine days in 1968 came close to bringing New York City to its knees.

Well, something similar happens to your cells as they age—except that in a sense it's *worse.* Rather than a temporary "interruption of service," aging cells undergo a progressive degeneration of their waste management infrastructure that would make the worst examples of inner-city decay look like models of sanitation.

Two chapters ago, in discussing the process whereby mutant mitochondria "clonally expand" to replace all of their genetically healthy cousins in the cell, I introduced you to the *lysosome*—an organelle that I called the cellular "incinerator." Actually, "recycling center" would be a more precise metaphor than incinerator, because a lysosome's job is not to out-and-out *destroy* cellular wastes, but to *break them down* at the molecular level into more basic components that can be used as raw materials for the biosynthesis of new cellular membranes, enzymes, and other important components of the cellular machinery. The incinerator metaphor is meant to convey the extraordinary *power* of the lysosome's molecular-level dismemberment of the materials that are thrown into it, as well as the *chemical* nature (burning is a chemical reaction, remember) of the lysosome's methods of breaking down waste into its fundamental components.

While the cell actually has a variety of mechanisms for reprocessing damaged cellular constituents, its lysosomes deal with some of the nastiest of them, including the waste materials that are still left over after the other cell waste-disposal systems have had a go at them but failed. In addition, when those alternative waste disposal units *themselves* become worn-out or damaged, it falls upon the lysosomes to break them (and, often, their semi-digested contents) down. This chapter is about what goes wrong with the cell's scrapyards of last resort and how we might avert that process.

Cleaning Up Life's Messes

Your cells, like your household, are constantly producing and consuming goods and generating wastes of various kinds in the process. One sort of waste is akin to packaging, or disposable pens, or tacky old bric-a-brac, the possession of which has come to embarrass you. It may have served a purpose at one time, but today you have no further use for it and wish it gone. Many cellular constituents are like this: Enzymes and signalling molecules are "disposable," produced for temporary or even one-time use in response to the immediate conditions in or around the cell, and they need to be degraded once they've served their purpose.

Another type of waste is more like something for which you would still have a use, except that it can no longer fulfill its purpose because it's been broken. Just as you can turn a piece of your grandmother's china into a jigsaw puzzle, or make your shirt unwearable at work by spilling dark red wine on it, so components of your cells—from the small (individual enzymes) to the very large (entire organelles, like a mitochondrion)—can become incapable of performing their vital cellular function after suffering molecular damage at the hands of free radicals and other products of the dirty underside of metabolism.

And a third type of waste is genuinely toxic material. Just as a useful substance (say, cottage cheese) can become a threat to your health through chemical changes (such as being taken over by mold), so normal cellular constituents sometimes become toxic to the cell through modification of their structure, or through being present in excess. As surely as, upon encountering the new ecosystem growing in your cottage cheese, you seal the container and drop it from shoulder level into the garbage, so, too, the cell needs to eliminate similar threats to its functioning.

Recycling's Dirty Details

All of this waste (except, again, that which is destroyed by simpler machinery) is directed to the cell's lysosomes. Functioning lysosomes ensure that it gets properly processed, removing toxic by-products of normal cellular machinery, returning usable molecular building blocks to the cell from the "slag" of the broken-down components, and making room for healthy, functioning cellular constituents.

So what exactly *are* these cellular incinerators? Lysosomes are membrane-bounded organelles packed full of a variety of enzymes, each of which evolved to target a "weak spot" in the chemical structure of a waste product that will accumulate in our cells and kill us if it isn't broken down. A lysosomal enzyme first binds to a waste product that carries the kind of chemical structure that it evolved to destroy, and then twists its shape like a tiny biological crowbar, physically tearing apart the target material's molecular joints. This is generally accomplished by a type of chemical reaction called *hydrolysis,* which is why such enzymes are called *hydrolases* (*hydro* being the Greek for water, as in "hydroelectric").

The exact chemical details of this process aren't terribly important for our purposes, but you should be sure to remember one key point. In order to break down a given waste product, the lysosome must have two things: the *right enzyme* for the job (one that targets a vulnerability in the structure of the *specific* waste product in question), and *enough acidity* in its interior for the relevant enzyme to function. This latter is required because different levels of acidity cause enzymes and other proteins to assume slightly different shapes, so that when the acidity in the lysosome is wrong, the enzyme becomes "bent out of shape" and thus no more able to do its job than is a flattened-out crowbar. Acidity is also required for the functioning of proteins that translocate some waste products into the lysosome in the first place, so that a mild neutralization of the lysosome's acidity prevents junk from even being delivered to the recycling center to begin with.

Lysosomal enzymes, like other cellular proteins, are created out of the blueprints present in the nuclear DNA. They are then shipped into the lysosome, though by machinery very different from the mitochondrion's

TIM/TOM complex. The extra protons that create the lysosome's acidity are actively pumped out of the main chamber of the cell and into the lysosome by an energy- (that is, ATP-) consuming pump located on its membrane (the *vacuolar ATPase*).

Incomplete Combustion

You won't be surprised to learn that bad things happen if your body fails to produce a lysosomal hydrolase that is needed to break down a waste product being produced in some cell type—or if it produces a defective form of the protein that doesn't do its job properly. In fact, this is precisely the description of a group of rare but well-established genetic disorders known as *lysosomal storage diseases* (LSDs).

There are about forty such diseases, but luckily only about one person in 7,500 is born with any of them. Victims of all of these diseases suffer from one or another type of failure of their lysosomal incinerators. Many of them completely lack the gene for a lysosomal enzyme, or bear a mutated copy of it, resulting in a misshapen and ineffective version of the hydrolase. In other cases, the problem is that one of the specialized transport proteins on the surface of the lysosomal membrane is missing or defective, so that the lysosome can't bring the junk *into* itself to break it down.

No matter what their origin in a given patient, the result of such mutations is a deadly degenerative disease. Which organs a given mutation affects, and how severely, varies from one LSD to another, depending on exactly which missing or malfunctioning enzyme is at the root of the problem. This is because different cell types produce different wastes at different rates, and each particular waste exerts a distinct pathological impact on the cell if it isn't degraded.

But in all cases, patients suffer pathology in major organs. In Gaucher disease, the spleen swells up and anemia develops. There are two inherited forms of Niemann-Pick disease. In the fast-acting version (Type A), the liver and spleen enlarge and the nerves degenerate starting at birth, killing its victims by age two or three. In the slower-acting variety (Type B), patients may develop fatty, yellow nodules on their eyelids, neck, or back, and an enlarged liver, spleen, and lymph nodes. And Hurler syndrome causes facial features to twist up and bone deformities to occur, along with enlargement of the spleen and liver, joint stiffness, clouding of the eye, early-onset dementia, and hearing loss.

The exact mechanistic links between the lack of effective waste disposal and particular pathologies have not all been worked out in detail, but the basic picture is clear. The undegraded waste material accumulates in the lysosome, causing it to swell up and take up too much room in the cell, impeding the traffic of other materials in the main cell body. Meanwhile, the acids and enzymes *within* the lysosomes are diluted, inhibiting their ability to both import and break down *other* wastes for which the cell *does* have the requisite enzymes, thereby setting up a vicious cycle.

There are also some cases in which it appears that toxic, undegraded waste accumulates in the main body of the cell. This can be either because it is not trafficked into the overburdened lysosome in the first place, or else because the failing organelle starts to leak or even bursts, spewing its toxic load—including the acids and enzymes that it carries, which are essential to lysosomal function but potentially deadly to the rest of the cell.

Lysosomal Limitations and the Deadly Dregs

In addition to the terrible, early-acting pathologies that ravage the victims of these genetic disorders, however, it's also long been known that undegraded gunk builds up in the lysosomes of all of us as we age. Called *lipofuscin* (LIP-oh-few-sin[1]) or popularly "age pigment," this noxious, accumulating goo is a chemical hodge-podge of fatty and proteinaceous materials derived from membranes, reactive metals like iron and copper, and a variety of other organic molecules. It is easy to see with a microscope because it glows red when exposed to light of a particular wavelength.

Lipofuscin is actually not a single, specific compound, but a catch-all term for the mixture of stubborn waste products that refuse to be broken down after they've been sent to the lysosome for degradation—materials so chemically convoluted that the normal complement of lysosomal enzymes just doesn't know how to deal with them. A combination of damage from free radicals and from *glycation* (the random sticking-together of different "branches" of a substance's proteins by reaction with the sugars in your blood and cells) twists their structure back on itself like some demented child's molecular origami, burying the vulnerable spots in their structure so that lysosomal hydrolases can't get at them to break them down.[2] As a result, these materials don't get properly degraded—and because lysosomes are in most cases unable to export them out of the cell, the material just accumulates, taking up more and more room in the lysosomes of long-lived cells like the heart and the brain.

First spotted in the nineteenth century, lipofuscin was largely ignored by biomedicine until it became a hot—and controversial—topic in biogerontology in the 1970s. At that time, it seemed just *obvious* to many researchers that lipofuscin must be bad for us: it slowly fills up our cells, (taking up as much as 10 percent of the total space in aged primates' heart muscle cells, for example), and the course of its accumulation tracks the cell dysfunction seen in aging animals (including people). Indeed, the *rate* at which lipofuscin accumulates in a given species' heart was found to be proportional to its rate of aging, so that adolescent, middle-aged, and old monkeys of two different species, with greatly different calendar ages but at similar stages in their life cycle, have roughly the same level of lipofuscin clogging up their cells. Many researchers outlined a mechanistic hypothesis of lipofuscin as a contributor to aging similar to what happens in the LSDs: lysosomal failure, waste accumulation, interference with cell trafficking, and the release of toxic enzymes and acidity from ruptured lysosomes.

But the point was a contentious one. Many scientists believed that lipofuscin was benign, in part precisely *because* it is so hard to degrade: With the reactive spots in their structure already balled up and stapled together by previous free radical and glycation damage, lipofuscin is chemically quite inert, so it doesn't interact with essential biomolecules the way that free radicals or other toxic chemicals do. Also, speeding up the accumulation of lipofuscin in experimental animals by denying them adequate vitamin E did not shorten their lives, as one would expect of any manipulation that accelerated a real cause of aging.

But those reports were strongly disputed by other experts in the field, because it was not at all clear that the material whose accumulation was increased by vitamin E deficiency actually *was* lipofuscin. Much of what gets referred to by this name in the older scientific literature is actually other, related substances (often called *ceroid*) that share many of lipofuscin's properties but are much easier for the cell to break down. It seemed likely that vitamin E deficiency was increasing the production of this relatively tractable material, while levels of "real" lipofuscin were unaffected.[3] Moreover, while ceroid accumulation was associated with a variety of diseases (diseases in which normally degradable substances are not degraded—one could think of them as nongenetic LSDs), it was not clearly related to "normal" aging.

And so, the debate went 'round. As with many such cases in the early days of biogerontology, the data were ambiguous, the definitions were imprecise, and there was little hope of a clear resolution.

I had, myself, remained an agnostic on the subject up to and including the writing of the first draft of my doctoral thesis. But that began to change in the spring of 1998, when I met Ulf Brunk, chair of pathology at Sweden's University of Linköping, at the Oxygen Radicals Gordon Conference in Ventura, California. After listening to his presentation of his recent results, I started taking lipofuscin more seriously as a potential nexus connecting the intricately orchestrated chaos of metabolism to the pathology of aging.

Brunk had done some first-rate work in assessing the role of lipofuscin in cultured heart cells—an important technical advance, especially because, in the body, heart cells never divide. When a cell divides, its load of cellular junk—including lipofuscin—is shared between the two daughter cells. Each now has half as much, on average, and this will continue with each new generation. If the junk is only being generated quite slowly, this dilution process will fully balance the rate of creation of new junk and an unmanageable level of junk will never accumulate. The quickly replicating skin cells that had been used in much previous work did just this—they tended to dilute out lipofuscin and other cellular junk, and so could not replicate the real effects of lysosomal buildup in critical nondividing cells like those of the heart and brain. This had long been recognised, but heart cells had been found to be notoriously hard to culture—in large part, it turns out, because of the high levels of oxygen in the atmosphere. Cultured cells are still too often grown under normal atmospheric air, despite the fact that our bodies ambiently contain only about *one seventh* of air's concentration of oxygen.

By growing the longer-lived heart cells under more physiological levels of oxygen, Brunk was able to show how increased oxidative stress—higher levels of free radical damage, in other words—increased lipofuscin formation. He also confirmed previous suspicions about how lipofuscin accumulation could impair the ability of the cell to recycle its used-up components. And on top of that, he was also able to show that older heart cells accumulate damaged mitochondria—which they would not do if the lysosome were operating properly, since the disposal of defective cellular power plants is one of their chief responsibilities in the cellular economy.

With his collaborator Alex Terman, Brunk outlined a "garbage catastrophe" theory of aging,[4] in which accumulating lipofuscin inside the lysosome dilutes the organelle's acidity and supply of enzymes. In this model, lipofuscin also wastes a lot of the enzymes that the cell body produces, by sucking them up without making effective use of them, thereby diverting them away from the other, still-functional lysosomal contents against which they could be put to effective use.

As the cell's lysosomes accumulate waste products that they aren't equipped to handle, they become ever less able to break down materials within them. As a result, the junk in question spends more time either out in the main body of the cell, or even trapped inside the lysosome, before being "incinerated." During that time, chemical alterations continue to occur in the structure of these wastes, mangling them further and further and making it more and more difficult for lysosomal hydrolases to reach the weak spots in their structure. As a result, even the standard cellular junk that lysosomes are, in theory, well equipped to degrade is no longer efficiently broken down, but instead accumulates—which then further dissipates the hydrolytic enzymes and acidity of the lysosome. In their culture experiments, Brunk and Terman showed that lipofuscin overload could even trigger cell death, as lysosomes become loaded with the stuff and rupture.

The data underlying this model were compelling, and I liked it at once, sneaking a short reference to it into my Cambridge biology thesis.[5] But I wasn't yet convinced that lysosomal failure was truly a significant contributor to aging, because if the theory were right you would expect to find evidence connecting lipofuscin to actual age-related disease, and no such evidence initially turned up when I went looking for it.

I quickly learned, however, that this seeming lack of data was more of a communication breakdown than an information vacuum. Researchers tend to get holed up in their narrowly specialized fields of study, and consequently they, too, rarely compare notes and observe the confluence of observations in different fields of science (or even subfields within those fields). I soon found that if I stopped specifically talking about "lipofuscin" and began asking researchers about the importance of lysosomal dysfunction in the diseases that they studied, I was suddenly inundated with evidence that the accumulation of junk that *should* be processed in the lysosome was at the heart of the matter—but that this fact was being obscured by the use of specialist jargon in referring to those wastes.

Making the Link to Pathology

Atherosclerosis

Just over a year after I was first exposed to Brunk's suggestive data, I was attending the biennial Gordon Conference on Atherosclerosis, at which I found myself listening to a review on the complex processes that lead someone from having too much cholesterol in the blood, to having fatty plaques in the arteries (atherosclerosis), to having diagnosable coronary heart disease and eventually a heart attack. As I quickly learned, researchers had been placing lysosomal failure at the core of the molecular events that underlie the formation of atherosclerotic plaques for years before I began looking into the issue—and they did so without ever mentioning "lipofuscin."

Most people visualise atherosclerotic arteries as being much like clogged pipes. Greasy gunk (whether it's bacon fat or blood cholesterol) simply accumulates on the inside of the tube, coating its surface and clogging it up, and blocking the passage of fluids—be those fluids the dishwater in your sink or the blood in your arteries. In fact, however, we've known for some time that the process is much more complicated than this.[6] Atherosclerosis begins with a microscopic problem in the blood vessel wall. Many things can cause or contribute to this, including friction from the passing torrent of blood flow, or the force of high blood pressure, or infection; most often, however, it's just the accumulation of our old friend LDL, low-density lipoprotein, which has a tendency to get stuck there. The body responds to this problem just as it does to any other injury: by secreting factors that inflame the site in order to attract immune cells called *macrophages*. Macrophages then infiltrate the damaged tissue to help it heal by cleaning up the debris.

I didn't tell you very much about LDL in Chapter 5; now's the time for more detail. Despite its bad reputation, cholesterol is actually a necessary component of cell membranes. In fact, the so-called "bad" cholesterol (LDL) in the blood is actually a *carrier particle,* designed by your body to bring needed cholesterol to cells, and those cells in turn have specialized receptors designed to allow them to ingest it for their internal use.

In order to reach most cells, the cells that comprise the walls of our blood vessels must allow LDL to pass between them, and beyond into the surrounding tissue. But when cholesterol is chemically modified—by exposure to free radicals (*oxidized LDL*) or reactions with blood sugar (*glycated LDL*), for example—it becomes more prone to stick together and thus more immobile. Because free radicals and blood sugar are (respectively) inevitable by-products of, and necessary raw materials for, some of the most fundamental metabolic processes in the body, they are ubiquitous. Hence, LDL particles are constantly being subjected to their chemically disruptive influence. Furthermore, enzymes that are designed to call in immune cells when the vessel wall is injured also alter cholesterol in ways that make it more toxic.

That's the main reason why having a high cholesterol level is bad for you. The more cholesterol there is in your blood, the more contact it has with these damaging agents, and the more toxic, modified cholesterol will be coursing through your body.

So, when macrophages are attracted toward an inflammatory signal, they find plenty of junk in need of removal. Initially, macrophages deal reasonably well with the gunk that they're internalizing, and they can often successfully remove the detritus. But when an already compromised blood vessel continues to be assaulted by high blood cholesterol levels, inflammatory signals created by excess body fat, or nasties from cigarette smoke, the problem persists and macrophages hang around for longer.

As they take in more and more waste—particularly an excess of modified LDL—macrophages begin to fall behind in their work. An increasing percentage of the load of junk is not successfully processed, but instead accumulates within the macrophages' lysosomes—or, just as bad, is puked out of the lysosomes without being properly detoxified, forming droplets of modified cholesterol in the cell body.

As this continues, macrophages eventually become the cellular counterparts to the obscenely bloated "Mr. Creosote" in the restaurant sketch in Monty Python's *The Meaning of Life.* If you've seen this movie, you will definitely remember the scene. Mr. Creosote comes into a fine French restaurant already stuffed with food and badly nauseous, but is plied with "moules marinières, pâté de foie gras, Beluga caviar, eggs Benedictine" and sauces "rich with truffles, anchovies, Grand Marnier, bacon and cream" by the perversely codependent and outrageously "French" maître d' (John Cleese).

Creosote becomes more and more ill as the meal progresses, but when he finally attempts to summon up the will to stop eating he is cajoled into having just one last "wafer-thin" after-dinner mint. When the spineless patron swallows the mint, a look of helpless horror fills his face; as the maître d' runs for cover, Creosote literally *explodes,* his innards and lunch splattering graphically over staff and guests alike.

Imagine your blood vessels to be such a restaurant, welcoming a steady stream of customers just like Mr. Creosote in the form of macrophages that come in to dine on modified cholesterol products. Imagine that they simply refuse to leave when you're trying to close up, but continue to stuff themselves until their "stomachs" (lysosomes) can't take any more—and then keep going until it kills them, and your restaurant (blood vessels) becomes their final resting place.

You now know, in essence, the genesis of the "foam cells" that accumulate in your vessel walls, forming "fatty streaks" as they become numerous enough to be seen with a microscope, and eventually developing into full-blown, unstable atherosclerotic plaques—the scabbed-over messes that ultimately form at the injury site, crammed with a miasma of clotting blood, inflammatory signal molecules, and dead foam cells. Once this happens, your days are numbered. It's only a matter of time until the pressures within and without the plaque cause it to rupture, spewing its contents into the bloodstream. This content is not a liquid but a horde of semisolid chunks, and these chunks are rapidly swept from their origin in the major arteries into progressively smaller vessels. They become stuck there, blocking off the flow of blood—sometimes into the heart (causing a heart attack), sometimes the brain (causing a stroke).

So we now understand that lysosomal dysfunction is the key step in the conversion of healthy macrophages into undead foam cells—and of healthy blood vessel tissue into an atherosclerotic time bomb. This fact is widely recognized, but unfortunately, nearly all researchers are pursuing old-school, ultimately *preventive* treatments for the problem. Existing anti-atherosclerotic drugs try to prevent macrophages from stuffing themselves so badly, either by reducing blood cholesterol levels or by reducing LDL's exposure to metabolically active agents (blood sugar, inflammatory enzymes, and free radicals). Drugs currently in the pipeline seek to approach the same problem from its flip side, by increasing the transport of cholesterol *out* of the blood, cells, or organs before it gets a chance to do its damage.

Meanwhile, basic researchers working in areas other than drug development are spending a lot of time trying to puzzle out exactly what causes macrophages' lysosomes to fail, with the idea that, if they understood the fine details of the process, they could design drugs that would interfere in the relevant steps in the metabolic chain. Unfortunately, the evidence is consistent with many interpretations, and the data are difficult to reconcile, which has more-or-less stalled progress in developing therapies based on this model.

For instance, some researchers focus on the fact that, in test tubes, oxidized cholesterol inhibits the necessary processing ("de-esterification") of normal (unmodified) cholesterol in the lysosomes, slowing it down enough to create a deadly backlog. Others think that modified LDL, like lipofuscin, is *itself* undegradable, and dilutes out the factors needed for the lysosome to degrade other materials as it accumulates. There is also evidence that something in modified LDL (or some metabolic by-product of it) is harmful to lysosomal *function*—such as the evidence (again in test tubes) that the oxidized cholesterol variant *7-ketocholesterol* (7-KC) interferes with the activity of the membrane-bound ATPase enzyme. When this enzyme is impaired, it can't maintain enough acidity in the cell's lysosomes to keep their hydrolases working properly, so there are those who think that this is where we need to look for a solution. And still others think that macrophages simply take in *too much* LDL, so that its sheer volume overwhelms their processing capacity; if so, slowing uptake might also slow down the development of the disease.

As yet, we don't know which school of thought is correct—and it's unlikely that we will resolve the question definitively any time soon, because the conditions under which the relevant studies are carried out are so unlike what happens in the body. While the scientific debates continue, vascular disease caused by atherosclerosis remains the number one killer in the developed world—and other problems that arise from the same failure to process cholesterol may be related to a wide range of other age-related diseases.

Fortunately, an intervention that came to me in a flash some years ago[7]—and that has since been worked out in greater therapeutic detail in collaboration with others[8,9]—offers a solution that sweeps away the need for this kind of detailed molecular map of the metabolic maze. This solution does not rely on such detailed understanding of what causes lysosomal failure in atherosclerosis. Instead, it provides a way for us to clean up the lysosome *itself,* rather than the metabolic processes that overload it—and in a manner that will work *irrespective* of what leads up to its initial failure.

But before we get into that, let's look at another fearsome disease of aging that has lysosomal dysfunction at its heart: the decay of the brain.

Neurodegenerative Disease

Except in the case of stroke—which I've covered above, and which is more of a one-off, traumatic injury than a degenerative process in itself—the brains of people suffering from *all* the major neurodegenerative diseases show evidence of inadequate lysosomal function. In most cases, the most obvious pointer is the presence of clumps of a distinctive aggregated protein material inside the brain cell: *Lewy bodies* in Parkinson's disease and the boldly named "Dementia with Lewy Bodies" (DLB), aggregated *hunt-*

ingtin protein in Huntington's disease, and *neurofibrillary tangles* (NFT), formed of aggregations of the protein *tau*, in Niemann-Pick and Alzheimer's diseases.[10] Yet, because these aggregates are not located *within* the lysosome, and are not themselves lipofuscin, the role of lysosomal dysfunction in these diseases has been obscured—so again, people specifically looking for a connection with "lipofuscin" can miss these data, obscuring the relationship.

In several cases, however, there is more direct evidence of trouble at the toxic waste dump. Some of the most remarkable such evidence has recently been found in the brains of Alzheimer's patients, in which the breakdown of proteins through another of the main components of the cell's recycling system (the *proteasome*) is badly impaired. In some victims, this may be because mutations in the gene for a protein called *ubiquilin* cause it to inhibit the activity of *ubiquitin*, a protein that "tags" proteins for breakdown in the proteasome. Both neurofibrillary tangles in Alzheimer's and Lewy bodies in Parkinson's disease are loaded with ubiquitin, yet the proteasome system seems incapable of picking these aggregated materials up.

The connection with the lysosomal apparatus is this: proteasomes that are not doing their job put more pressure on the lysosomal system as the defective proteasomes (and the material they have failed to destroy) are sent to the lysosome, increasing lipofuscin formation.[11] At least some of the waste that the proteasome fails to pick up—along with damaged proteasome units themselves—is ultimately sent to the lysosomes: this phenomenon has been definitively observed in the case of aggregates normally degraded by the proteasome in Huntington's disease, and is probably what's responsible for the finding of a lot of ubiquitin inside the lysosomes of the neurons of Alzheimer's patients.

But the most dramatic hallmarks of abnormal trash disposal in Alzheimer's disease are the signs of malfunction in the lysosomal system itself. As a bit of background: One of the main ways in which cellular rubbish gets delivered to the cellular recycling center is through a process called "macroautophagy," in which the waste in question is swallowed whole by a membrane structure called an *autophagosome* or *autophagic vacuole* (AV), which then hooks up with the lysosome and fuses with it. (If this term rings a bell, that's probably because I briefly mentioned it in Chapter 5 as the way in which damaged mitochondria are delivered to the lysosome.) The result, in effect, is a bigger lysosome, with a single combined membrane that surrounds both the contents of the AV and the hydrolytic enzymes (and acidity) of the original lysosome to digest that contents.

Recent studies show that this aspect of lysosomal function is in a very bad way in the brains of Alzheimer's victims.[12] It has been known for some time that the lysosomal system in the Alzheimer's brain is, like the proteasome, apparently both hyperactivated and inactivated: it's as if the neuron were an unthinking driver of a car with a worn-out engine, trying unsuccessfully to compensate for its misfiring cylinders by pushing down harder on the gas pedal. The new work suggests one major reason why they fail: Their brain cells—and especially the cells located in areas of the brain that are most badly affected by the disease—are full of multilayered AV-based structures that are a lot like Russian nesting dolls, with one AV contained within another, larger one, which in turn is sequestered inside another, still larger AV.

Some of these structures seem to form when AVs fail to fuse with lysosomes, and hang around in the cell long enough to begin to take some damage, ultimately becoming so badly degraded as to be recognized as junk—at which point they are swallowed up into *another* autophagic vacuole. Then, the cycle repeats itself, as the new AV itself fails to fuse. In other cases, it appears that the AVs *have* fused with a lysosome, but that the lysosome is so weak—or perhaps so immature—that it can't degrade the AV contents.

It's a picture that reminds me of nothing so much as the infamous *Khian Sea,* a ship hired by the city of Pennsylvania in 1986 to haul its incinerator ash to an artificial island in the Bahamas for disposal. Unfortunately, the Bahamian government had not given the operators of the *Khian Sea* permission to dump its waste there. And so began a fourteen-year world cruise of garbage, in which the ship traveled from port to port, attempting to dispose of its load in different countries all over the world—first back up the east coast of the United States, then back down south to the Caribbean and South America, and ultimately wandering as far afield as Indonesia and the Philippines.

Ultimately, the *Khian Sea*—renamed and reflagged—relieved itself of its toxic burden by illegally dumping it into the Atlantic and Indian Oceans. Sooner or later, peripatetic AVs can only be expected to discharge *their* hazardous contents, too.

Scientists all agree about the basic facts: the major neurodegenerative diseases are characterised by the presence of aggregated proteins and lysosomal dysfunction in the brain, and it's clear to everyone involved that there is *some* kind of connection between the clear failure of the cell's waste disposal systems to deal with the aggregates and the diseases in which these disruptions occur. The question is just *what* that connection is. Intuitively, it makes sense that the aggregated junk sitting around in our brain cells must be bad for them. Most scientists in the field share this intuition, and indeed it's easy to show, in relatively crude test-tube experiments, that these substances work mischief in brain cells to which they are added, including the initiation of a vicious cycle in which the accumulation of aggregates disrupts normal neuronal function, leading to further lysosomal dysfunction and protein aggregation.

But others have a different take on these phenomena. Surprisingly, some scientists think that protein aggregates may in some sense be *protective*. The idea is that while the aggregates themselves may in the *long* term interfere with cell function by blocking cellular traffic with their sheer size, the soluble, highly reactive units that *make up* the aggregate are a much more immediate threat to the health of the cell. By handcuffing these units together into a single cellular chain gang, the cell can keep them from attacking other cellular apparatus in their environment, preventing a deadly *short-term* threat to cellular health.

And then there are those who view the aggregates as being more of an epiphenomenon: a sign that something is wrong with the cell, but not an actual contributor to pathology. In this model, undegraded protein deposits are more like gunsmoke than actual guns or the bullets they fire: in themselves they are more-or-less harmless, but their presence is an unmistakeable signal that you're in a crime scene. Perhaps, for instance, some *other* contaminant is building up in the lysosome, preventing it from properly incinerating cellular garbage, so that the aggregates build up—but the aggregates *themselves* aren't the source of the problem or a major contributor to cellular pathology. This is still a bad thing to have happening, of course, because cells rely on a functional lysosome—both to break down benign cellular constituents that are past their useful life in order to make use of their building blocks for future cellular construction projects, and to destroy genuinely toxic wastes. But the source of the problem is to be found elsewhere than the obvious piles of trash cluttering about the main body of the cell.

For example, Alzheimer's patients may have more defective mitochondria in need of recycling than healthy people do, putting demands on the lysosome that it just can't satisfy; once the lysosome fails, other components may form the observed aggregates, but it's still the dysfunctional mitochondria that started the ball rolling downhill. But again, it's pretty hard to escape the conclusion that the resulting protein clumps constitute cellular "speed bumps" that must eventually cause the cell some serious problems of their own.

Unfortunately, there is substantial evidence—both in neurodegenerative disease and in aging—to support each of these positions. "Unfortunately," I say, because I feel it is paralysing researchers in their quest for cures. Researchers spent much of the 1990s in entrenched holy wars between the "BAPtists" (named for "Beta-Amyloid Protein") and the "Tauists" (named for the tau-based neurofibrillary tangles or NFTs), each of which expended considerable effort in trying to prove their favored candidate to be the primary problem in Alzheimer's disease. ("What's beta-amyloid?" I hear you cry. You'll learn plenty about that in Chapter 8.) Today, there is a similar feud simmering over the different interpretations of the role of protein aggregates generally in neurodegenerative disease. And in old-school thinking—in which the goal is to find drugs that will shut down the metabolic processes that lead to a disease outcome, or at least perturb that aspect of the pathway that causes the most harm—issues of this kind must be definitively resolved in detail before we can even *begin* to design treatments for humans, since interfering with metabolic pathways is a risky business that can only lead to harm if the process that you're blocking turns out to be an innocent bystander.

Even more so than with atherosclerosis, then, traditional medical approaches to neurodegenerative disease are, with respect to protein aggregates, at a standstill because of inadequate understanding of the link between the junk in question and the disease itself.[13] Again, however, I have a solution in mind that sweeps aside the need to resolve these ambiguities.

Macular Degeneration

While I don't wish to tease you, I do want to go over the critical role of undegraded aggregates in a third important aspect of aging before finally revealing my proposed therapy for all diseases involving lysosomal failure—including aging itself. This third age-related problem is *age-related macular degeneration* (AMD).

There is some relief from suspense in this section, inasmuch as there is no controversy about the involvement of aggregates in AMD. This is a classic case of how biochemical cycles with which we absolutely cannot dispense lead to the destruction of the systems in which they are embedded. Vision, like all of life's processes, is ultimately mediated by a carefully controlled, complex chemical chain reaction, and our conscious perceptions correspond in a one-to-one fashion with the particular electrochemical phenomena that this cascade triggers in our brains. In order to perceive an object, the energy from the light that reflects off it and into the lens of our eyes must be translated into the chemical signalling language that corresponds to our subjective "sight" of the object.

For our purposes, the important step in this translation process—important because fatal to the cells that suffer it, and hence to our eyesight—is the (nearly) perpetual cycle of a derivative of vitamin A between two forms.[14] The rods and cones of your eyes contain the "storage" form of this compound (11-*cis*-retinal), which is chemically transformed into an "activated" derivative (all-*trans*-retinal) when it absorbs energy from incoming light. This activated form is used as a signal to turn on the electrochemical firing of the optic nerve, which carries the signal to your brain; then, normally, an enzyme converts it back into its "storage" form, readying it for the next burst of incoming light.

But any system that relies on chemically unstable components always runs the risk that their reactive chemistry will spill over the tight controls of the system they're meant to serve. In this case, all-*trans*-retinal can react with some of the fatlike molecules that make up the cell membrane, leading through a complex series of steps to the formation of a stubborn end product called *A2E*. This compound is completely resistant to digestion within the lysosome, so it's a major source of undegraded junk in the lysosomes of these cells. Over time, so much A2E is produced and absorbed into the lysosome without being degraded that it can take up as much as one fifth of the *total cell volume* in the cells that accumulate it. These unfortunate cells make up the *retinal pigmented epithelium* (RPE) of the eye—a part responsible for maintaining the function of the light-sensing areas of the retina.

But again, because of the specialist terminology in use (A2E, rather than "lipofuscin"), the role of lysosomal inadequacy has been—and you will pardon the unfortunate pun!—*obscured*.

Toxic Waste Problem—Toxic Waste Solution

By the morning that I was throwing some clothes into a duffel bag for the 1999 Society for Free Radical Research meeting in Dresden, I had come to see lysosomal inadequacy—and the resulting accumulation of cellular waste products—as perhaps *the* key step linking the mitochondrial mutation-driven rise in oxidative stress with age with the actual pathology of aging. Remember from Chapter 5 that, by then, I had a scheme for how mitochondrial mutations in a few cells could propagate toxins to mitochondrially healthy cells elsewhere. What I didn't explain in Chapter 5, not least because when I developed the Reductive Hotspot Hypothesis I didn't know it, was just how these "toxins" are toxic—what harm they might do to the cells that ingest them.

But now, a year later, this mystery was beginning to resolve. It was clear, once one got over one's attachment to the term "lipofuscin," that the failure to dispose of specific waste products was at the root of the most terrible diseases that accompany biological aging: atherosclerosis, age-related macular degeneration, and neurodegenerative diseases like Alzheimer's. It was just that the *kind* of waste that was linked to a given disease was specific to the cell type and the particular diagnosis.

As it happened, Ulf Brunk was again presenting his data in Dresden. As I listened to his talk and contemplated his slides—the telltale red glow of lipofuscin choking cells, his computer-generated diagrams illustrating his and Terman's "garbage catastrophe" theory—I saw that it was a waste of time to argue about whether lipofuscin contributes to "aging" in the narrow sense. Clearly, we needed a way to solve this problem if we were to protect our bodies from age-related pathology. But I also became convinced that it would not be enough to try to prevent the accumulation of this junk "upstream" by obviating mitochondrial mutations. We were also going to have to deal with the junk *directly*.

But how? With the recalcitrant materials in question being so multifarious, and with the metabolic pathways, chemical identities, and even specific role in pathology of these materials being still largely unknown, it seemed that no classic "magic bullet" approach—one small molecule to match one therapeutic target—would work. And simply putting the lysosome into overdrive wouldn't really solve the problem: While, as later animal studies would show,[15,16] simply souping up lysosomal activity or topping up their existing enzyme supply could slow down the progression of lysosomal storage diseases, such approaches could not ultimately stop those diseases. It is the very nature of the problem that the body simply *does not have the enzymes* to degrade the really ugly junk—and thus, that it will choke up your cells, steal your mind, blind you, and clog your arteries later, if not sooner.

Ulf's talk ended, and with a hundred other scientists I got up and streamed into the outer hall for the coffee break. The red glow of lipofuscin on the slides had gotten me to thinking of the stuff as the toxic waste that it is, and of the aging cell as a tiny contaminated environmental site equipped with a woefully inadequate waste management system. So the job of restoring the cell to health was really a kind of environmental cleanup job, and what was needed was a biomedical superfund project to develop new remediation technologies capable of dealing with materials that had so far evaded the lysosome's capacities.

Superfund!

It suddenly occurred to me that this was more than a metaphor. (If you've never heard of Superfund, be patient—I'll explain shortly.) There *were* actual land sites all over the planet that *should* be very badly contaminated by lipofuscin, because their soil has been seeded with the stuff for generations. I speak, of course, of *graveyards.* Think about it: hundreds of bodies put into the ground—sometimes *en masse,* as happened throughout Europe during the horrors of the Plague, and more recently following acts of genocide in Rwanda and elsewhere. These soils should be chockablock with aggregates from their inhabitants' decaying bodies.

Yet, to my knowledge, there was no accumulation of lipofuscin in cemeteries—and if there were, we certainly ought to be aware of it, because lipofuscin is *fluorescent.* Months later, when I was discussing the issue with fellow Cambridge scientist John Archer, he would put the disconnect succinctly: "Why don't graveyards glow in the dark?"

Soil microorganisms struck me as the most likely explanation. Bacteria, fungi, and other microbes normally play a role in turning our remains into compost, of course, but it was not so immediately obvious that they would be able to digest something so resistant to enzymatic action as lipofuscin. And yet, I recalled, we'd known for decades that soil microbes display an astonishing diversity in their choice of food.

Scientists became interested in this phenomenon in the 1950s, when it was noted that the levels of many hard-to-degrade pollutants at contaminated sites were present at much lower levels than would have been expected. A big part of the explanation turned out to be the rapid evolution of quickly reproducing organisms like bacteria. Any highly energy-rich substance represents a potential feast—and thus, an ecological niche—for any organism possessing the enzymes needed to digest that material and liberate its stored energy. The presence of high levels of such a material therefore creates a powerful evolutionary "pull," driving the evolution of the necessary enzymes in microorganisms that come into contact with it.

And this is especially so if the substance is not easy to break down, because then the chances are good that most of the *other* organisms in the vicinity will *not* have enzymes capable of this degradation.

It was proposed in 1952 that these forces might well be so strong as to guarantee that, given enough time, evolution would find a way to create microbes with the capacity to digest *anything* we throw at them that is both carbon-based and rich enough in energy to be a worthwhile fuel source. This was given the immensely memorable name "the microbial infallibility hypothesis." While it has turned out to be a *bit* of an overstatement—no one has yet discovered the microorganism that can eat Teflon, for instance—studies over the next few decades tended to confirm the general principle. U.S. Geological Survey scientists collected case studies showing that microorganisms were breaking down significant amounts of a variety of organic chemical pollutants in wastewater. Oil spills, chlorinated solvents, pesticides—you name it: soil bacteria learned how to digest almost anything that was thrown at them, leaving only harmless residues like carbon dioxide and water.

Scientists' first attempts to harness this power failed, because they were trying to *invent* organisms to order, *imitating* what nature was already doing very well. But eventually, researchers realized that they just weren't as smart (nor, more accurately, as *fast*) as the forces of nature. Out of these observations developed *bioremediation*: the exploitation of evolution's ability to generate novel digestive capacities in microorganisms for the intentional cleanup of contaminated environments. "Superfund" was the name of a U.S. government initiative to stimulate and commercialize bioremediation research.

Sipping my coffee in Dresden, my mind brought this train of thought full circle, feeding back into my original musing on the lysosome as an inadequate toxic waste disposal system. The lysosome *already* deals with the cell's waste products using enzymes to break them down into their constituents. But it is not equipped with the capacity to deal with *every possible* waste material. This is just what you'd expect from evolutionary theory. Remember, again, as we discussed in Chapter 3, that evolution only designs your body to last as long as your environmental niche will allow it to last. In the Paleolithic environment in which we evolved, that meant about three decades—far less time than it takes lipofuscin or atherosclerotic cholesterol aggregates to build up to life-threatening levels.

For this reason, evolution has never bothered to equip the lysosome with enzymes designed to deal with these wastes, because it has never had a good reason to do so. But, as we've seen, it seems very likely that evolutionary forces *have* pushed soil microorganisms to develop these capacities in order to exploit a new fuel source—an issue, for them, of day-to-day survival. Not only do evolutionary theory and the "microbial infallibility hypothesis" predict this, but it also seems to be confirmed by the absence of large accumulations of lipofuscin in mass grave sites: were it not so, *all* such locations would have an eerie glow about them, instead of such a phenomenon being confined to cheesy horror flicks.

Suddenly it came together. The imaginative spark of metaphor had fallen upon the very concrete fuel of data in the oxygen-rich environment of evolutionary theory, and a fire began to burn in my brain. These two observations implied that we could perform a sort of *medical bioremediation,* in which we would identify the soil bacteria that already clean up our undegraded junk *after we have died,* determine the enzymes that allow them to do it—and then deliver these enzymes into the lysosomes *of people who were still alive to benefit from it.* **Figure 1** gives a graphical depiction of this cycle.

These enzymes would give new powers to our cellular recycling centers, allowing them to process materials that presently go undegraded within us—not just preventing, but *reversing* their pathological accumulation. Our brains would be cleared of neurofibrillary tangles; the dying macrophages in our arteries would gain new life, letting them clear out the oxidized LDL toxins and allowing the necrotic vessel tissue to finally heal; *the blind would see.* And aging cells all over our bodies, choking on their own filth, would become clean and new again.

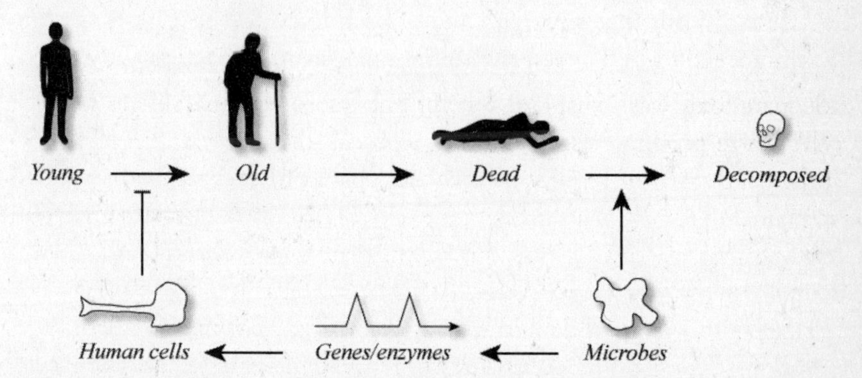

Figure 1. Medical bioremediation, exploiting the microbial enzymes that turn dead people into decomposed people, may retard many of the processes that turn young people into old and eventually dead people in the first place.

Normally, when new ideas come to me, I give myself a few days to try to punch holes in them before bouncing them off anyone else. But this time I felt supremely confident. Taking my nose out of my coffee, I scanned the hall for Dr. Brunk. I spotted him across the room: chubby, graying, earnest but with a mien of compassion that made you think of him as an aging social crusader. In a few purposeful strides I was confronting him.

"Listen, Ulf," I said quickly, "I've just had the most fabulous idea . . ."

A Quick-and-Dirty Test

I was a little disappointed by Brunk's reaction, though I was not sure how much of what I was seeing was a reflection of his assessment of the feasibility of the entire scheme versus its inherent audacity or my jumbled, hot-off-the-fire delivery. Perhaps it was just Nordic caution. Whatever it was, it was clear that, while not dismissive of the proposal, Brunk was clearly not experiencing spontaneous combustion from the white heat of my brainwave.

Still, I pressed him for his thoughts on possible ways to perform preliminary tests of the idea. The first thing, we agreed, would be to test the sturdiness of the very foundation of the castle that I had just constructed in the air: the idea that soil microorganisms are, indeed, routinely digesting lipofuscin in corpses. Fortunately, there was a reasonably straightforward way to execute such a test: collect some soil microorganisms from a site likely to be "enriched" in human remains, and then see if they could break down lipofuscin in a test tube.

It turned out that what you might think to be the easy part—getting the lipofuscin needed to test the abilities of graveyard bacteria—was actually almost impossible to pull off in the real world. There are only small amounts of lipofuscin in most of our cells, and the tissues in which there's more (such as the heart) are not so readily available, so to get a useful quantity of the stuff would be a challenge. But Brunk said that he could whip me up a batch of an excellent substitute: the *synthetic* lipofuscin used by people working in his field, which is created by merely exposing mitochondria to enough ultraviolet radiation to induce cross-linking of their membrane proteins. The resulting recalcitrant gunk has the same fluorescence spectrum as the real thing, and seems to also have the same physical and chemical properties—which is as you'd expect, since most experts think lipofuscin largely *is* the remains of unsuccessfully degraded mitochondria, damaged by the effects of free radicals and left festering in the lysosome.

The next thing would be to gather microorganisms from soil that had been exposed to a large supply of lipofuscin, to look for the ones that, in my hypothesis, had been responsible for breaking the stuff down. My mind had already leapt ahead to the fact that John Archer—the man who would later make the "glowing graveyard" quip—was working on bioremediation at Cambridge. As such, he was well-versed in the techniques used by scientists in the industry to isolate and culture bacterial strains capable of digesting classical toxic waste materials, and to identify and clone the genes involved in producing the enzymes responsible for that capacity. If I could enlist his help, we could do the same thing for lipofuscin-digesting hydrolases.

Tomb Raider

Fortunately, John was immediately fascinated by the whole idea, and agreed to give it a go. So it was that his graduate student would find herself, like a good mad scientist, at Midsummer Common in the twilight of a late summer day, digging into the soil of an ancient mass grave with a trowel, seeking not bodies but tiny, mysterious creatures imbued with the power to turn the most stubborn, aggregated junk in our bodies into compost.

The sense of being immersed in a Gothic horror novella lasted only a moment. Having scooped up the soil and brought it back to the lab, John and his student isolated the microbes and put them in petri dishes with the synthetic lipofuscin as their only potential food supply. Then, we waited to see if the force of natural selection would reveal the existence of strains that were capable of surviving on a diet of pure lipofuscin.

Almost immediately, the microbes that we had isolated began to give off the characteristic red glow of lipofuscin under the fluorescing light of the specialized microscopes. This was not yet a success, because all it meant was that the microbes were *engulfing* the material; they weren't necessarily *digesting* it. But it was not long before clear differences began to emerge among different strains. Most of the colonies of microbes were in a state of growth arrest, failing to thrive for lack of nourishment. But a few of them were clearly enjoying a ghoulish feast: their numbers were expanding rapidly as their hydrolytic enzymes slowly broke the stubborn goo down into usable components, tearing apart its complex organic chemical bonds to release the stored energy. Within short order, we had a sample of microorganisms equipped with enzymes that could digest lipofuscin in the same way that the enzymes in your stomach digest a steak.

The hypothesis had been confirmed. The next challenge would be to move those enzymes into our own lysosomes. No one has done this yet, and indeed there is a sense in which the task sits just where it was when I finished my work with John Archer. Fortunately, however, we do not have to create an entirely new field of medicine in order to get going with this idea (which I will from here on call "LysoSENS"). That's because the fundamental biotechnology required to pull it off is already in clinical use. Pioneering physicians have been introducing "foreign" lysosomal enzymes into patients for several years—not in aging, but in the lysosomal storage diseases.

Cleaning Out the Drain

Lysosomal storage diseases, the syndromes that we now know to be the result of mutations in genes that code for our *normal* complement of lysosomal enzymes, had been known for decades before researchers figured out what was causing them. Once their origins became clear, however, a way to treat most LSDs became apparent: *enzyme replacement therapy* (ERT—don't confuse the abbreviation with *estrogen* replacement therapy). In individuals lacking an enzyme for some common metabolic waste, undegraded cellular waste products build up within the lysosome (and also outside it in the main body of the cell), and cellular dysfunction inevitably results. Therefore, it was reasoned, if the right enzyme could be delivered to the lysosome, the cellular recycling center would return to normal function, the piled-up garbage would be broken down, cells would return to health, and victims would be able to lead normal lives.

After a few decades of work, victims of three of the most common LSDs are now being successfully treated with such therapies. There are, for instance, about four thousand people now living normal lives despite having Gaucher's disease, thanks to regular injections of the lysosomal enzyme that their cells are unable to produce for themselves. The drug development process has been reasonably clear, although technically challenging. In one disease after another, scientists have identified the enzyme whose absence causes the disorder; modified it in various ways to allow it to be injected, taken up by cells, and delivered to the patient's lysosome, where they function exactly like the same enzyme does in the rest of us when it is produced by our own cells; and watched as symptoms have disappeared, lives have been extended, and victims have been enabled to live the life that the rest of us take for granted.

Of course, the same fundamental problem faces all of us in the case of diseases of *long-term* lysosomal failure: we will all ultimately suffer from *age-related* "lysosomal storage diseases" (such as age-related neurodegenerative disease, macular degeneration, and atherosclerosis), even though only a tiny proportion of the population is stricken with the currently recognized *congenital* ones (Gaucher's disease and the like). While the exact *origins* of the two kinds of LSDs are different (rare genetic mutations in genes for lysosomal hydrolases that are otherwise part of the species' standard evolutionary legacy in the congenital LSDs, versus never having evolved the enzymes needed to break down neurofibrillary tangles, A2E, etc., in the age-related diseases), the *molecular natures* of both congenital and age-related LSDs are essentially the same—and as anti-aging bioengineers, that is quite sufficient to let us do our job, which is to clean out the accumulating molecular damage. To arrive at that destination, we will need to address a series of specific challenges. Fortunately, in all cases we have options available with which we already have experience, or for which the solutions are clearly in sight and under development by researchers in other fields of biomedicine.

First Challenge: Identifying Suitable Enzymes

Our "grave robber" raid on Midsummer Common proved that the enzymes exist to degrade the highly cross-linked remains of mitochondria—believed to be the single largest contributor to lipofuscin. However, we still don't know exactly what enzyme or series of enzymes is doing the job. Moreover, enzymes that degrade this synthetic lipofuscin will not be enough: we also need to identify other enzymes that will deal with wastes clogging up lysosomes in a variety of tissues and associated with various disease states.

This doesn't necessarily mean enzymes that will degrade any known aggregate, such as neurofibrillary tangles. As we discussed above, the messes that we see are not necessarily the ones causing the problems: they may be the gunsmoke rather than the gun itself. For instance, it's possible that some *other* junk is actually responsible for backing up the system, and thus that the aggregates that pile up like so much trash in the street are simply what results when the lysosome can't keep up with its normal load. In fact, the problem might not even be the presence of undegradable materials: in several diseases, some researchers have presented evidence to suggest that the problem is a substance (for instance, A2E in macular degeneration) that directly inhibits the activity of the pump responsible for keeping lysosomes acidic enough for their enzymes to work.

Fortunately, we again don't need to trouble ourselves with this. Once again, our job is not to tease apart the minutiae of metabolism, but to clean up age-related damage. To do that, we can follow the example of the bioremediation experts: throw enzymes at the problem until the problem is solved (normal lysosomal function is restored), and then identify the enzyme that did it. That's a simple task in principle—but when the bioremediation field first got going thirty years ago, it was a long slog to actually narrow down which of hundreds of enzymes in a strain of microorganisms were responsible for breaking down the wastes of interest.

One of my reasons for optimism, therefore, is the fact that so much progress *was* made in those days—and today, we have much more sophisticated molecular tools available to do the job. One is a method called *molecular fingerprinting,* but that term is a little bit misleading. It suggests a process of finding a clear, unique identifier of an individual—like a fingerprint—and then finding the individual that bears that same identifier. Instead, molecular fingerprinting is based on the fact that the members of a closely knit family of organisms tend to carry genes with broadly similar sequences, and that (similarly) genes for a range of enzymes with broadly similar functions within such a community also tend to have a similar stretches of code.

This allows us to winnow our way down to the genes (and, therefore, enzymes) of interest from either of two angles. One option is to focus on a *class* of enzymes for whose encoding genes we are searching (in this case, hydrolase enzymes), and then to look for genes that match the overall pattern and are expressed in large amounts when the parent organism is feasting on the contents of the dysfunctional lysosomes. The other option is to identify, within a community of organisms, which specific ones are thriving best when only given those contents as nourishment—and which therefore carry the genes encoding those enzymes most effectively tearing their contents down.

Another powerful tool at our disposal is *DNA microarrays,* or *gene chips*. These are tools that identify, in real time, which genes in an organism are being actively expressed at a given moment. So, if we can isolate strains that are doing well on a diet of lysosomal detritus, we can sequence their genetic libraries, and then test which of those genes are being used intensively when they are feasting on the stuff.

We can also use techniques that allow us to "knock out" (simply *remove*) specific genes in such strains, and then retest them. When we knock out a given gene in a strain of microbe and see that the mutants starve on a diet that was previously their version of a gourmand's wet dream, we can infer that the gene in question encodes a protein that is critical to the sequence of processes involved in digesting such materials. We can then identify these genes, and see whether the enzymes they encode are the crucial ones that will fill in the weak spot in our human hydrolytic arsenal.

Second Challenge: Getting Them to the Cells

Once we have enzymes in hand that will do the job, we'll have to find ways of getting them into the cells that need them. Not every cell type will be confronted with the same kinds of waste: as we've seen above, particular disease states are characterized by specific aggregated waste products, and (as is especially likely in the case of A2E in macular degeneration) this is the result of the particular metabolic pathways that produces them. How much we have to do to address this challenge will depend on exactly how we're going to get them into the body at all—for which there are again multiple options.

Right now, for instance, doctors treat LSD patients by intravenously injecting modified forms of their missing lysosomal enzyme. Of course, the enzyme does the patient no good when it's just floating around in the bloodstream, and it might even do some harm if it were active (since it might start attacking functional proteins), so the enzymes are tweaked in ways to ensure that they go where they're needed. First, they are targeted to the right cells. In Gaucher's disease, for instance, *macrophages* are especially vulnerable to the lack of the enzyme that causes the disease. Therefore, the enzyme is hitched to targeting molecules that are already recognised by macrophages as passports to entry. The same trick might, therefore, be used to target enzymes needed to clear away the substances that cause macrophage lysosomes to fail in atherosclerosis.

This method has the advantage of being relatively simple to implement in the short term, and indeed of already being in use for a recognized disease (so that we have a large body of practical, clinical experience with the basic technique on which to draw). It does, however, face a variety of limitations. Most notably, there's a big difficulty in using it to move enzymes past the protective *blood-brain barrier,* which is a highly effective shield designed to guard your brain against exposure to the many potentially toxic substances floating around in your blood. Obviously, injecting enzymes that cannot reach the brain will seriously limit their benefits—and represents an almost complete barrier against their use for age-related neurodegenerative disease. Even today, some Gaucher's patients develop neurological complications as a result of their enzyme deficiency, and injected hydrolases are of little help to such patients.

Fortunately, scientists are making progress in coming up with ways to move proteins across the blood-brain barrier—and in the future, we can expect to have much more powerful delivery systems. In the relatively short term, we should be able to develop a form of *cell therapy,* involving seeding the patient with cells that produce the needed enzyme and secrete it into the bloodstream or into the fluid bathing surrounding cells.[17] This would therefore act like a biological version of the nicotine patch, providing a continuous dose of the enzyme. This might be extremely useful: right now, LSD victims rely on regular injections of heroic amounts of their needed enzyme, and it's possible that the sheer quantity of (several) enzymes required to combat all types of lysosomal failure due to aging or age-associated disease might make injection impractical. One reason why so much enzyme is needed is that some of these enzymes are proteases—that's to say, they break down proteins—but enzymes are proteins, so proteases in the lysosome actually destroy themselves and each other.

And, of course, the ideal would be to modify our own cells using somatic gene therapy, introducing DNA to instruct the relevant cells to produce the very enzymes that they need to stay healthy. This, and the cell therapy option too, are still a long way away from the anti-aging clinic—but again, fortunately, the major hurdles will be tackled first for a variety of better-recognized diseases, from sickle cell anemia to the severe combined immunodeficiency disease that creates "bubble babies." Thus, we can expect to ride on their coattails to a certain extent. Indeed, once somatic gene therapy is available for use in treating relatively common genetic disorders, it will doubtless be seized upon by LSD researchers as a way to replace genes for the lysosomal hydrolases missing in their patients. Again, the specific use of gene therapy to provide better solutions for LSD victims will be a useful source of information and collaboration to develop a gene therapy version of the LysoSENS project.

Third Challenge: Getting Them to the Lysosome

This is similar to the second challenge, above: lysosomal enzymes do us no good—and might conceivably even cause us some problems—if they wind up (or, in the case of gene therapy, are synthesized) inside the cells where they're needed but are not then localized in the lysosome where the junk accumulates and where the acidity is available to let them do their work. Again, one potential solution is already in use in the LSDs: the use of molecules of the sugar *mannose 6-phosphate,* which is recognized and taken up—along with its cargo—by the lysosome.

But again, we are also in the process of learning to hide a few other cards in our sleeves. We might be able to use a backdoor solution, by turning a targeting system that is currently used to ensure *wastes* get delivered to the lysosome into a method of sending *enzymes* into the heart of the same system. (This system is called *chaperone-mediated autophagy.*) Lysosomes would take the enzyme up just as if it were any of numerous classes of cellular detritus, but would instead incorporate a hydrolase capable of preserving and restoring its normal functioning.

We might also be able to take advantage of targeting systems already in use in the organism from which we originally isolated the enzymes in question. Bioremediation typically uses bacteria as the microorganism of choice because they digest their food quickly. Fungi, by contrast, are usually viewed as too slow-acting and slow-growing to provide viable solutions for oil slicks or contaminated chemical spill sites. But in a slowly accumulating, low-volume toxic waste problem like age-related lysosomal failure, these issues would be less of a problem. The advantage of using fungi is that they—like us, and unlike bacteria—have a lysosome-like structure of their own, called the *vacuole,* which shares many of the key characteristics of the human equivalent (including, for instance, the need for an acidic internal milieu to work properly). Enzymes taken from such sources, then, might come already equipped with a range of features that would be useful for the LysoSENS project in humans.

Fourth Challenge: Potential Side Effects

Even once we have ways to deliver useful enzymes to the lysosomes of affected cells, we will still face the key challenge of preventing the intervention itself from causing us harm. One potential issue is that the enzymes in question might, as suggested earlier, also be active elsewhere than in the lysosome. One reason to expect that this won't pose a major challenge is the fact that, as already mentioned a few times, lysosomal enzymes typically require a very acidic environment to function properly, so they will probably be nearly inactive in the main body of the cell.

We might also, however, further modify the enzyme so that it only becomes active after it has been taken up by the lysosome. One possible such modification would be to attach an extended sequence of amino acids that would prevent the enzyme from being active, but that would be cleaved off by enzymes present and active in the lysosome, liberating the active form for duty at its destination. This concept sounds really tricky, but it is already used by the cell for the safe delivery of some members of the standard human lysosomal enzyme complement, so the technique should be adaptable without overmuch heroics.

Another potential worry is that the enzyme might cause an immune reaction, just as any "foreign" protein might. But experience with the LSDs suggests that this will not be as big a problem as one might at first expect. Remember that, for a person who was born unable to produce the hydrolytic enzymes that the rest of us take for granted, these proteins are every bit as "foreign" as the microbial hydrolases will be to all of us. Normally, we learn to be tolerant of the proteins of our own bodies because our immune system is exposed to them early on in our prenatal and childhood development, allowing it to recognize them as "self." Having never been exposed to such proteins in early life (because without the gene, the protein can't be constructed), LSD victims lack immune tolerance to them. And in such patients, immune reactions do occur. But, reassuringly, they are always mild, and they taper off with time. This seems to be because enzyme replacement therapy delivers enzymes to the lysosome in a way that does not allow the cell to hack them up and to display them on its cell surface, alerting a suspicious immune system to their presence.

Moreover, even if the experience with the newly introduced enzymes is not the same (for instance, if we find reasons to use gene therapy and chaperone-mediated autophagy rather than ERT), we aren't necessarily stuck. Dampening down an excessive immune response is a necessary part of many medical procedures, from organ transplants to over-the-counter allergy medications, and we are getting better at it all the time. We might also eventually be able to produce the protein within the bone marrow, as has already been done with some lysosomal enzymes; this might also help to induce tolerance, because of the role of bone marrow cells in immunity.

Inside, Outside

As you can see, there are quite a few hurdles to be overcome before we will be able to use novel hydrolytic enzymes to clear out the junk in our cells, preventing or reversing many of the most debilitating health problems of old age. But, as I've shown, perfectly plausible solutions to all of these problems seem to exist that are either already in use in treating the recognized (congenital) LSDs, or else have clear routes to implementation that are the subject of intense study by researchers the world over. Identify the enzymes we need, and a first-generation therapy might look much like ERT for lysosomal storage diseases today: expensive, inconvenient, and limited in its scope, but lifesaving. And as we go on, we will progressively improve the therapy, making it more comprehensive and advancing its safety and efficiency in lockstep with the advance of gene therapy and other enabling technologies that will also be exploitable in LSD treatment.

As in previous cases, the pursuit of this solution will depend on an interdisciplinary synthesis of research performed in areas that have, ostensibly, little to do with aging, and original work done by scientists dedicated to the goal of adapting existing technologies to the novel problems associated with the aging process. What is clearly needed is to get private and public capital devoted to the latter half of the equation, which suffers from a serious lack of investment of dollars and brainpower, and without which the greatest killer of all in the modern world will continue to cripple, torture, and kill our fellow human beings in enormous new cohorts every day.

Let me now turn from the junk *within* our cells to some of the aggregated junk *coating* our cells, exploring how it is harmful, what can be done about it, and how its threats to your health—and its therapeutic solutions—are intimately tied up with the lysosomal failure problem that we've been exploring here.

8

Cutting Free of the Cellular Spider Webs

Our cells—and thus our bodies—are progressively damaged by protein-derived junk that gathers over the years in the space between cells. Alzheimer's disease is perhaps the best-known condition associated with this, but there are others that are equally fatal. However, there is a way forward for medical science and our health: recent and very promising research demonstrates that science can turn our own immune systems against this dangerous material.

In the previous chapter, I talked about the junk that accumulates *inside* our cells with age—how it contributes to the biological aging process, and what can be done to get rid of it. In this chapter, the focus is on the garbage that accumulates on the *outside* of our cells and tissues, enmeshing them in webs of damaged proteins, impairing their function, and contributing to aging and age-related disease.

Most of the junk that we'll be discussing is *amyloid* of one kind or another. When I say "amyloid," of course, almost everyone thinks of *beta-amyloid protein* (also called "amyloid beta"), which accumulates as the waxy "senile plaques" that cluster around the brain cells of people with Alzheimer's disease. But many other, less-well-known diseases (*amyloidoses*) are also rooted in abnormal protein aggregates of this type. Most amyloids are cell-snaring chains of molecules that begin their existence as healthy proteins already present naturally in our blood, or in the fluid bathing our brains. A wide range of proteins can become amyloids under the wrong circumstances, including *immunoglobulin light chain,* a key component of the antibodies in your immune system; the protein *transthyretin,* which is responsible for carrying around thyroid hormones in your blood; and a small protein (*islet amyloid polypeptide,* or IAPP—also referred to as *amylin*) that helps your body regulate its blood sugar levels in association with insulin.

What turns these proteins into snares that squeeze the life out of cells and organs is how they are *folded.* Misfolded proteins are just what they sound like: proteins that have become twisted out of their proper configuration in ways that cause them to undergo toxic interactions with each other, or with other constituents of the cell. The ones that cause amyloid diseases contain sites within their structure that, if exposed, readily stick to other proteins of the same sort, causing them to link together one after another in a sinister, self-assembling daisy chain. These sticky sites are normally kept safely tucked away *within* the complex folding of the protein's three-dimensional architecture, precisely to prevent such interactions from happening. Misfolding exposes such sites, initiating the spinning of a cell-choking web.

Many of the amyloid diseases result from the victims having faulty genes that produce defective versions of these proteins. In some such disorders, the underlying mutation introduces fatal flaws into the structure of the protein itself, causing it to open up at inappropriate locations, exposing the critical "sticky" site in its structure. Others involve mutations in enzymes that normally chop up the protein into functional units as it emerges from the cell's protein-assembly machinery. These mutations cause the enzymes to cut too close to the critical site, again unleashing it from the restraining influence of the rest of the protein's normal conformation. Another route to congenital amyloidosis is errors in "chaperone" proteins whose job it is to direct the emerging (and potentially amyloidogenic) protein to assume a safe, nonamyloidogenic final shape.

But in addition to these *inherited* protein-misfolding diseases, there are also *universal* amyloidoses—ones that are the result not of mutations, but of the fundamental vulnerabilities that proteins face in the course of their critical jobs in the molecular maelstrom of cellular biochemistry. With free radicals, sugars (sugars? Oh yes. See Chapter 9) and vibration constantly acting on them, proteins are bound to get bent out of shape now and again in ways that open them up to becoming the seed of an amyloid fibril. Once one such protein is formed, it can sometimes twist other proteins out of shape as it grabs at them, exposing another site and forming the nucleus of an ever-expanding fibril chain. One example of this happening in fast-forward is seen in people with kidney failure, when the body ceases to pass out *beta-2-microglobulin* in the urine. Beta-2-microglobulin is normally a perfectly safe protein that helps the body to distinguish its own cells as "self" from "nonself" cells of bacteria or other organisms. But without regular excretion, levels of this protein begin to climb to abnormal levels, and they eventually reach so high a concentration that they start to spontaneously glom together, forming amyloid deposits.

Indeed, Cambridge professor Chris Dobson, who has spent his academic life looking into protein misfolding diseases, says that "conditions could be found in which seemingly *any* protein could form amyloid fibrils [emphasis mine] . . . although the propensity to form such structures under given circumstances can vary greatly from one protein to another."[1] Over time, these fibrils build up to potentially pathological levels, coiling around our cells and organs, choking them off like so much bindweed.

Mind-Forg'd Manacles

Most researchers now believe that the horrors of Alzheimer's disease can mostly be traced to abnormal processing of an otherwise healthy molecule called *amyloid precursor protein* (APP). Everyone's brain produces APP, and it is required for *some* essential function in our bodies. Ironically, in fact, properly processed APP actually appears to be required for many of the key activities of healthy neurons, such as their ability to rewire themselves in response to new learning and to grow out the branching "electric cables" (*neurites*) that allow them to talk with one another.

When things go right, APP is produced in the main body of the cell and then sent for further processing to *alpha-secretase,*[2] a type of enzyme called an *endoprotease*. The result is the creation of two molecules, one of which remains in the membrane of the neuron's neurites, while the other is released into the fluid inside the cell. APP *cannot* form the evil beta-amyloid when it is processed by alpha-secretase. After this, one of the fragments is chopped further, by a distinct enzyme named *gamma-secretase*.

APP only become dangerous when, instead of being trimmed down by alpha-secretase, it is mistakenly cut up by a different, but related, enzyme called *beta-secretase.*[3] Beta-secretase, like APP, is not a villain: it has a proper place in the cellular "factory," as part of *another,* distinct cellular assembly line from the one that handles APP. In that assembly line, beta-secretase makes essential trims in the structure of *other* proteins that bear some molecular resemblance to APP itself. But if beta-secretase performs the same action on APP, it chops it in the wrong place. This distorts the protein's shape and creates a molecule with a totally different action within the cell.

It's as if beta-secretase were an overly helpful laborer who, while crossing the factory floor on the way back from lunch, had seen some APP lying on a stopped assembly line and mistaken it for a part that he or she normally works on. Seeing no alpha-secretase around, and presuming to know what the half-finished product needs, beta-secretase steps in to do alpha-secretase a favor by taking care of a little bit of its workload. After giving it a few whacks of its molecular hammers, beta-secretase tosses the APP fragment—now subtly misshapen—back onto the line, where it eventually reaches gamma-secretase. And because gamma-secretase is a busy enzyme, it's too caught up in its work to notice the change, and proceeds to splice and dice the distorted APP fragment just as it would if alpha-secretase had made the *proper* modifications. Beta-amyloid is the product of this mistaken sequence—the sequential cleavage by beta- and gamma-secretase rather than by alpha and gamma.

When processed appropriately, the middle APP component (between the sites of cleavage of alpha- and gamma-secretase) assumes a shape similar to a stretched-out coiled spring—a conformation called an *alpha helix.* But thanks to beta-secretase's molecular meddling (and gamma-secretase's unwitting cooperation), this fragment loses its normal shape—and, just as would happen if you were accidentally to cut into a tightly drawn spring with a pair of wire cutters, the improper slicing of APP causes the fragment to jump backwards on itself, creating a shape more like a bent hairpin (a *beta-sheet*) that gives beta-amyloid the fatal molecular *stickiness* that characterizes amyloid proteins.

Once released by gamma-secretase, individual fragments (*monomers*) of beta-amyloid initially float free in the brain. But they soon come into contact with other monomers, and their "stickiness" causes them to glom together into larger—but at this stage still free-floating—units called *oligomers*. These fibrous strands, in turn, get stuck to one another to form still longer strands, which eventually grow so large and complex that they can't stay dissolved in the brain's fluids, but precipitate out into the spaces between neurons to form the notorious plaques. Under a microscope, these mind-snaring webs can be seen extending to the neurons' caretaking support staff (the *glial cells*), and down to the neurites (the wiring system that I mentioned earlier).

Some people produce unusually large amounts of beta-amyloid because they have inherited mutations that either cause their bodies to produce too much of APP itself (thus increasing the odds that the problematic enzymes will come across molecules of it and mistakenly hew their structure), or else encode defective secretase enzymes that are not so good at doing their selective jobs as the more common varieties. But because *everyone* has both APP and the enzymes that can sometimes turn it into beta-amyloid, we *all* produce beta-amyloid, and given a constant output of the stuff, some fraction of that amyloid precursor is bound to get snipped in the wrong way now and again. Once that happens, it's only a matter of time before enough of it builds up to form Alzheimer's-type plaques—and indeed, all of us have at least some plaque in our brains by the time we reach late middle age.

Thus, like other aging damage, beta-amyloid plaques simply accumulate over time, and it's reasonable to think that neurological impairment occurs when a critical threshold is reached. This is probably why most cases of Alzheimer's disease are not inherited, but instead occur sporadically in the population: the underlying biochemistry is just part of the kind of organisms we are, living in the kind of universe that we do. Lifestyle risk factors and most genetic predispositions merely determine how early on in our lives the process begins to impair our intellects and identities.[4] This is also why, apart from a very small number of inherited, early-onset cases, almost no one in early middle age or younger gets Alzheimer's . . . and why the prevalence of the disease *doubles* every five years beyond age sixty-five, so that victims pile up with age like the grains of rice on the Emperor's chessboard in the old fable. Our brains are slowly being enmeshed in beta-amyloid plaques—it's just a matter of when we reach the threshold beyond which our brains can't keep up sufficient function to carry on the lives and identities that we have spent so many years creating. Barring some radical new therapy, *each and every one of us* will be struck down by Alzheimer's dementia if something else doesn't kill us first.

"O Captive, Bound and Double-Ironed . . ."

Beta-amyloid also causes brain damage and death in many people who never develop Alzheimer's disease. This is because, in addition to building up in the *neurons* of the brain, beta-amyloid also clings to the interior surfaces of its *blood vessels.* The resulting condition, called *cerebral amyloid angiopathy* (CAA), is a crusting-up of these pipelines of oxygen and nutrients, weakening them and reducing their ability to flex in response to the surging flow of the pulse. This leaves them vulnerable to bursting open in a bleeding stroke.

CAA is certainly more common in people with Alzheimer's (about a quarter of all patients have it as a complication), but as we age it becomes an increasingly serious issue in people *not* struck by the latter disease. Just 5 percent of us have CAA in our seventies, but after the age of ninety over half of us are suffering from the disease, and it is responsible for about 15 percent of all bleeding strokes in people over the age of sixty.

But there's more: beta-amyloid is just one mangled protein among many. Less well known, and less recognized as causes of death and disability, are a variety of other age-related amyloidoses that also don't seem to be related to an *inherently* malformed protein, but to the *healthy* version being damaged in the rough-and-tumble of its biochemical environment. *Senile cardiac amyloidosis* is one example. As you might guess, this disorder is most clearly characterized by amyloid fibrils building up in the heart, although it damages the lung, liver, and kidneys, too. This buildup interferes with the regular beating of the heart, and can cause heart failure. The fibrils are made up of *transthyretin*—the rickshaw driver for thyroid hormones that I mentioned earlier on—and while it *can* arise from a mutated version of the protein, it can also result (at a slower rate) from damage to the form that most of us carry.

As the name implies, senile cardiac amyloidosis is a strongly age-related disease—first showing up in people over the age of seventy, and found at pathological levels in about a quarter of people over ninety. As in the case of Alzheimer's disease, if we all lived long enough without something else killing us first, each of us would wind up with the lives squeezed from our hearts by this form of aging damage. The disease is known to be a common contributing cause of death in the "oldest old," such that about half of people over ninety years old have diagnosable senile cardiac amyloidosis at autopsy.

Much earlier on in life, nearly everyone gets some degree of amyloidosis of the *aorta,* the main blood vessel leading out of the heart. Two different proteins are involved, one of which builds up in the innermost layer of the aorta in an astounding 97 percent of people over the age of fifty, while the other accumulates deeper into the middle of the vessel wall in about a third of these cases. This amyloidosis is not currently recognized as a cause of specific pathology or death, but again it seems that this is only a matter of stepping over a fatal threshold that we don't reach in a normal lifespan today because we die of other things first.

Add them all up, and amyloid deposits of any of several misfolded proteins in the heart are significant contributors to death in the elderly, causing abnormal heartbeat, weakening of the muscle of the heart, "blackouts" of the electrical activity that keeps it beating, and heart failure.

And those are just deposits of the cardiovascular system. *Every one of us* becomes riddled with microscopic amyloid deposits across multiple tissues in the body by the time we hit our eighties. Its toll isn't widely appreciated because the very old are autopsied so rarely, and you just can't *see* the deposits without opening a body up. This lack of curiosity about death in the very old of today is just another example of our routine acceptance of the massive toll of aging processes in people who have enjoyed only—yes, only—a few score years of life.

Moreover, while the evidence is still preliminary, amyloidosis of one organ system or another appears to become an increasingly critical factor in the snuffing-out of people at the extremes of the current "natural" longevity range. Some of this evidence comes from Japan, where the presence of a few centenarian "hotspots" has made it an exception to the widespread pattern of voluntary ignorance about what kills the oldest old. An autopsy study carried out at Aichi Medical Center in Japan from 1989 to 1995 found brain-wide CAA in 16 of their 19 centenarian patients.[5] Unfortunately, this study was restricted to the central nervous system, so it did not provide any information on what other amyloid diseases might have riddled these long-lived humans' bodies or to what extent such diseases may have contributed to their deaths.[6]

Even more suggestive is the early evidence coming out of the important effort by the newly launched *Supercentenarian Research Foundation* (SRF)[7] to autopsy as many "supercentenarians" (the extremely rare people who live beyond the age of 110) as can be identified and convinced to donate their remains to science after their deaths. Of the six who have thus far been examined, *four* were felled by some form of amyloid disease (the other two deaths were cancer victims).[8]

Again, we don't yet know what the pathological consequences of many of these deposits may be, but it seems awfully likely that they are doing us harm—so by the engineer's definition, they're aging damage, since they're not found in the young. Hence, you can bet your life that I want to clean them up along with the ones that *have* already been exposed as culprits in specific age-related diseases.

Alzheimer's, Amyloids, Aging

In a perverse way, then, the fact that Alzheimer's is such a widespread and obviously terrible disease has aided the cause of general anti-aging biomedicine. The attention to this specific age-related curse—together with the widespread professional and public belief that amyloid beta deposits are a major factor in its development and progression—has driven scientists to attack this particular form of amyloid as a therapeutic target in its own right, and that work has opened up the strong possibility that a similar strategy can form the basis of foreseeable therapies for amyloid-type extracellular damage generally. As with other cases that we've discussed in previous chapters, the existence of a *recognized* disease that is caused by aging damage has "legitimized" research into ways to clean it up—and, as this research bears fruits in new cures for these diseases, anti-aging biotech will be able to hitch a ride to develop treatments for aging damage itself.

Alzheimer's is an especially good example of this phenomenon, because it is both so utterly fearsome *and* so common in our parents and grandparents (in contrast to rare and rapidly fatal disorders like the mitochondriopathies or the lysosomal storage diseases). As the sheer number of Alzheimer's victims explodes as the population's biological age creeps upward, victims' families and loved ones have organized politically. There are now thousands of people in the United States and elsewhere who are *demanding*—and getting—enormous government investments of intellectual and financial capital into the quest for a cure. (Indeed, for better *and* for worse, Alzheimer's research now consumes over half of the National Institute on Aging's budget.)

Once the view that beta-amyloid was the key to the disease became dominant, Alzheimer's specialists (possibly because they were not biogerontologists?) began thinking about this particular form of extracellular junk along the same damage-reversal lines that underlie the engineering approach to age-related damage generally. All that currently available treatments for Alzheimer's disease can do is improve the *symptoms* of the disease: sadly, *no* existing therapy has the power to check the ongoing *degeneration* of the brain itself (see sidebar, "Alzheimer's treatment today"). This does mean that users of these drugs are better off at any given point than they would be without them, but the underlying progression of the disease continues unabated with every passing day. Functionally, even modern Alzheimer's drugs perform the way ibuprofen and antidepressant medications do for diabetics—in providing merely superficial relief from the nerve pain that often accompanies the disease. While the pain may indeed diminish, diabetics' nerves themselves continue to be savaged by the "caramelization" chemistry of that disease. (See Chapter 9 for SENS's main answer to late-onset diabetes.)

ALZHEIMER'S TREATMENT TODAY

Currently, the most widely used treatments for Alzheimer's disease are the *cholinesterase inhibitor* drugs, such as donepezil (Aricept), rivastigmine (Exelon), and the herb galantamine (Reminyl/Razadyne), which attempt to bolster brain functioning by boosting levels of some of the signalling molecules involved in some of the flagging aspects of memory. Of course, such a treatment is purely palliative, with no effect on the underlying disease process. A recent study[9] suggests that these drugs are even less effective than is implied by their inability to prevent the brain-ravaging effects of the disease: It appears that while the drugs boost scores on standardized tests of some aspects of brain function, they have no effects on the kind of *real-world functionality* whose loss forces families to institutionalize victims.

There was hope, for a while, that a more recently introduced drug called memantine (Namenda) would at least *slow down* the progress of Alzheimer's disease, by shielding neurons against the damaging effects of another signalling molecule (*glutamate*) that can kill brain cells when present in excess. A recent trial[10] suggests that this isn't so. The study compared people who had already been on memantine in a six-month placebo-controlled trial, and who were then allowed to *continue* taking it for an additional six months, to people in the same trial who had originally been taking the placebo but who were given the real deal for those subsequent six months. If memantine were really slowing down the underlying disease process, you'd expect that people who started on the drug earlier would have been in better shape than those who'd had to wait through the first six months on the sugar pills, because they would not have been suffering the full effects of brain degeneration for the first six months and thus would have more intact brains later on. But instead, it was found that the patients who started taking the drug later quickly caught up, in terms of their improvement over their baseline condition, with those who had been getting it all along. This suggests that memantine's effects are only on the *immediate* symptoms of the disease.

That's good news, in a sense, for those who start taking memantine at a more advanced disease state, because it means that they haven't lost anything by waiting to get started. However, the bad news is that *no one* taking memantine can expect that it will actually stop their minds slowly dying from the tangled mess in their brains.

It's actually not clear that blocking glutamate's effects on neurons would be an entirely good thing in any case. As with so many things in the finely tuned network of metabolic pathways, glutamate is a molecule with two faces. While it can stimulate brain cells to death when it's present in excess, it's also a key chemical signalling molecule in the brain, required for the normal storage and retrieval of memories. This suggests the possibility that memantine may cause problems in the laying down of *new* memories, even as it preserves the brain cells that store the old ones. There is no direct evidence of such an effect yet, but it still isn't clear that the drug *helps* much, either: While this trial did seem to show a benefit from memantine, there were *no statistically significant* benefits compared to sugar pills in the two other major trials of the drug.

In any case, even a drug that could slow down the *rate* at which brain cells are lost would be unable to *prevent*—let alone *reverse*—the degeneration of the brain, since the underlying damage that has already occurred is left unrepaired.

As the evidence supporting a central role for beta-amyloid in the development and pathology of Alzheimer's built up, a new hope emerged. Scientists began talking seriously about the idea that, by making beta-amyloid itself the target for new medical interventions, they would be able to develop new treatments that would *treat* the disease instead of merely providing crutches to a crippled—and rapidly deteriorating—mind. Once researchers had the tools they needed, in the form of engineered mice whose brains produced variations on human beta-amyloid that led to the formation of brain plaques and dysfunctions of brain and memory, they could start work on testing therapies that would target beta-amyloid directly.

New Targets, Old Rifles

But simply believing that that beta-amyloid is the main villain in the Alzheimer's story doesn't tell you what to do about it. Thus, it's no surprise that academic labs and pharmaceutical companies around the world have been working on quite a variety of anti-amyloid strategies, each hoping to make a Nobel-quality breakthrough or market a blockbuster new drug with a desperate—and ominously, inexorably expanding—"target market."

Predictably, however, when scientists began thinking about how to tackle the beta-amyloid plaque problem, many of them first turned to classically *preventative* strategies typical of the old-style gerontological approach to aging. Recall that beta-amyloid, like other amyloid proteins, is formed from an essentially healthy protein—amyloid precursor protein (APP). That protein is found in long strands woven through brain cell membranes, and while its exact function is unknown, it's at least *harmless* as long as it remains intact and in place. But occasionally, the *beta-secretase* enzyme mistakenly latches onto the APP protein and chops it at an unintended place; gamma-secretase innocently follows suit, not recognising the fatal flaw in the misprocessed APP; and beta-amyloid—with its exposed, sticky binding sites—is released to wreak its havoc in the brain.

With this in mind, one of the first ideas for an anti-amyloid therapy was to create drugs that would dampen down the activity of these enzymes, thereby cutting down the production of beta-amyloid. This in turn would reduce plaque formation, and thus either slow down or prevent the emergence of the disease.

First out of the gate was a drug that interfered with gamma-secretase activity. Animal studies showed that even a single dose of the drug could reduce levels of soluble, pre-plaque beta-amyloid in both the brain and the plasma, and it was duly moved through the development pipeline into the "Phase II" trials that are designed to give preliminary evidence of a drug's efficacy and safety in a moderate-sized population of people with the disease.

That was in 2001. I'm writing this in early 2007, and to date there has been total silence about the results of the trials of this first gamma-secretase inhibitor. We may never know what happened, but we may be able to guess. Even as the drug was being tried in humans, greater understanding of the role of gamma-secretase in the body emerged. A question that had long hung over the enzyme-inhibition approach was what, exactly, the enzyme was *supposed* to be doing in the body.

Many harmful mutations lurk in isolated pockets of the human family, but we *all* have gamma-secretase in our brains—and, as I discussed in Chapter 3, evolution does not design us to suffer horrible diseases. Although gamma-secretase has the unfortunate long-term side effect of beta-amyloid production, scientists always had in the back of their minds the acknowledgement that it had to serve some *useful* purpose, too. And, sure enough, researchers discovered while the trial was still progressing that gamma-secretase operates on several proteins in the body—including *Notch receptor 1* (NOTCH1), a protein with critical functions. *By* this I mean *really* critical: They include the activation of stem cells that renew damaged muscle tissue, the growth of new blood vessels, and the maturation of some kinds of immune cells.

So, what happens to these important biological processes, when you start interfering with an enzyme that's needed to keep them going? Animal models, using either a "knockout" of the gamma-secretase gene or an alternative gamma-secretase inhibitor drug, showed that dampening down the enzyme clearly prevented immune cells from developing in both the bone and the thymus, reducing the number of these cells and causing pathology in the gut. It seems highly plausible that the reason for the stony silence on the human trials is a similar profile of side effects.

However, some researchers are still chasing after therapies based on the same basic strategy. In 2002, researchers at Eli Lilly presented animal data showing that LY450139, the code name for a new gamma-secretase inhibitor, lowered beta-amyloid levels in Alzheimer's mice without interfering with NOTCH1. As this chapter was being drafted in 2006, the results of an early human trial in seventy Alzheimer's patients came in, showing that the drug reduced levels of beta-amyloid by 38 percent without apparently causing any serious side effects. By April 2006, the company had partnered with several universities and hospitals and was gearing up to perform a larger clinical trial to see if it could actually affect the disease. Elan Pharmaceuticals is also continuing to research a gamma-secretase inhibitor.

But deactivating NOTCH1 is far from the only potential concern with these drugs. Remember, gamma-secretase is also an essential partner in the *normal, non*-amyloidogenic metabolism of APP into products that appear to be essential to the functioning of neurons. It's unlikely that you can get away with ratcheting its normal activity down by force with no negative impact on the very brain function that researchers are desperately trying to preserve. It may just take longer than the brief six weeks over which the new drug's first safety test was performed. The new trials are just starting to recruit patients at this writing; we'll see how they fare in the clinic, and keep our fingers crossed for the people taking them.

Other scientists are pursuing a slightly different version of the same basic strategy. Some are developing drugs that inhibit *beta*-secretase, or that turn *up* the activity of *alpha*-secretase, with whose normal processing of APP beta-secretase interferes. Because such drugs are still in the early stages of development, we don't yet know what their side effects will be, but again it seems unlikely that the activity of an enzyme produced normally throughout the body—and especially in the brain—can be altered without cost. For instance, one specter that already looms over the beta-secretase inhibitors (based on animal studies) is that they may make users more vulnerable to some psychological disorders. While very young animals with their beta-secretase genes knocked out seem to be *physiologically* more or less normal, they are timid and don't like to explore their environment, and seem to run through *serotonin* (the chemical messenger whose metabolism is modulated by drugs like Prozac) abnormally quickly.

Freeing the Prisoners . . . or Letting Loose the Inmates?

Other scientists are pursuing an alternative approach that is, superficially, more in line with the engineering principles that I have advocated. Namely: to ignore the formation of beta-amyloid itself, and instead focus in on the process whereby beta-amyloid becomes aggregated into neuron-enmeshing plaques. A surprisingly large number of drugs and even herbal concentrates—from extracts of the spice turmeric (used in curries) to custom drugs with the highly marketable moniker *beta-breakers*—either interfere with the glomming-together of beta-amyloid into plaque fibrils, or even break existing aggregates apart . . . in a test tube. Most of these compounds never worked out once they got beyond the petri dish and into a living, breathing organism, but a few have been shown to reduce the plaque burden in animals genetically engineered to produce large amounts of beta-amyloid, and a few of *those* are now going into clinical trials in people with Alzheimer's disease.

Here, however, we once again run into the problem of over-reliance on our hypotheses about what biochemical processes "cause" the disease. In the other amyloidoses, the connection between fibril and pathology is pretty obvious. Indeed, you can dramatically extend life expectancy in patients with several kinds of amyloidosis by "simply" *replacing* the amyloid-strangled organ with a transplanted one free of fibrils.

But the reality is that, despite a decade of intense research into the "amyloid hypothesis" of Alzheimer's disease, we still don't understand the mysterious metabolic underpinnings of the disease. Even among the majority of researchers who *are* convinced that beta-amyloid is the key to Alzheimer's, the consensus around the detailed mechanistic role of the protein in the disease is as shallow as it is wide. Controversy continues to rage about what exactly links it—the protein itself, and/or the plaques that form from it—to the decay of brain and body that we see in its victims. Thus, the premise that beta-amyloid plaques *cause* Alzheimer's—or link the underlying metabolic defect(s) to disease—is still not enough in itself to tell you what should be done about them.

If anything, the balance of evidence is that it may not be the plaques *themselves* that impair neurological functioning the most, but the soluble beta-amyloid *oligomers:* short chains, made up of just a few single beta-amyloid molecules ("monomers") linked together in the same way that plaque fibrils are, but whose small size allows them to remain dissolved in the fluid bathing the cells of the brain rather than precipitating out into deposits. In cultured cells and in experimental animals, beta-amyloid oligomers derived from human nerve cells clearly disrupt neuronal function and interfere with normal memory, in ways that are not observed with either the beta-amyloid plaques that they form or the lone "links" of beta-amyloid monomers of which they are composed. After being microinjected with human-derived oligomers, rats become confused and forgetful. Beta-amyloid monomers don't have the same effect, and while giving animals or brain cell cultures chemicals that clear *all* forms of beta-amyloid (oligomers and monomers) out of the fluid bathing the neurons prevents the negative effects of exposure to a mixture of oligomers and monomers, an enzyme that selectively degrades free beta-amyloid monomers while leaving the oligomers intact provides no protection.

Likewise, the rats' memory function is largely recovered a day after introduction of the oligomers, when they have been cleared out by the animals' natural protective systems. This again suggests that the oligomers, rather than the plaques that they form, are the guilty parties in cognitive function. And indeed, preliminary studies suggest that the oligomers act as the molecular equivalent of dust in the eyes of neurons, interfering with their ability to receive signals from other neurons and pass the signal on to their internal machinery. Adding to the uncertainty, one group have reported a mouse model in which abundant plaques form but no neurological deficits result.

Another series of experiments strongly suggests that the plaques are the *result,* rather than the cause, of the widespread death of neurons in the Alzheimer's brain.[11] This study looked in detail at the location of both soluble amyloid species and plaques in the brains of people who had died of the disease. Soluble amyloid was found accumulated in still-intact cells within their lysosomes (the cellular garbage disposal units discussed extensively in the last chapter). Areas with very high levels of amyloid showed evidence of the rupturing of neurons, with beta-amyloid and the lysosomes' digestive enzymes dispersed outward from a central locus in a pattern that suggests nothing so much as a bomb blast. And wherever plaques were found, the researchers *also* found the remnants of a destroyed neuron's cell nucleus in the debris.

The strong suggestion was that the cells had been trying to dispose of the toxic amyloid oligomers by dumping them into the lysosome. Remember, again from the last chapter, that the Alzheimer's brain shows clear evidence of dysfunction of these organelles. Moreover, a lot of the beta-amyloid found inside brain cells is actually clustered in and around lysosomes, and the aggregates themselves are in fact naturally taken up and degraded in *microglia* (the immune cells of the brain), but at a rate that is too slow to keep up with plaque formation.[12]

This suggests that the burden of amyloid may eventually overcome the capacity of lysosomes to dispose of it, leading to the death-spiral of dysfunction outlined previously—and eventually to the death and rupture of the cell, during which the beta-amyloid and lysosomal enzymes spew forth from the dying neuron, creating plaque deposits like so much slag from a bombed-out building.

However, the innocence of plaques in Alzheimer's disease can't be asserted any more confidently than their guilt. The beta-breakers that have been shown to be effective in animal models (as opposed to just test tubes) *do* restore memory function as they break up the plaques. And again, a mere glance at the snarling webs of amyloid-enmeshed cells of the

Alzheimer's brain defies the observer to accept the notion that the plaques are harmless.

One theory that may reconcile these conflicting conclusions is the idea that, at least in the short term, the plaques' most damaging effect on the brain is to act as *reservoirs* for the beta-amyloid oligomers. You may have done simple experiments involving solutions in junior high-school science class, in which a substance is dissolved at very high concentrations in a glass of water. Eventually, levels reach so high a concentration that the water can't hold any more, and a crystal precipitates out of the solution.

But the teacher may have explained to you—or even demonstrated—that the crystal is not a static entity, but exists in a state of "dynamic equilibrium," with some dissolved material continually precipitating out onto its surface to build it up even as some of its existing surface molecules are continually dissolving out into solution. The volume of crystal will remain constant at a given solution concentration, as the rate of dissolution *out* of the crystal remains equal to the rate of precipitation *into* it. But of course, if you add more water or dissolved material into the solution, the equilibrium shifts accordingly.

The same thing, more or less, may be happening with amyloid plaques. As the concentration of beta-amyloid monomers and oligomers increases, they aggregate into plaques, which keeps the level of dissolved oligomers lower than it might otherwise be, thus effectively reducing their potential toxicity to local neurons. But when the level of oligomers dissolved into the fluid is reduced, the aggregated oligomers dissolve back into solution, maintaining their signal-jamming influence in a toxic steady state.

This potentially creates a real dilemma for therapies designed to deal with beta-amyloid. Using drugs like beta-breakers to prevent amyloid fibrils from forming, or to break existing plaques apart, would free up beta-amyloid oligomers that would otherwise have been sequestered in the plaque mass—which would actually expose neurons to *more* oligomeric interference than just leaving the plaques alone or even letting them build up.

In fact, both things would happen at once—and this could well have parallels in the Alzheimer's brain. It seems very likely to me that beta-amyloid plaques play—or eventually *would* play—a similar role in the brain of the Alzheimer's patient to what, say, transthyretin deposits do in the hearts of people struck by senile cardiac amyloidosis: you can't look at the mess of a victim's brain and not suspect that the plaques have been choking neurons to death. While the evidence is strong that any intervention which results in an increase in neuronal exposure to soluble oligomers of beta-amyloid can be expected to damage the brain, simply leaving the plaques to grow larger and larger as the diseased neurons continue to produce beta-amyloid (and/or to rupture and spill unprocessed beta-amyloid out of their lysosomes) surely sets the brain up for an even greater problem further down the road.

These intriguing theoretical questions are the kind that excites the curious minds of scientists; moreover, answering those questions seems to many such scientists to be the natural way of identifying and developing treatment for this terrible disease. As for other kinds of aging damage, however, I believe that this assumption is flawed. Let me say it again: We do not need to understand in detail how aging damage accumulates, or by what mechanism it wreaks its havoc, in order to *undo that damage.* However exciting an *intellectual* challenge it may be to sort out the exact pathway that leads from the fatal nips and tucks on APP, to beta-amyloid formation, to plaques, lysosomal dysfunction, cognitive impairment, and neuronal cell death, the bottom line in terms of the *biomedical* challenge is that we have here a material that is clearly accumulating and altering the composition of our aging and diseased bodies. When I see that, I say: *it must go.*

You can probably guess, by now, what sort of anti-amyloid therapy I prefer. While the remaining concerns about the exact role of plaques and oligomers in the disease process make simply *breaking apart* the aggregated beta-amyloid a potentially risky strategy, that doesn't rule out a solution based on *removing the plaques in whole cloth.* That would eliminate the source of the problem, *no matter* what step in the formation, metabolism, or aggregation of the constituent material is in fact the key to its toxicity. Such an intervention would be classic anti-aging engineering as I've envisaged it—if it could be done.

Fortunately, we should know very soon. A potential solution is already undergoing clinical trials.

"Immune" from Plaque: The Beta-amyloid Vaccine

As I mentioned earlier, researchers found evidence some time ago that *microglial cells*—the immune cells of the brain—slowly eat up and digest away beta-amyloid deposits from nerve cells. Unfortunately, it was clear that the rate of clearance was not nearly high enough to keep up with the pace of deposition in Alzheimer's patients. But researchers guessed that this natural defense mechanism could be stimulated to greater throughput. The obvious way to do this would be as we do for other targets of the immune system: with a vaccine.

The vision: inject patients with beta-amyloid, and the silent sentinels of the immune system would be roused up in defense, seeing the protein as a foreign invader. The same forces that your body marshals against chicken pox or influenza would be mobilized for an all-out war on the brain-choking protein, churning out antibodies specific to it and inducing the brain's microglial cells to go on a search-and-destroy mission against brain beta-amyloid.

This was an even more exciting idea than it might at first appear, because an anti-beta-amyloid vaccine would be expected to have an even greater impact against Alzheimer's disease than had the vaccines used against earlier epidemics. Such vaccines had allowed us nearly to wipe out diphtheria, polio, and measles in developed countries by *preventing* new cases from emerging. The promise of a vaccine targeting beta-amyloid was that it would actually *cure* all but the most advanced cases of the disease. Armies of activated microglia, their appetites for beta-amyloid whetted, would actually *consume* and thereby *remove* the *existing* beta-amyloid deposits. Once those choking fibrils were removed, the brain's normal structure—and thus, function—would be restored.[13]

Critically, this approach should work no matter what theory of the link between beta-amyloid and memory loss turned out to be correct. (It would not be the whole solution if other hallmarks of Alzheimer's, such as the intracellular neurofibrillary tangles, were also contributing to disease progression—but SENS incorporates solutions to those other aspects too, as you've seen.) The vaccine cleared out the plaques—and also, it shortly turned out, the more soluble oligomers, even *inside* the nerve cell itself[14]—not by breaking them apart into their constituent elements, but by causing immune cells to internalize and digest them.[15] Because no soluble beta-amyloid is released by such an approach, there is no risk of doing new damage as levels of the soluble form rise: beta-amyloid just *goes away,* clearing neurons of its malign influence no matter what its exact mode of toxicity might be.

It turned out to be a relatively easy matter to test this concept in animals engineered to develop a version of Alzheimer's disease. While vaccines against viruses must be carefully modified to ensure that they are close enough to the real thing to set off the immune system's alarms, but yet different enough that they don't actually infect people with the disease, no grand feat of molecular engineering was used in the first test of the concept: mice were simply injected with aggregated human beta-amyloid.

It worked smashingly well from the outset. Plaques quickly regressed from the mice's brains. The swelling and dysfunction of the *neurites* (the branches that transport chemical messengers from one nerve cell to the next) faded away, leaving healthy, functional units. The dense, inflammatory overgrowth of supporting cells around the neurons retreated.[16] And memory function—evidenced by the animals' abilities to find hidden platforms in flooded mazes—became more like that of younger, healthy animals.[17,18]

The company coordinating this work, Elan Pharmaceuticals, obtained results in many different models of engineered mice, each with a different mutation in the processing of APP. Best of all, the treatment appeared to be quite safe. Contrary to what had been feared, the immune attack on the beta-amyloid *around* the animals' brain cells did not cause collateral damage to the fine network of supporting cells onto which the plaques were glued. Likewise, some had feared that the immune attack on beta-amyloid would be so aggressive that it would punch holes in the protective shield that protects the brain from the toxins in the bloodstream, triggering a flood of foreign substances into the brain from the rest of the body; but little evidence of such an effect was found. The FDA was so impressed with the results that it quickly gave Elan the nod to move their vaccine—codenamed *AN-1792*—into placebo-controlled clinical trials.

Nearly 400 patient volunteers were recruited, of whom 300 received the amyloid vaccine and 72 were given injections of saline solution as a placebo control. Their baseline condition was determined, their mental state and functionality were assessed on a battery of neuropsychiatric and clinical tests, and a regimen of periodic injections was instituted.

Twelve months later, disaster struck.

A Fire in the Brain

Just a few months after this trial had begun, some patients started exhibiting serious side effects. Of more than 300 patients recruited from 28 clinical centers across Europe and North America, about one in 15 developed *meningoencephalitis,* a life-threatening swelling of the brain, apparently as a result of an overreaction of the immune system inside the brain itself.[19]

As soon as the side effect was discovered, the trial was halted, sending researchers scrambling to try to figure out what had gone wrong. The problem came as a complete shock. The vaccine had been tested in mice with a wide range of genetic abnormalities, each leading to Alzheimer's-like plaque formation from a different defect in the synthesis or metabolism of APP, and no such side effect had been observed—this despite the fact that scientists had been much more aggressive in their treatment protocols with mice than they would dare to be with human patients.

How such a crisis could occur, after such careful preclinical testing, has been documented extensively in the media and the academic literature, and I think it's too much of a digression to describe here. The important point is that researchers rapidly homed in on the problems with the first vaccine—and, as we shall see, on how those problems can be overcome.

Snatching the Ore from the Smelter's Ashes

The first trial proved catastrophic for some patients, and it was terminated before science could assess the full range of effects of the vaccine—positive and negative. But despite—and in a way, *because of*—the constraints imposed by this serious adverse reaction, researchers were desperate to squeeze every available bit of data that they could from the trial, to make up for its human and financial costs.[20,21,22] Through careful sifting of the data, scientists managed to collect some preliminary information suggesting that, despite the horrors of inflamed brains in a few patients, immunization with beta-amyloid fundamentally does *work* as a therapy in humans. And when combined with further animal studies, the findings also suggesed ways to avoid this side effect (and others) in future vaccines—some of which are either under development, or even in clinical trials, already.

In the immediate aftermath of the study's sudden termination, information about the patients was still scattered throughout the twenty-eight independent clinical centers at which the patients had received treatment. Preliminary assessments of the readily available information were not promising: researchers saw little evidence of improvement in the memory or other cognitive functions of people in the trial. But as the dust settled and researchers began to collect and analyze volunteers' full medical records, a new pattern emerged.[23] A given vaccine (any vaccine) does not do the same to all its recipients: some people create a stronger immune response to it than others. By separating out those participants whose blood work showed that they had mounted a substantial antibody response to the injected beta-amyloid by the time the study was shut down (fifty-nine volunteers) from the rest (who had not), scientists were able to show that those who had responded well to the vaccine *had,* apparently, fared better.

This finding took time to emerge because it was initially obscured by the statistics. When the researchers looked *individually* at each of the cognitive tests that volunteers had been administered, they could find no differences that passed the test of statistical significance. However, an integrated analysis of the *whole battery* of tests suggested that people whose immune systems had responded to the challenge suffered less decline during the study period than had people administered the placebo—a difference that was only beginning to become clear at the twelve-month mark when the trial was halted. Of particular interest was the fact that the emerging difference was most apparent for a composite score on the memory tests. And most suggestively, there did seem to be a sort of "dose-response" effect, with the greatest improvements in scores on overall memory, immediate and delayed memory, a nine-component memory score, and possibly a test of "executive function" (that is, higher-order brain functions involved in governing ourselves in a goal-directed fashion in the beginning and over time) in those whose antibody responses were strongest.

These results were all the more impressive considering the likelihood that most of the active responders were suffering from low-level brain inflammation and "ministrokes" triggered by the vaccine—a possibility suggested not only by autopsy findings and animal studies, but by the fact that headaches and confusion were reported as side effects so much more often in subjects who had received vaccination than in those who had received the placebo. Indeed, studies in breeds of Alzheimer's mice that had beta-amyloid deposits on their vasculature, and were thus vulnerable to microhemorrhages in response to the vaccine, *still* showed some cognitive improvements, despite the direct damage their brains suffered from the tiny bleeds. If your brain function is being preserved in spite of a quiet, chronic attack on it by your own immune system, the implication is that something *else* about your underlying clinical condition has improved *even more.* A vaccine that would clear out beta-amyloid (as this one appears to do) *without* causing the inflammatory side effects would therefore be expected to yield a much more robust improvement in the workings of the mind.

Another interesting finding came when researchers looked at the levels of the protein tau, which is the major constituent of neurofibrillary tangles, in the fluids bathing the central nervous system (the *cerebrospinal fluid,* or

CSF). Even though they only had about ten subjects in each group with good enough data to compare, the medical teams did observe that the vaccine responders had lower levels of tau than did the patients receiving placebo. While it's very indirect, this *might* be a sign of reduced rates of brain cell death, since high levels of tau in these fluids are associated with the death of neurons in people with the disease.

What about evidence on the *intended* effects of the vaccine—its ability to actually clear out beta-amyloid plaques? Unfortunately, it's not yet routine for scientists to look into the brain to see the plaque burden in people's brains before they've died, although new imaging techniques to do just that have been developed and are being tested for accuracy now. What *can* easily be done is to analyze levels of beta-amyloid in the cerebrospinal fluid. In animal studies, vaccination against beta-amyloid almost always results in an *increase* in CSF levels of beta-amyloid—a finding usually interpreted as a sign that the vaccine is helping the animals to clear the stuff out of their brains and then transport it elsewhere in the body for disposal. This effect was *not* seen in the small number of trial volunteers for whom scientists had samples that they could compare. But we have another, more direct source of evidence on the subject: the three vaccine responders who died over the course of the trial. While one must again be cautious in reading too much into the results observed in just three people's brains, autopsies of these volunteers by independent groups did show a *dramatic* reduction in levels of plaques in key regions of each of them as compared with control subjects.

Moreover, pathologists examining these brains found microglia in close proximity to many of the plaques that remained, suggesting that the vaccine was working as scientists had always hoped: the activated immune system had successfully mobilized microglia to clear out the deposits of beta-amyloid. This conclusion was further bolstered by a study—only completed *after* the end of the human trial—showing that old Caribbean green monkeys given beta-amyloid vaccination exhibited huge (66 percent) reductions in beta-amyloid levels in the brain, and a *complete absence* of plaques, and that the dense tangling-up of neurons' supporting *glial cells* usually seen in human Alzheimer's patients was considerably reduced.[24] This is an important piece of supporting evidence, because these monkeys develop some Alzheimer's-type pathology *naturally* as they age, and are much closer relatives to us than any mouse. The researchers who reported this finding are currently working to develop ways to assess the monkeys' cognitive function.

The Next Beta-amyloid Vaccine

All of this tantalizing, albeit inconclusive, information has once again erected the banner of beta-amyloid vaccination as a true *cure* (or at least a major component of a cure) for Alzheimer's disease. We've learned enough about both the potential of the vaccine and the reasons for the deadly brain inflammation to develop a variety of *new* approaches to the basic strategy, one or more of which will almost certainly prove effective without putting its recipients at risk.

The key to designing a safe vaccine is, of course, to avoid enraging the immune cells that attacked the original subjects' brains in the first trial even as the microglia were loyally cleaning out the amyloid plaques. There are several ways that this might be done, and each of them now enjoys support in animal models of the disease—supporting their efficacy while providing evidence that they will not have a deadly immune side effect.

One relatively straightforward approach that has already entered clinical trials is something called *passive vaccination*. Unlike a conventional *active vaccine,* in which the patient receives the offending agent *itself* (in this case, beta-amyloid) in order to signal the immune system to produce its *own* antibodies in response, passive vaccination involves directly providing the very antibodies that are desired, bringing out the immune response that the same antibodies elicit when they are produced by the body. The advantage of this approach is that it would allow scientists to choose *which* antibodies would be circulating throughout a patient's body. These antibodies might be chosen, or even custom-made, to activate the type of response that sends the microglia out to break down and digest the amyloid deposits[25] *without* the risk of eliciting the undesired antivasculature response.

The main disadvantage of this approach is that it would not induce the kind of semipermanent immune vigilance we've come to expect from vaccination against diseases like mumps or polio, but would instead require patients to receive regular reinjections of the antibodies to keep their amyloid-fighting supplies topped up. But even this apparent inconvenience has an upside: because the recipient's immune system isn't put into a long-term state of vigilance against beta-amyloid, physicians can stop treatment of an individual patient—or even stop an entire trial—at any time in case of side effects, without the fear that the body will remain in a destabilized autoimmune-like state of the kind elicited by the original vaccine.

Another approach is to manufacture an active vaccine composed of only *part* of the beta-amyloid molecule. When scientists looked at the immune response to active vaccination with the whole beta-amyloid molecule, they found that a mixture of different antibodies was being produced. Only a small number of those antibodies were of a type that harms the vasculature—and, significantly, these antibodies were only observed in *humans* administered the original vaccine, and not in the mice or monkeys that received the same vaccine and had not suffered the tragic assault on the brain. These antibodies were sensitized to only one segment of the total beta-amyloid protein, located in the middle of the molecule. By contrast, in mice, monkeys, and humans, *most* of the antibodies were of types that would mobilize the immune system against a completely different segment of the beta-amyloid molecule, located on its far end.

Since it's reasonably clear that the antivasculature response is not only unnecessary (since it isn't seen in the animal models with active or passive vaccination, yet these procedures dramatically clear out amyloid and improve memory) but is extremely harmful (due to its role in brain inflammation), it seems highly likely that a vaccine based on the above principles will work fine and be free of the risk of brain inflammation so long as it reacts only to this key subsection of the protein. Such a vaccine would induce the desired immune response on an active basis, without sending the body's immune system on a misguided mission against the very brain in whose defense it was aroused.

To date, several vaccines have already been developed based on this principle. They work by binding the key segment of the beta-amyloid protein to other proteins or antigens, or by combining it with appropriate immune stimulants, or by joining several such segments together into a sort of beta-amyloid molecular pincushion, all in an effort to maximize the antibody response to the vaccine without initiating the overreaction. These vaccines have all been shown to reduce levels of beta-amyloid, and often of plaques—some of them using very convenient delivery systems, such as transdermal patches similar to the ones widely used for nicotine treatment, or else the kind of nasal spray now used for quick decongestant relief of stuffed-up noses.

And even this has not exhausted scientific creativity. As this book was in preparation, for instance, researchers at the University of Texas' Southwestern Medical Center reported that injecting animals with the DNA for the most toxic form of the beta-amyloid proteins, smuggled under the skin inside tiny gold microparticles, led to production of the protein and to the body responding with a rigorous and apparently safe antibody response. The result: after receiving eleven injections over the course of several months, the Alzheimer's mice enjoyed a 60 to 77.5 percent reduction of plaque burden.[26]

As I mentioned, the most well-studied passive vaccine is already in mid-stage clinical trials orchestrated by Elan Pharmaceuticals. And while this vaccine was the first out of the gate, the race to develop a clinically effective and safe vaccine to defeat beta-amyloid is very much on. Indeed, the question today seems to be not so much *whether* vaccination against amyloid will work, but *which* of these many ingenious strategies will ultimately prove to be most effective and safe. Surveying the progress made so far, we can say with a high degree of confidence that we should soon be able to harness the ancient powers of the immune system to cut our way through the mind-stealing webs of beta-amyloid, redeeming the captive minds that they bind.

Amyloid Vaccination: Beyond Alzheimer's

I've spent so much time discussing the Alzheimer's beta-amyloid vaccine that you may well have forgotten how I got started: by noting that beta-amyloid is just one famous example of a *class* of extracellular aggregating proteins ("amyloids") associated with aging and with age-related diseases. And no, I'm not repeating the mistake of the National Institute on Aging in this regard, which is to throw nearly *half of all of its resources* every year into Alzheimer's research, at the expense of its real mandate, which is (or ought to be) to find ways to treat aging *itself* rather than one particular age-related disease. Instead, I've been guiding you through the vaccine's progress simply because it is in such an advanced state of clinical development. There's every reason to think that we will be able to exploit the same sort of immunological strategy to tackle most of the other age-related disorders caused by coatings of extracellular junk.

Next to beta-amyloid vaccination for Alzheimer's disease, the immunological amyloid-buster in the most advanced state of development is a vaccine for *systemic AL amyloidosis,* also known as *primary amyloidosis.* I briefly mentioned this form of amyloidosis early on in this chapter. This is the most common form of amyloidosis in the United States and some other industrialized countries—it strikes two to three thousand Americans annually. It's the result of overproduction by a specific cell type, *plasma cells,* of *immunoglobulin light chain* (L—thus "AL," for "Amyloidosis Light-chain"), a component of a class of antibodies.

However, I'm not going to spend much time on AL amyloidosis here, because it's not an *age-related* amyloidosis and the amyloid involved may differ in important ways from ones that are age-related. The main difference is that it's laid down so extremely fast that it may not have the same degree of problematic cross-linking as age-related amyloids. The main thing I do want to tell you about AL amyloidosis concerns an immunotherapy protocol for it that shows promising signs of being transferable to age-related amyloidoses.

The existing therapies for AL are decidedly inadequate. Until recently, the standard intervention was a regimen of high-dose chemotherapy designed to kill off the originating plasma cells, often combined with bone marrow transplants to replace some of the *other* blood cells that are destroyed in the process. More recently, a new chemotherapeutic drug called I-DOX has been used, after the serendipitous discovery that it could accelerate removal of systemic AL amyloid plaques through an as-yet unknown mechanism that apparently does not involve suppressing plasma cells. But even this new therapy tends to cause blood disorders, and both these treatments are only helpful for people with soft-tissue deposits, extending no benefit to the more serious cases involving the heart or kidneys. They're also not terribly successful at saving lives, with a mortality rate of about 40 percent.

At the turn of the millennium, scientists in Dr. Alan Solomon's lab at the Human Immunology and Cancer Program at the University of Tennessee Graduate School of Medicine developed a new animal model for AL, created by simply injecting mice with human light-chain amyloid extracted from the livers or spleens of patients who had died of the disease. The AL quickly began forming amyloid "tumors" in the animals, the size of which varied with the amount of material the animals were administered: at higher doses, the amyloid masses on their backs grew so large that they could be felt by hand.[27,28]

As they explored the effects of the disease on the animals, Dr. Solomon's group demonstrated that antibodies against a segment of the unique "beta-sheet" conformation of the light-chain amyloid fibrils were partially breaking down the deposits, making them more susceptible to immune attack. When they took amyloid tumors out of mice, subjected them to such antibodies and then returned the tumors into the mice, the animals cleared them out twice as quickly as new tumors of similar size. Clearance was also speeded when the original, unaggregated amyloid extracts were pretreated with antibodies before being injected. Even immune-deficient mice cleared out the deposits more quickly when they were given antibodies along with the amyloid extracts.[29] One such antibody—an *immunoglobulin G1 antibody* that they named *11-1F4*—was found to have the strongest "homing instinct," rapidly converging on tumors composed of either of the two major classes of human light-chain amyloid fibrils, whether in a test tube or in mice bearing them. Just as important, the antibody was found to target the amyloid specifically, without infiltrating tissues anywhere else in the body or samples of human tissue. And it also worked on *pre-formed* AL amyloid.

A Jack-of-All-Amyloids?

As I mentioned, the reason I'm telling you about 11-1F4 is not because of its effects on AL amyloidosis. The one disadvantage of Solomon's model of systemic AL amyloidosis was that it's an imprecise model of the human disease, with little spread of the deposits to major organs. Noting this, his group sought ways of testing it against *other* amyloid diseases, such as *amyloid protein A amyloidosis (AA*—also termed "inflammatory" or "secondary" amyloidosis), the most common amyloid disorder outside the United States.[30] AA has the advantage of being easy to induce as a full-blown disorder in many strains of mice: you simply inject them with chemicals like silver nitrate that induce a strong inflammatory response, resulting in overproduction of *serum amyloid A* (SAA), a protein produced by the liver during times of inflammation. Like immunoglobulin light chain, SAA is only partially degraded by the body's macrophages, resulting in the release of sticky, half-digested SAA fragments that tend to clump up and accumulate in the kidney and liver. This is the genesis of the disease in both mice and humans, so the amyloid disease that results when mice are given these inflammatory chemicals very closely mimics its human counterpart.

Surprisingly, the University of Tennessee team found that the 11-1F4 antibody also reacted to AA amyloids from mice—and that it cleared them out of affected mice.[31,32] In fact, it was very nearly as effective against AA deposits as it was against the original AL target, with the average organ amyloid burden dropping by over three-quarters in liver and spleen alike! This may be because the molecular architecture underlying the different fibrils' stickiness is similar, leading to a similar antigenic profile. It may also be related to the long-established fact that extracts of AL amyloid deposits accelerate the development of AA in response to inflammation in mice. If so, maybe the aggregating properties of the different amyloids are such that they can interact, with one serving as a sort of crystallizing center around which other amyloids gather. Think of a small deposit of cooked-on food on the surface of a pot. If the stain isn't immediately scrubbed out, it becomes increasingly stubbornly imbedded on its surface by future cooking, and then begins to catch food particles from subsequent meals, slowly expanding into a larger and larger stain that is ever more difficult to remove.

Even more remarkably, the University of Tennessee team then went on to show that 11-1F4 could also react with, and remove, amyloids based on *transthyretin* (TTR), the thyroid-hormone transporting protein whose aggregation causes senile cardiac amyloidosis, and which is a cause of death in so many of the oldest humans among us today. If passive immunization with 11-1F4 can reverse the course of both of these major forms of amyloidosis, we are well on our way to clearing out some of the most important sources of extracellular junk in the population at large—as well, potentially, as other forms of amyloidosis that are less common causes of death only because our lives are currently so brief.

This is exciting research, and the next step for the antibody is clearly to test it out in humans with the corresponding amyloidoses. For this to be done, the antibody must first be "humanized." Remember, while the light-chain amyloids that the antibodies remove are derived from humans, the agent that's doing the removal is a *mouse* antibody, which would probably not interact well or safely with the human immune system. To overcome this problem, Dr. Solomon's team "chimerized" the antibody, combining its antigen-seeking "business end" with a human "handle." The resulting antibody still recognized both light-chain amyloid and amyloid protein A aggregates, and even cleared them out of mice just as the original vaccine had done. The results are so promising that the National Cancer Institute's Drug Development Group has arranged for large-scale pharmaceutical production of the new antibody, with the intention of moving it into preliminary human clinical trials.

Open Possibilities

There is every reason to believe that this kind of immune-based, vaccination approach to amyloids, demonstrated in animal models of Alzheimer's disease and three human amyloidoses (and now in clinical trials for the former), will also work in other cases of cellular bindweed. Take, for example, amylin, or "islet amyloid polypeptide," whose amyloid-inducing properties I briefly mentioned early on in this chapter. Amylin aggregates accumulate on insulin-producing *beta cells* in the pancreases of nearly all people with type 2 (late-onset, non-insulin-dependent) diabetes. Either the aggregates or the soluble oligomers of which they're composed appear to play some role in the gradual dying-off of beta cells that occurs as the disease progresses,[33] leading to the failure of the body to produce enough insulin to keep up with the incessant surges of sugar that accompany every meal.

No one has yet tried to develop a vaccine to remove these deposits, but the feasibility of such an approach is suggested by the fact that amylin fibrils have been identified inside macrophages harvested from areas adjacent to the amylin deposits, where amylin accumulates without being fully degraded. Moreover, amylin fibrils are engulfed by and accumulate within macrophages exposed to them under test-tube conditions.[34] All of this suggests that that the immune system mounts an attack against *this* form of extracellular junk just as it does against amyloid beta and the junk responsible for secondary amyloidosis—in which case, there is every reason to think that this attack could be strengthened with a vaccine similar to those currently in the pipeline for those other amyloidoses. The therapeutic promise of such an approach would be even greater if it were combined with a souping-up of the macrophages' lysosomes with enzymes that are more able to digest the amylin fibrils—a job that cries out for the use of the LysoSENS approach that I discussed in the last chapter.

Other forms of amyloidosis could also fall before an infusion of targeted antibodies or other vaccines. And while the focus for drug development today is on treatments for specific amyloid-based diseases, this same research can be bootstrapped into the SENS agenda. Once it has proven its efficacy in Alzheimer's disease, senile cardiac amyloidosis, and type 2 diabetes, the spinoff technology will allow for the rapid development of vaccines for more obscure amyloid deposits that today may go nearly unnoticed except in people with a hundred candles or more to illuminate their birthday cakes.

The fact that these therapies have moved so quickly from the laboratory into clinical trials (remember, results in mice with the first beta-amyloid vaccine were reported in 1999, and it was in clinical trials by 2001) suggests that we will be able to move even more rapidly in the future, when the first anti-amyloid vaccines have passed through clinical trials and have been successfully used in doctors' offices all over the world.

Eventually, I envisage a protocol to keep our bodies clear of extracellular junk in which we might take a regular sequence of anti-amyloid vaccines, not unlike the standardized series now given in regular succession over the course of our childhood. The timing and frequency of administration of a given vaccine would depend on how quickly its target builds up to levels that impair function: we would get a "booster shot" of some every few years, while others would be administered only a few times in each century of a greatly expanded lifespan. Each time we took one of these vaccines, our cells and organs would once again live and function free of a specific species of molecular bindweeds, returning them to the literally *unbound* potential of youth.

9

Breaking the Shackles of AGE

Year after year, ongoing chemical processes are shackling the structural proteins of your body together, holding them back from their vital jobs. Eventually, this leads to a familiar (and ultimately fatal) range of age-related disabilities and diseases—especially in the kidneys, heart, eyes and blood vessels. What if we could break these chemical shackles, and thereby allow those proteins to get back to work, as they did when you were young? Scientists are making progress toward drugs that could achieve just such a goal.

You're in the last hours before the big holiday feast, and the atmosphere is heavy with the smells and emotional charge of the season. It's been a long day in the kitchen—the oven running continuously, the matron of the house trying to keep cool by leaving a window left slightly ajar—and at last the hurry and stress are giving way to a more expectant, eager sort of tension. The potatoes are mashed, the cranberry sauce has been spooned into serving dishes, the sweet potatoes are being kept hot in the oven, pumpkin pie is cooling on the windowsill . . . and now, a single component of the meal dominates the cook's attention and the appetites of her family.

Every fifteen minutes, like clockwork, for the last hour and a half, the turkey has been lovingly basted with its own fat, and perhaps a little honey; now, to perfect the feast, the broiler is flipped on, to glaze its surface.

All the time that the turkey has been in the oven, complex chemical processes have been imperceptibly proceeding—and now they accelerate. Down at the molecular level, the high heat causes the sugars and fats to attack the proteins in the bird's skin. Molecular bonds are forged; new chemical products arise and are broken down; neighboring proteins are tied together in shotgun marriages, tightening the outside surface of the turkey and coating it with thick, gooey chains of linked proteins, fats, and sugars.

Finally, the deed is done. Mom flips the oven off and dons her oven mitts as she calls to Dad to get the carving knife. The family looks on her handiwork with eager eyes, gazing with hunger and appreciation at the darkened, crispy, sticky, slightly *toughened* surface that the chemical maelstrom has made of the turkey's skin. Dinner is served.

I'm sure that you and your family have played out a similar script at Christmas—or at Hanukkah, browning the latkes or *sufganiyot.* But, in a profound sense, you have as much in common with the *feast* as you do with the *family.*

Every day of your life, the same processes that are involved in the browning of meats and other glazed or fried foods are insidiously at work in your body. In your arteries. In your kidneys. In your heart, your eyes, your skin, your nerves. At this very moment, in all your tissues, the sugar that provides your body with so much of its energy is also performing some unwanted chemical experiments, *caramelizing* your body through exactly the same processes that caramelize onions or peanut brittle. Slowly but steadily, unwanted bonding by sugars and fats handcuffs your proteins, inactivates your enzymes, triggers unhealthy chemical signals in your cells, and damages your DNA. Aging you.

Make that: AGEing you. And I'm not just reminding you of my nationality by adding that final e.

The Way We AGE

The body relies on sugar as a key energy source. But, like any fuel, sugar can only be "burned" by our cells because it is chemically reactive—and, again like other fuels, that volatility can make it dangerous to work with. *Advanced Glycation End-products* (or "*AGE*s," as they're appropriately called) are the end results of the complex chemical processes through which the structure of proteins is warped by sugars and other fuels. This same chemistry is the cause of the "browning" you see when you roast a turkey, caramelize a sauce, or pop a slice of bread into a toaster. AGEs accumulate in your tissues, leading to gradual loss of function, then disease, and ultimately an early grave. AGEs transform the supple grace of youth into a "crusty" old age, through exactly the same chemical processes by which they form the crust on a loaf of bread.

The many chemical reactions, intermediates, and stable end products of AGE chemistry have been the subject of an enormous amount of research, first in the food technology and chemical sectors and more recently in biomedicine. Scientific study began with work by a food chemist named Maillard in the 1910s and '20s, but it took until the 1980s for role of AGEs in the complications that ravage the diabetic body to become a hot topic in diabetes and aging research. Even now it's clear that we've only begun to understand the furious promiscuity of this biochemistry and its impact on the aging or diabetic body.

Though the details needn't concern us here, you will need to grasp the outlines of how AGEs form if you are to understand the various strategies that have been employed in the search for a way to shield us from their fossilising influence. In the best-understood pathway (the main stream of the *Maillard reaction*—see **Figure 1a**), a molecule of sugar opens its structure and glues onto ("glycates") a protein molecule, forming a *Schiff base*. This structure is relatively unstable, so the Schiff base will often spontaneously fall apart. Sometimes, however, it will collapse into a more stable structure called an *Amadori product*. Amadori products are much longer-lived than Schiff bases. (This fact has long been exploited in a lab test that measures levels of *glycated hemoglobin* or *HbA1c,* an Amadori product in red blood cells, as an indicator of the average amount of sugar that has been present in the blood over the course of the previous few weeks.)

Relatively stable though they may be, however, Amadori products are still subject to the biochemical hurly-burly around them. They can therefore be put through any number of further chemical transformations, such as rearrangement or degradation of their basic structure, forcible insertion of water molecules or removal of amino groups, or attack by free radicals. Many of these changes lead to the formation of even more stable structures, either directly or via highly reactive intermediate compounds such as *oxoaldehydes*. These structures are stable enough, in fact, to be called "end products"—they are the advanced glycation end products, or AGEs.

For our purposes, the important outcome of these processes is the

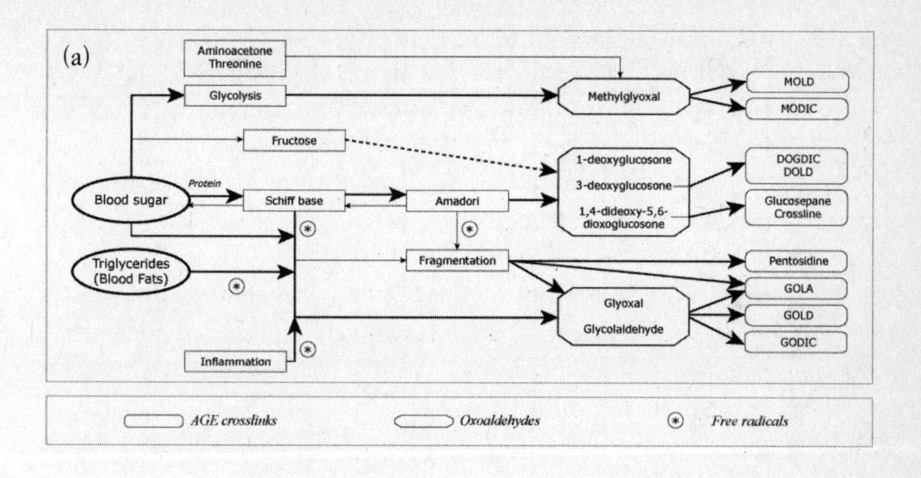

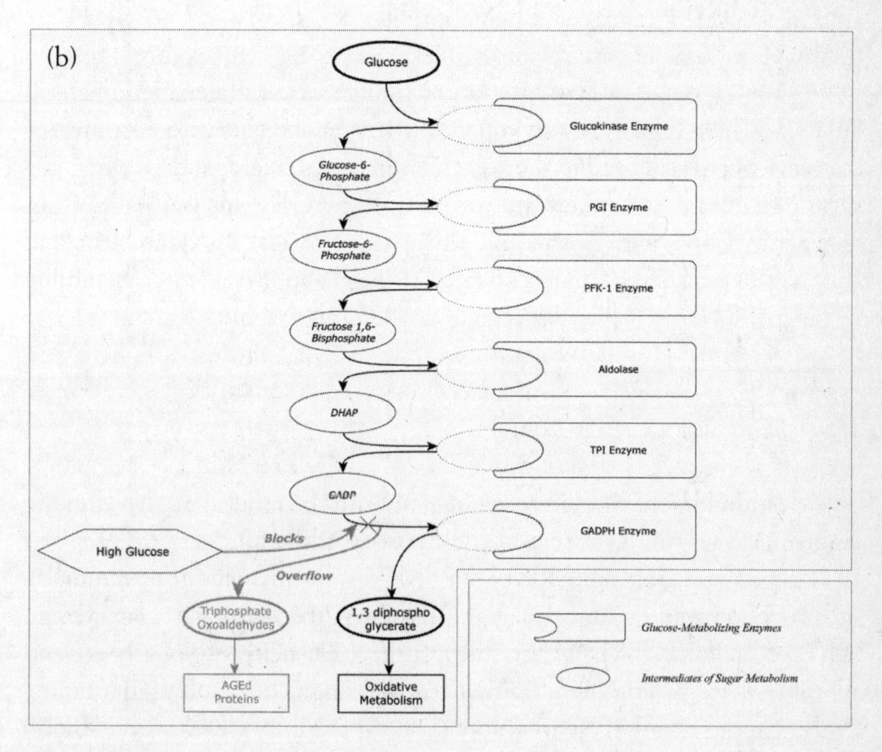

Figure 1. The ways we AGE. (a) The "chemical" (Maillard) pathway; (b) The "metabolic" (triosephosphate) pathway; (c) Sources of methylglyoxal.

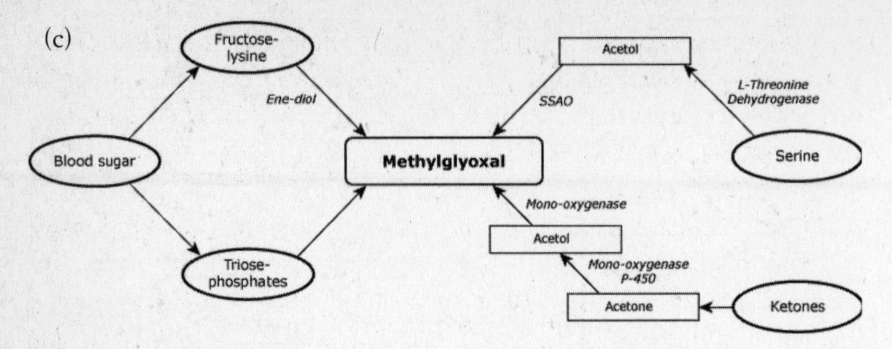

Figure 1. (continued).

formation of AGE *cross-links*, a subset of AGEs in which proteins that are already working with one arm tied behind their backs because of glycation become shackled to a second, neighboring protein.

AGEing happens much more quickly in people with diabetes than in the rest of the population, partly for the simple reason that diabetics' blood sugar levels are higher: In any chemical reaction, a higher concentration of an active agent will tend to increase the rate of its interactions with its targets, provided those targets are plentiful. But AGE cross-links also accumulate in people with normal blood sugar levels, and it's quite clear that they are responsible for much of the pathology and increased vulnerability to the insults of daily life that accompany "normal" aging.

Browning to Death

The cross-linking of proteins is similar, at both the molecular level and the functional level, to the processes that cause windshield wipers to lose their flexibility. For people who don't have diabetes, the most life-threatening locus of the ensuing stiffening of the tissues is the cardiovascular system. AGE cross-links slowly impair the youthful elasticity of your heart and blood vessels, making them rigid and unyielding. The resulting hardening of the arteries is in large part responsible for the increase in systolic blood pressure that everyone suffers with age. (Systolic pressure is the first of the two numbers that you get from a blood pressure reading, like the "110" in "110 over 80.") Meanwhile, the AGEing of your heart impairs its capacity either to *contract* to pump blood through your body, or to *expand* in order to fill up with that blood in the first place. The combination of these two factors increases the workload on the heart, ultimately leading to one of several forms of heart failure *if nothing else kills you first*. The same lack of plasticity also means that your blood vessels become less able to withstand the constant surges of blood that course through them: they become brittle and eventually break under the pressure like old rubber bands, one potential result of which is a bleeding stroke.

And the damage caused by AGE cross-links extends well beyond the cardiovascular system. They shackle proteins all over the body, accumulating with age in tissues as diverse as the tiny blood vessels in your eye and the supporting *myelin sheaths* of your nerves. Everywhere they occur, AGE cross-links impair the functioning of those proteins, contributing to age-related dysfunction, disability, and death. In your eyes, they accumulate on the *crystallin* proteins that make up the structure of the lens. AGEd lens proteins stop allowing light to pass through them, leading to the brown pigmented spots in the lens that we know as cataracts. The combination of this browning with several effects at the cellular level is why age and diabetes are the major risk factors for this, the single greatest cause of vision loss worldwide.

And that isn't the only way in which AGEs contribute to vision loss. Elsewhere in the eye, AGEs contribute to *diabetic retinopathy* (vision loss in diabetics linked to damage to the fine blood vessels feeding the light-absorbing tissues at the back of the eyeball), to age-related macular degeneration, and possibly also to open-angle glaucoma.

The kidney, too, suffers badly from AGE assault—again, especially among people with diabetes. Diabetic damage is the single biggest cause of kidney failure in the United States, and a third of all patients who find themselves in the dialysis ward got there because of their diabetes. Indeed, the severity of kidney disease in diabetics tracks the level of renal AGEs, which cross-link the proteins of the kidneys' biological filter material and trigger an inflammatory process that leads the body to overcompensate by growing too much replacement tissue, in a sort of out-of-control wound-healing response. The net effect of these two processes is a buildup of something similar to scar tissue in the kidney, which accumulates to levels that literally *squeeze* the tiny blood vessels where filtration is supposed to occur, reducing the amount of filtering surface available and leading to inefficient screening of materials in the blood—as if you had glued a coffee filter paper back on itself before running the machine, leading to a ground-filled mess when the water starts to back up.

AGEs also contribute to *diabetic neuropathy,* the debilitating damage to the *nerves* that is suffered by so many diabetics. The severity of this disease can vary, but the most common symptom is an unremitting version of the experience one has after a temporary, pressure-induced reduction in blood flow to the hands or feet (i.e., when the extremity is said to be "asleep"): a sensation of "pins-and-needles," pain, or numbness in the affected limbs, along with some loss of control or clumsiness in their use. People with diabetic neuropathy also lose some of the *unconscious* control by their nervous systems of functions such as the regulation of the heart's rhythm, the digestive process, the bladder, and erectile function; they also often suffer dizziness, and nausea that may extend to vomiting. Whether AGEs play any role in similar, more subtle defects in nerve function with age in otherwise healthy people is unclear, but it seems likely.

Comparisons of the rates of accumulation of cross-links in the tissues of slower- and faster-aging species, and of slower- and faster-aging individuals *within* a species, suggest that AGEing plays an important role in aging per se, not just in specific diseases or the complications of diabetes. Both the rate of age-related buildup of one of the more easily measured AGEs (*pentosidine*) and the related toughening-up of the proteins in skin or tail tissue are inversely associated with the maximum lifespan of different mammalian species. This means that the more slowly a species ages, the more slowly its collagen is stiffened by AGEs (see **Figure 2**). Likewise, calorie restriction—which is, as I've mentioned in previous chapters, the best-studied way to slow down aging in mammals—slows these processes down; and in fact, higher rates of tissue AGEing have been shown to predict early death in individual calorie-restricted animals.[1] In our own species, studies show that even within the "normal" range (i.e., at values well below those typical of people with diabetes), higher blood levels of either glucose itself[2] or of the Amadori product HbA1c[3] are associated with a higher risk of death from all causes.

A drug that would slow down or reverse the accumulation of AGEs would thus help people with a wide range of diseases and disabilities. It could potentially improve, or even cure, problems as wide-ranging as the gradual increase in blood pressure over a lifetime; the terrible kidney,

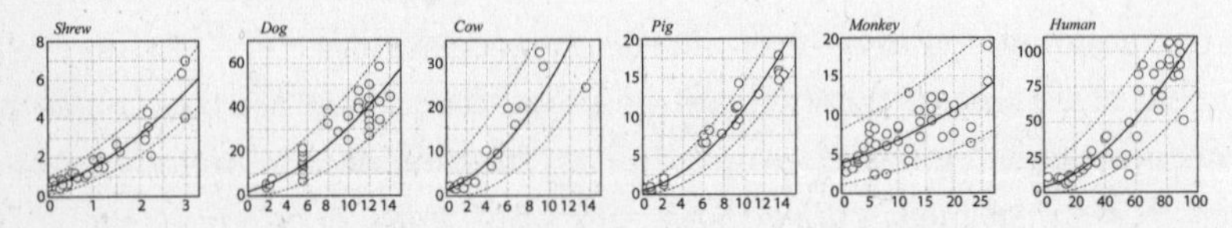

Figure 2. Maximum life span as a function of AGE formation rate. Redrawn.[4]

nerve, and visual complications of diabetes; and several forms of heart failure. And it could also help us to address a major contributor to aging itself.

This idea didn't just pop into my head recently, of course: a variety of schemes to reduce the tissue AGE burden have been explored over the years, and many of them have even reached relatively advanced clinical trial status. Yet, despite years of work, none of these treatments has been shown to be safe and effective enough to find its way into the drug arsenal of any developed country. The obstacles that have plagued their development and limited their usefulness represent yet another case study in the problems of trying to deal with age-related damage by tinkering with the complex biochemistry of life.

Listening to Parmenion

"Sugar Pills"

The fact that AGE cross-links are often ultimately the result of sugar molecules acting like a glue, gumming up our tissue proteins, immediately suggests one possible solution to the problem: just lower people's blood sugar levels, and you'll reduce the formation of Schiff bases (see **Figure 1**) in their bodies and thereby lower their AGE burden. Of course, this has long been the major focus of diabetes management, and in the 1990s, two massive and widely cited scientific studies—the *Diabetes Control and Complications Trial* (DCCT) and the *UK Prospective Diabetes Study* (UKPDS)—were hailed as the clearest proof yet of the effectiveness of this strategy when taken to its limits. These two studies showed that when diabetics take strict steps (aggressive use of blood-sugar-lowering drugs and regular feedback in the form of frequent blood sugar testing) to keep their blood sugar under very tight control, they are at greatly reduced risk of developing the major complications of the disease. The DCCT, in particular, showed that—as compared with the standard of care at that time—a regimen of intensive blood sugar control could reduce a diabetic's risk of developing nerve disease by close to two-thirds, diabetic kidney disease by about half, and diabetic retinopathy by an astounding *three-quarters*.

The results of these two studies were trumpeted around the world—by their government sponsors, by patient advocate organizations, and by pharmaceutical companies looking to boost sales of glucose-lowering drugs. The plan was to encourage doctors to prescribe these drugs to patients whose blood sugar control was in the range that made them safe by previous standards but demonstrably at risk based on the new data, and also to increase the doses taken by people with worse control who were already using the drugs.

The benefits that would accrue to patients as a result of such a surge in drug use seemed to be clear-cut: people with diabetes all over the world would enjoy miraculous improvements in the quality and length of their lives through dramatic reductions in their risk of blindness, nerve damage, and kidney failure. But when scientists actually assessed the *overall* quality of life of people who had undergone the intensive therapy regimens in the trials, the results were surprisingly gloomy. Despite the fact that more aggressive treatment had reduced the risk of all major diabetic complications, the intensive-therapy patients enjoyed *no improvement* in their net well-being as compared to people who had been assigned to standard care.[5,6]

Many factors probably contribute to the lack of clear-cut benefits from aggressively lowering blood sugar levels. While diabetic complications clearly have a negative impact on quality of life, the drugs used to lower blood sugar also come with costs that are not included in the sticker price. People on such medications tend to gain weight, which reduces their quality of life—both directly, and by increasing the risk of other diseases such as osteoarthritis. Many patients also find that sticking with the rigid schedule of injections and finger-prick tests required to keep up with these regimens imposes real restrictions on their lives, which some studies report contributes to depression, frustration, isolation, or troubles at work.

And finally, constantly trying to push blood glucose even into the "normal" range carries with it the risk that blood sugar levels will drop *too* far, leading to a "hypoglycemic crisis" whose consequences can range from dizziness to a coma. This is of particular relevance to normal aging. If pushing blood sugar levels down is a mixed bag for *diabetics,* you can see that it would be a decidedly dubious solution to the AGE problem for the rest of us, in whom the wiggle room between our normal blood sugar levels and a hypoglycemic crisis is much smaller, making the potential benefits more limited and the risks higher.

And even if we could safely bring our blood sugar down to the lowest possible safe level, we'd be quite far from a complete solution to the AGE problem. All of us must maintain *some* level of glucose in our blood as an energy source, and some percentage of that glucose will *always* wind up reacting with tissue proteins, leading to cross-linking.

And on top of that, not all AGEs are even derived from glucose. Blood *fats* (*triglycerides*) can also cause the cross-linking of proteins, particularly if there's a high level of oxidative stress; this is the chemistry that underlies the browning of a turkey skin as it roasts, even without a sweet, syrupy slather on its surface. As with blood sugar, diabetics usually have high triglyceride levels, and even many nondiabetic people would benefit from having their triglyceride levels brought down; but triglycerides also resemble blood sugar in being indispensable to normal function, so there's only so far that such a strategy can be safely pursued.

Less Is More . . . Is Worse

And that's not all: attempts to control levels of both these early precursors of AGEs, even by nonpharmacological means, can have perverse metabolic consequences.

For instance, one established effect of very low-carbohydrate diets of the Atkins type is to bring down both triglyceride levels and the body's total exposure to carbohydrates, so some advocates have hypothesized that these diets would reduce a person's AGE burden. Unfortunately, it turns out that the metabolic state that these diets induce (the notorious "ketosis") has the unfortunate side effect of causing a jump in the production of the oxoaldehyde *methylglyoxal,* a major precursor of AGEs that is also, ironically, produced within the cells of diabetic patients when they are forced to take in *more* glucose than they can immediately process (see **Figures 1b** and **1c**). A recent study tested the size of this effect in healthy people who successfully followed the first two phases of the Atkins diet for a month, and who had the ketones in their urine to prove that they were sticking to the diet. These previously healthy people suffered a *doubling* of their methylglyoxal levels, leading to concentrations even worse than those seen in poorly controlled diabetics.[7] Like other oxoaldehydes, methylglyoxal is far more chemically reactive than blood sugar (up to *40,000 times* more reactive, in fact), and is known to cause wide-ranging damage in the body, of which AGE cross-links are but one example. This potentially makes the Atkins diet a recipe for *accelerated* AGEing, not a reprieve from it.

"Radical" Proposal—Lukewarm Results

Even before the counterintuitive results of the DCCT came out, it was obvious that a blood-sugar-lowering strategy would not be a complete solution to the AGE problem. The body *needs* blood sugar and fats as fuels, and yet *no* level could be so low as to eliminate all cross-link formation: at best, lowering the concentration of glucose and triglycerides would delay the inevitable. So some scientists turned their attention elsewhere, to cross-link-inducers whose harmful role in the body is less ambiguous.

One such AGE-prevention strategy is the use of high doses of antioxidants to bring down free radical levels. As you can see from **Figure 1a**, free radicals can accelerate the conversion of some AGE precursors into certain specific full-blown cross-links—a phenomenon called *glycoxidation*. Based on the effect of adding free radicals to proteins and sugars in test-tube experiments, glycoxidation can be predicted to hit diabetics with a double whammy, because in addition to the excessive levels of blood glucose and fat, the impaired metabolic state of diabetes also causes an overproduction of free radicals in victims' cells. Put the two factors together and you have a potentially synergistic interaction. Also highlighting the importance of free radicals in AGE formation is the fact that birds have sky-high blood glucose levels that would rapidly kill a human, yet generally live around ten times *longer* than mammals of the same size; part of how they get away with this is probably by having really good control of oxidative stress.

If glycoxidation *were* a major reason for diabetics' high levels of AGE, then sopping up their excess burden of free radicals with antioxidants might considerably reduce the cross-linking of their tissue proteins, resulting in longer life expectancy and reduced risk of crippling complications. And many studies carried out in laboratory rodents have supported this expectation: dosing them with various free radical quenchers typically reduces the cross-link burden in their tissues considerably, cutting back on the incidence and severity of diabetic kidney, nerve, and even retinal damage.

When antioxidants were tried as an anti-AGE therapy in humans, however, the results were disappointing. The effects on AGE levels and symptoms were minor or nonexistent—and even when benefits *were* observed, the effect was almost exclusively confined to the most severe cases of the disease, with more typical diabetics getting no relief.[8,9,10] We now know that there are a couple of major reasons for this. First, human diabetes causes a much less severe increase in oxidative stress than is suffered in the rodent version of the disease, as can be seen by comparing the levels of molecules damaged by free radicals in the two species' skin. This lower free radical load makes glycoxidation a less important factor in human diabetics' AGE chemistry, and thus weakens the potential benefit of *reducing* its impact.

But another, more general reason for the lack of efficacy of antioxidants as an anti-AGE therapy is the sheer riotous promiscuity of the highly reactive precursors of AGE cross-links. One highly revealing animal study[11] illustrates the point. Diabetic rodents were given diets fortified with different antioxidant supplements (vitamins or green tea extracts), and the impact on the animals' AGE burden was assessed by comparison with both healthy animals and diabetics given unsupplemented chow. To tease out the biochemical pathways involved, scientists measured levels of substances produced at several different steps across the spectrum of the glycoxidation process, from initial glycation events to the creation of specific AGE cross-links—some of which are created by glycoxidation, and others by straightforward glycation, i.e., without the involvement of free radicals.

As previous rodent studies had shown, antioxidant treatment did exert some benefits on diabetic complications. Equally predictably, the treatments had no effect on the *initial glycation* of proteins, since the window of opportunity for free radicals to work mischief in AGE chemistry opens up further along in the process (**Figure 1a**). But the researchers got a surprise when they began looking at actual *cross-links*. Antioxidant supplements had no effect on the levels of those AGEs whose formation doesn't require free radicals, of course—but the intervention actually *increased* the levels of the two glycoxidation-derived AGEs, so that diabetic animals receiving green tea extracts actually wound up with *more* total cross-linking than those who simply suffered the "natural" course of the disease.

This remarkable result yet again illustrates the hopeless complexity of the tangled skein that is metabolism. The precursors of these AGE cross-links don't simply disappear when they aren't hit by free radicals—they have to go *somewhere*—and when much of the excess oxidative stress was relieved by antioxidant supplementation, these precursors began to build up until they spilled over into one of the alternative pathways of cross-link formation. It was the same effect you see in traffic jams when a main traffic artery is cut off: a few drivers may indeed just turn their cars around and go home, but most of them turn off onto the local side streets, creating secondary traffic congestion in hitherto sleepy residential neighborhoods.

Collateral Damage

The best-understood pathways of AGE cross-linking are fundamentally random events, not too far removed from what happens in the browning of food, or in a test tube. The fuels of metabolism, dissolved in your blood or in the fluid inside your cells, randomly bump into tissue proteins; depending on factors like temperature, concentration, and the presence of transition metals and free radicals, a series of chemical events may occur; and if they happen in just the right order, an AGE cross-link will form.

But some AGEs result more directly, from the *regulated* activity of metabolic processes. One recently identified example is the enzyme *myeloperoxidase,* which is used by macrophages to kill bacteria by generating toxic *hypochlorous acid.* It has been shown that hypochlorous acid, in the presence of the protein building-block *serine,* can itself induce AGE-type cross-linking, independent of the usual fuel chemistry of sugars and fats.[12]

If myeloperoxidase were only ever activated to kill bacteria, it might be a relatively unimportant source of AGEs in people living in the developed world who don't have chronic infections (although the number of such people is much higher than is generally appreciated). But, as we saw in Chapter 7, macrophages don't just attack bacteria: they also become aggravated—and crank up their myeloperoxidase activity—in their shortsighted efforts to clear trapped cholesterol from your arteries. Some scientists now believe that myeloperoxidase is probably a major contributor to the high levels of AGE found in the atherosclerotic foam cells of nondiabetic people.

While reducing excess myeloperoxidase activity might be desirable at the sites of atherosclerotic plaque, we probably could never lower its activity pharmacologically without also impairing our ability to defend ourselves against bacteria. As people with AIDS know, when your immune system is suppressed, you're not just at risk from relatively rare bacterial killers like tuberculosis: you can be felled by infections that most of us shake off before we have even the beginnings of symptoms. Moreover, and surprisingly, one study found that animals bred to produce something like human atherosclerosis, but lacking the ability to produce myeloperoxidase, showed *more* severe atherosclerosis than animals with normal activity, again illustrating the frustrating complexity of metabolic processes.[13]

The Drug That Failed

Okay, so trying to lower the levels of the ultimate AGE precursors, like glucose and fat, is difficult—and also unsafe, because they are essential biological fuels. Soaking up free radicals and sequestering transition metals is of limited effect because there are so many alternative routes to AGE formation.

But an overview of **Figure 1** may suggest a much more attractive target: oxoaldehydes. For one thing, these reactive compounds are present at much lower concentrations than blood sugar or triglycerides, meaning that you would only have to knock out relatively few molecules to lower the total level in the body by a significant proportion: methylglyoxal, for instance, is several *thousand* times less concentrated in the blood than glucose. On top of this, oxoaldehydes are very virulent molecules (as I mentioned earlier, methylglyoxal is up to *40,000 times* more prone to attacking tissue proteins than glucose is), so that each molecule you take out of circulation is much more likely to translate into the prevention of a cross-link in the waiting. Oxoaldehydes also play their role in cross-link formation relatively late in the process, leaving fewer alternative pathways by which an AGE might form if they could be soaked up. And in contrast to sugars, which are *essential* molecules for which there is a limit beyond which lowering their concentration in the blood becomes life-threatening, oxoaldehydes are fundamentally *toxic* molecules, so that one should be able to reduce their concentration drastically without doing any harm to the body.

Thus, if trying to lower AGE formation with antioxidants or blood-sugar medications is like launching a wide-sweeping crackdown on an entire crime-plagued neighborhood, a drug that mops up oxoaldehydes would be like springing a carefully targeted police raid on known members of a brutal criminal gang.

For a long time, a drug called *aminoguanidine* (trade name *Pimagidine*) seemed poised to fulfil this promise, revolutionizing the treatment of diabetes and perhaps landing the first serious punches on the Mike Tyson that is biological aging. The drug enjoyed a lot of buzz in the scientific literature on diabetes, as well as in some of the commentary on life extension in popular magazine articles and Internet discussion groups, because its clearest mechanism of action was precisely its ability to mop up oxoaldehydes.

Over the course of many years, researchers put aminoguanidine to the test—first in test tubes filled with AGE precursors and mixtures of catalysts, then in cell cultures, and eventually in animal studies. And at nearly every juncture, hopes for the drug continued to rise. In diabetic rats, it lowered AGE formation in the cells of the retina and reduced the maladaptive overgrowth of the blood vessels feeding them. In dogs, it prevented the loss of the retinal blood vessel cells and the associated accumulation of dead blood vessels through which blood had stopped flowing. It also kept both species' hearts and blood vessels more flexible.

Somewhat less consistently, aminoguanidine showed promise against other complications of diabetes. It reduced the total level of kidney tissue that was so cross-linked as to be indigestible by strong acids, and prevented much of the thickening of the kidney's filtration machinery that accompanies diabetes in rats—although it was unable to affect the course of the disease in dogs. Further, diabetic rodents (but not baboons) given the drug exhibited less loss of blood delivery to nerves, and improved ability to conduct nerve impulses.

Most important for those of us looking for interventions against AGEs' role in the degenerative processes of aging, aminoguanidine even seemed to reduce heart and kidney AGE levels (and the resulting loss of those organs' function) in animals that were suffering from purely age-related AGE accumulation rather than diabetes.

After a few small human studies designed to test for any obvious toxicity seemed to go well, the company that had patent control over aminoguanidine for use as a drug for diabetes attracted the attention of the biotech giant Genentech, which partnered with them to launch two full-scale clinical trials. Each one was to involve about six hundred people with the beginnings of diabetic kidney disease, in medical centers spread out all over North America. The first trial (ACTION I) recruited people with type I (autoimmune) diabetes; the other trial (ACTION II) was to involve patients with the more common type II (late-onset) diabetes, which usually develops as a result of lifestyle, albeit sometimes overlaid on genetic vulnerability. Ambitiously, patients in both trials were to be well medicated to control both blood sugar and blood pressure before being put on aminoguanidine, so that differences between the groups would be entirely the result of the test drug's direct anti-AGE effects.

But when ACTION I was completed in 1996, the best spin that one could put on the results would be to say that they were disappointing. On the positive side, risk factors like blood pressure, LDL ("bad") cholesterol, and triglycerides went down in people who had received the drug. And some crunching of the data suggested that the drug might improve some indicators of kidney function. Plus, in a tiny subgroup who had been tested before and after the trial, diabetic damage to the retinas *seemed* less serious in patients taking aminoguanidine than those taking placebo—though these observations were suspect, because they were made as an afterthought after the trial had been shut down rather than being part of its original design.[15]

What the study was *supposed* to show was a direct effect on kidney health, as measured by a standard laboratory test for kidney function—and the data just weren't strong enough to support that conclusion. The raw numbers looked, on their face, better in aminoguanidine users than in the placebo group, but the difference was so small relative to the number of patients in the trial that it seemed likely to be a statistical fluke, like getting "heads" in six out of ten coin tosses instead of the expected five.

Worse: while the *benefit* attributable to aminoguanidine was dubious, the *risks* associated with the drug seemed undeniable. Along with signs of an overactive (and possibly damaged) liver and strange flu-like symptoms that went away when they stopped using the drug, a few people taking aminoguanidine developed signs in their blood of an autoimmune disorder, which—in three patients taking the higher dose—was associated with a form of highly inflammatory kidney disease that leads to complete loss of kidney function in a matter of just weeks or months. Two of the three patients who developed the disease progressed to end-stage kidney failure. Fortunately, this apparent side effect was caught early in the trial, and the safety committee accordingly introduced a monitoring program, after which no one was allowed to progress into clinical signs of the disease.

A FATAL RESEMBLANCE

Even now, we still don't know for sure what caused aminoguanidine's severe toxicity, which had not been observed in animal studies. But there's a good guess—and if it's correct, it makes aminoguanidine yet one more case study in how trying to repress the dark side of metabolic processes so often has repercussions.

What made aminoguanidine a promising AGE-blocking drug was its mechanism: the sopping-up of oxoaldehydes. These substances are in a class of chemical compounds called *carbonyls:* organic molecules with a carbon atom double-bonded to an oxygen atom. This structure makes many carbonyls highly biologically active, which is why oxoaldehydes are such relentless shacklers of bodily proteins. Of course, metabolism relies on the harnessing of highly active compounds to carry out the biochemistry of life—and so it's hardly surprising that many essential biological molecules *also* feature prominent carbonyl groups.

The problem is that aminoguanidine can't necessarily tell one carbonyl-bearing molecule from another. That might be expected to cause it to sop up some *essential* carbonyl-bearing molecules along with toxic ones like oxoaldehydes. In fact, we *know* that it does so in at least one case: vitamin B6. As a result, animals given aminoguanidine can easily develop a deficiency of this vitamin indistinguishable from simply putting a subadequate supply of it in their diet.

Damningly, a blood pressure drug called *hydralazine*, which brandishes the same carbonyl-trapping *hydrazine* surface that aminoguanidine's business end uses to neutralize oxoaldehydes, is well known to cause a lupus-like autoimmune disorder, the first sign of which is the appearance in the blood of the same antibodies observed in the aminoguanidine users in ACTION I.

If a drug as promising as aminoguanidine can't safely prevent enough AGE damage to improve the health of diabetics, you can be sure that it won't do much for the basically healthy among us. Because the concentrations of blood sugar and fats are much lower in people without diabetes, the buildup of AGE is much slower, and thus harder to slow down to a degree that causes a measurable change in their health. Thus, it would take a lot longer for any potential benefits to accrue, whereas the risks remain at the same high levels for each individual year of use.

Indeed, a study published after aminoguanidine's withdrawal from clinical development[17] appears to show that even the initial reports of a reduction in age-related AGE cross-linking in nondiabetic *rodents* were specific to the strain of rat used in early studies (which is particularly susceptible to kidney disease). Other strains showed little or no benefit from lifelong aminoguanidine administration.

These are just a few illustrations of the known or anticipated ways in which the mechanisms underlying cross-link formation undermine our ability to prevent AGEing. This nightmare of biochemical complexity is so elaborate as to cause even the most dedicated puzzle enthusiast to snap pencil in two and go to bed in frustration; it should raise serious doubts about the wisdom of continuing to invest resources in seeking new ways to interfere with such a poorly understood, multiply-branching network of pathways (**Figure 1**). In the pell-mell of the body's biochemistry, a certain quantity of AGEs is simply inevitable, and trying to prevent *enough* cross-linking from happening to have a real impact on the stiffening of our tissues, without somehow disturbing essential metabolic processes, may ultimately be futile.

If you've read the preceding chapters of this book, you probably have a pretty good idea of the sort of strategy I'd like to see used to deal with the problem of AGEs, whether in diabetics or in "normal" aging. Don't mess with blood sugar. Don't try to block free radicals. Don't go chasing after ways to outsmart metabolism. Don't try to *prevent* AGEs from forming at all. No, the anti-aging engineer's solution should be to allow metabolism to proceed in its infamously messy way, and then to remove *full-blown AGEs themselves* before they build up enough to impair tissue function, robbing us of youthful flexibility of heart and sinew and increasing our risk of death and disability.

In this case, however, I am not playing the role of visionary, so much as of cheerleader. At least two companies have developed such drugs and tested them in animals. One of them has undertaken several clinical trials already.

A SENS Serendipity

In the decade since Drs. Tony Cerami and Peter Ulrich had first suggested that the cross-linking of proteins by glycation might be the link (pun intended) between high blood sugar levels and the complications of diabetes, they had spent a lot of their time working on ways to do something about it.

They had played key roles in the development and early testing of aminoguanidine, but well before the failure of ACTION I they knew that much stronger molecules would be required to help two specific groups of people with quite different AGE-induced disabilities. On the one hand, diabetics whose disease had progressed so far that they had already suffered a lot of cross-linking would be quickly approaching the threshold at which their *total* level of cross-linking would begin to result in disability and death, and would therefore require much more effective interventions than would people only in the early stages of the disease. And on the other hand, many people who suffer with AGE-derived diseases such as hypertension and heart failure have *normal* blood sugar levels. In these people, the precursors of AGEs are present at much lower levels and are thus are much harder to intercept: it's like trying to shoot down a single bird in the sky, whereas taking on AGE precursors in diabetics is like sighting one of the incredible flocks that once blacked out the sun in the early days of European colonization of the Americas, into which hunters could fire off buckshot without even bothering to aim.

So in late 1991, Cerami arranged for a summit at their labs at the Picower Institute for Medical Research in Manhasset, New York. The meeting brought together himself, Ulrich, several other Picower staffers, and scientists working for a company Cerami had helped form named Alteon, to brainstorm on new ways to inhibit AGE formation.

Analysis of what was believed about the chemistry and products of these reactions had already led many researchers to conclude (correctly) that *reactive carbonyls* (like oxoaldehydes) would be important potential sources of AGE cross-links, and thus targets for anti-AGE drugs. This was exactly the rationale for the development of aminoguanidine. Ulrich saw that, theoretically, a lot of the body's AGEs might be formed from a class of reactive carbonyls known as *Amadori diones* and the related *Amadori-ene-diones*. These molecules would form when Amadori products broke down: carbonyl groups would join hands across the gulf of adjoining proteins, resulting in an *alpha-dicarbonyl* link—specifically, a type of alpha-dicarbonyl called an *alpha-diketone* link. On chemical grounds, such a link would not be expected to remain intact for long—but it wouldn't just disappear, either. Most likely, Ulrich thought, it would rearrange into a more stable, final structure—a molecular marriage which only death would part.

If that were true, Ulrich could see one potential path to the development of novel anti-AGE interventions. The body has enzymes that break down at least some kinds of dicarbonyl compounds, and many of these enzymes share in common the incorporation of the vitamin *thiamine*. Research by Ukrainian scientists in the mid-1980s had shown that molecules in the same chemical family as thiamine (called *thiazolium* compounds) break linkages of the same chemical type, albeit embedded in organic chemicals rather than as AGEs in tissue proteins. The inclusion of thiamine in so many of these enzymes, combined with the mechanism of other thiazolium compounds, suggested that thiamine was the essential feature to all of them, like the common head shape of different brands and sizes of Phillips screwdrivers. The active core of the incorporated thiamine would get an electrochemical grip on the carbonyls in the enzyme's target molecule, whereupon the enzyme would twist its shape, opening up and tearing the bonds apart.

Ulrich wanted to design a new molecular "tool" that would do the same splitting job with the dicarbonyl bonds in Amadori diones, eliminating their cross-link-forming potential. Starting with the concept of a thiamine-like "business end," the assembled scientists began throwing out suggestions on how different kinds of molecular "levers," "swivels," and "sprockets" might behave, predicting their interactions with Amadori diones from their structures. Ulrich stood at a blackboard, drawing out their proposals.

Finally, they came up with a basic template of a class of molecules that might be expected to cleave the kinds of bonds present in the Amadori-diones that they believed would probably be found in the body. Then they threw together a variety of specific variations on the theme by tacking on various "limbs" to the core "backbone" structure.

At the end of a marathon session, the Alteon scientists took the results of their work back with them for preliminary testing. At Alteon, Dr. Jack Egan assigned several junior scientists to synthesize test quantities of each of their various candidate molecules, and also to cook up large batches of some model Amadori diones. From there it was a straightforward series of simple experiments: pipette small quantities of the candidate molecules into test tubes full of the AGE precursors, and see if they could inhibit their conversion into more permanent structures.

As it turned out, however, "straightforward" did not mean "quick." After having run dozens of tests at different concentrations and still not having nearly exhausted the range of experiments they wanted to do, Egan was looking for a faster way to run the experiment. In collaboration with scientist Sara Vasan, he devised an alternative method. He wasn't sure this new method would work, but they would certainly save a lot of time and effort if it did.

At first, the new procedure seemed to work fine: a lot of the work was moved from the old protocol to the new one, and soon they had accumulated a broad enough sample of data to expect that the answers they needed would be buried somewhere inside their mountain of notes. Vasan gathered the results together and began writing them up for internal analysis and possible publication.

And at first, it seemed that they had their results: the test tubes contained varying amounts of AGEs, suggesting that the compounds had inhibited their formation from their precursors to varying degrees. But a few of the results seemed to be wildly out of step from the main body of work, with the levels of the expected reaction products being well in excess of what could be accounted for by the small variations in concentrations and other factors as compared with other, similar tests using the same compound. The chemistry just didn't make sense.

Embarrassed at having to ask her supervisor, and afraid that she had simply overlooked something or that she or one of her coworkers had improperly performed the experiments, Vasan showed the results to Egan. He agreed that the results didn't make sense, and they began going back to the original lab notebooks to double-check the results.

It didn't take long to see that the outliers were coming quite consistently from experiments using the new, faster protocol. Egan and Vasan went back over the protocol, looking for a flaw—the kind of "garbage in, garbage out" error that makes an accounting program tell you that you owe twice your annual income in taxes. Eventually, they found a mistake in the last few steps in the initial production of the model Amadori diones themselves. At one key point, the correct procedure is to put a halt to the reactions occurring in the test tube, preserving the compounds that have been produced. Instead, the protocol was allowing further reactions to occur, generating alpha-diketones instead of freezing them at the precursor phase. They had, in effect, been "overcooking" their biochemical soup, generating mature alpha-diketone linkages and leaving few or no intact Amadori diones available against which Ulrich's carbonyl-busters could be tested.

But while this was clearly *a* mistake, Egan and Vasan doubted that it was the *only* one, because it couldn't fully account for the anomalous results of the inhibition tests. The results of those experiments had initially looked right, because after the inhibitors were mixed in with their test compounds, their quick-and-dirty assay methods had detected the remnants of shattered Amadori diones floating like molecular flotsam in the test tubes. But how would such chemical debris have been produced if there had been no Amadori diones present for the thiazolium inhibitors to destroy?

It was then that it hit them. The explanation was staring them in the face; indeed, the chemistry would've been obvious, if they hadn't walked into the experiment with a preconceived understanding of the reactions that they would be observing. Because they could see the broken carbonyl groups that would be expected to be left behind after the thiazolium compounds had torn apart the Amadori-diones that they *thought* were present in the test tubes, Egan and Vasan had been assuming that the presence of those broken bonds meant that the inhibitors were doing what they had been designed to do. But what if they were doing something else entirely? What if the carbonyl groups that they were detecting in the final samples were the remnants of alpha-diketone links that had been produced in error during the generation of the test compounds, *and then torn apart by their model drugs?*

Egan felt no "Eureka!" moment of insight, however—and not only because he still hadn't figured out how those alpha-diketones could have persisted long enough to be attacked by Ulrich's model drugs. No, his mood was neither intellectual satisfaction nor ongoing curiosity, but a sinking recognition that they had been wasting their time. The Amadori diones would have to be resynthesized, probably using the original, time-consuming protocol, and the inhibition assays run over again.

There was no sense in covering things up. Egan contacted Cerami and explained the situation, apologizing for the wasted time and emphasising that it was all going on Alteon's bill.

Egan initially thought nothing of the questions that Cerami asked about the experiment: scientists are nothing if not inquisitive. But it *did* begin to seem odd as the doctor pressed him for more and more arcane details of the procedure, the reasoning underlying his conclusions, and even speculations about how the thiazolium compounds might have reacted with the alpha-diketones. Unfortunately, Egan had no real idea how such an interaction might have led to his observations: he was a bench scientist, not a medicinal chemist. *What a relief,* he thought: *Cerami's curiosity seems to have gotten the better of his frustration with this setback.*

Cerami hung up the phone and leaned back in his chair, his mind racing. Was he missing something? Did he dare to believe what Egan was telling him—or to accept the implications?

Trying to calm his trembling fingers, he dialed Peter Ulrich. Impatiently, he waited for his partner to pick up the phone. Finally, the click of a lifted receiver. "Ulrich," said the familiar voice at the other end.

"Peter?" Cerami said, keeping his voice steady. "Can you explain to me how one of these compounds could break an AGE cross-link?"

Reverse-Engineering Serendipity

Over the course of the next few weeks, Ulrich worked backward from Egan's protocol and Vasan's results, developing a tentative scenario under which mature AGE compounds based on an alpha-diketone linkage could be broken by their new thiazolium compounds. Finally, he thought he had the chemistry right.

If the result held, then they were really on to something. Thanks to the laboratory flub, Picower and Alteon were at the center of not one, but *two* breakthroughs in AGE biochemistry: one theoretical, and one of enormous potential medical significance. First, the result implied that alpha-diketone AGEs might be stable enough to persist in the body long enough to contribute to tissue stiffening *without* any further chemical alteration. And second: they had unwittingly designed a class of molecules that would not merely *prevent* AGE from forming, but actually buzzsaw their way *through* them.

The biomedical implications were startling. Imagine being able to take patients whose bodies were already extensively riddled with cross-links, and to give them a drug that would *break the AGE apart.* AGEd tissues would be rejuvenated. Arteries would dilate outward in response to the pulsing tide of blood; hearts would fill with the incoming flow; even skin could become flexible again. It was just the solution that one would dream of for advanced diabetic cases, or for people whose AGEs had built up because of *time,* not high blood sugar. The market would be enormous.

The hard-nosed chemist in Ulrich brought him back from this vision to the steps that lay between him and its realization. For starters, Alteon would have to run the previous experiments again, monitoring the reactions at each step to provide evidence to support the theoretical chemistry that he had outlined to explain the original result. Additionally, everything that they had done thus far was with an AGE that had been cooked up in a beaker: he still didn't know whether *any* alpha-diketone AGE (let alone the specific molecule that Vasan had accidentally produced) ever actually formed in the body at levels sufficient to impair tissue function. And then there was the question of whether his test compounds would be able to reproduce in human subjects what they were doing under glass: the body's detoxification machinery might metabolize them into inactive forms, or it might be impossible to take enough of the stuff to have any effect.

The first few questions were answered by a more careful, intentional repetition of Egan's and Vasan's initial experiments, which seemed to confirm all his hopes. The results of these studies were consistent with the hypothesis—that the predicted AGE was in fact formed from the model Amadori diones, and that it persisted long enough to react with his thiazolium compounds, which did indeed appear to sever the cross-link at the alpha-diketone bridge. And based on the observed results, one particular thiazolium compound—a chemical known as *N-phenacylthiazolium bromide* (PTB)—was an especially effective wrench with which to pull these AGEs apart.

But the fact that PTB severed a bond in an artificial AGE didn't prove that it would break any of the cross-links that actually tie up the arteries, hearts, and other organs of aging and diabetic humans. At this point, it was time for Cerami, the more medically minded member of their tag-team, to get more actively involved. The two decided to put PTB through a graded series of increasingly challenging tests using more and more lifelike model systems, working their way step by step from single cells up to functional and molecular investigations in living, breathing animals.

The Manacles Fall Away

These necessary studies were again farmed out to people working under Jack Egan at Alteon, whose lab scientists first confirmed PTB's ability to cut through AGE using isolated, cross-linked proteins and tissues. With each successful jump over an experimental hurdle, their optimism grew, until they were ready to move their work into the living laboratory of diabetic laboratory rodents. When Egan's team injected the animals with their new compound, the results were again positive: levels of glycated proteins bound to the animals' red blood cells dropped by over a third in the first week, and kept dropping, going down to half of the original level after three weeks and to just 40 percent by the end of the month. It really looked as though they were on to something.

With that evidence to hand, Alteon scientists began giving PTB injections to rodents with hearts, kidneys, and arteries hardened by AGEs, accumulated over a normal healthy lifetime or in the fast-forward mode of diabetes. Here the real excitement began to build, as PTB continued to live up to expectations, *restoring* supple performance to cardiovascular systems that had previously lost their youthful flexibility, instead of just *slowing down* an inevitable decay as aminoguanidine had done. Structurally, the tissues of treated animals were softer and more elastic, stretching out like rubber bands fresh out of the pack, and readily melting away when doused with digesting chemicals; functionally, their hearts were expanding to fill with incoming blood like new balloons, and blood coursed through their arteries without the large backward-rippling "echoes" of pulse that are characteristic of old blood vessels.

They did have one problem, though, which was that PTB is too unstable to succeed as a drug for human use: by the time a pill had made its way through digestive system and the complex chemistry of the body's drug-metabolizing processes, too little would be present to have a meaningful therapeutic effect. But Ulrich was not going to give up on so promising an agent, and with a bit of work he and the Alteon chemists were able to develop a variant on its basic structure that was not only more stable but more active: *4,5-dimethyl-3-(2-oxo-2-phenylethyl)-thiazolium chloride*. For convenience, Alteon first shorthanded this mouthful to *ALT-711* (because it was *ALT*eon's *711*th compound); later, the compound would be rebranded to the more marketable *alagebrium*.

A drug with the ability to cleave AGEs that had already formed in the body would have applications in diabetes as well as in a wide range of diseases of aging, but regulators only approve drugs for one specific indication at a time. Wanting to carve out as exclusive a niche for the drug as they could, Alteon strategists set their sights on developing alagebrium for conditions that were not already being successfully treated with existing medicines, and that would be expected to respond uniquely well to their new treatment.

One of these diseases was *isolated systolic hypertension* (ISH), the kind of high blood pressure in which a person's *systolic* reading (again, this is the first of the two numbers that you get from a blood pressure cuff, like the "110" in "110 over 80") is high, even though their *diastolic* pressure (the second number) is fine. Systolic pressure is a measure of how much pressure is applied to the artery wall by the surge of blood into the vessel as the heart contracts, whereas diastolic pressure is the baseline pressure in the arteries at rest (technically, at "diastole."). Hormonal and other factors can *actively tighten up* the blood vessel, keeping the pressure inside the artery high even during diastole; such effects raise blood pressure irrespective of the intrinsic flexibility of the artery as a tissue. But when systolic pressure is high despite a normal diastolic pressure, it's a sign that the *vessel itself* has become stiff, unable to expand to accommodate the incoming rush of blood from the heart.

This nonatherosclerotic "artery hardening" is not just a concern in people with a diagnosis of isolated systolic hypertension. As people push past middle age, arterial stiffness becomes an increasingly powerful predictor of heart disease and heart attack, and indeed comes to override many conventional risk factors like cholesterol and blood pressure when it comes to risk of actual cardiovascular *events* (heart attacks and strokes). FDA and other regulators don't recognize this "normal" effect of the aging process as a "disease" for which they'll approve a drug, so Alteon knew that they could never get official sanction for the use of alagebrium to treat these people; but they also knew that, once the drug had been proven to buzz-saw through the constricting manacles of AGEs in the artery, restoring flexibility and opening up the vessels to the systolic flow, they could vastly expand the market for it by quietly encouraging its *unapproved* ("off-label") prescription to untold thousands of aging people with age-related arterial stiffening.

Another disease whose victims don't get much benefit from existing drugs and would be expected to respond more specifically to an AGE-breaker is *diastolic heart failure* (DHF). The more common, *systolic* form of heart failure occurs when the heart's lower, pumping chamber loses the strength to push out enough of the blood that it receives from the upper chamber to keep the body supplied with oxygen and nutrients. But about a third of heart failure patients have a perfectly normal capacity to *pump* blood; their problem is that the same chamber can't *expand* sufficiently well to take in the required volume of blood in the first place, so that the body's needs remain unmet even after it squeezes out nearly all of the load that it first takes in. The *result* is the same—the body's tissues are starved for blood—but the *cause* is different, and treatments that admirably address systolic heart failure leave the bodies of DHF patients still crying out for critical fuels. While the underlying loss of the heart's filling capacity can be the result of a variety of factors, many cases of the disease are associated with AGE stiffening of the heart. Again, an AGE-breaking drug would be uniquely suited to restoring healthy functionality to these people, and trials showing that it could restore the elasticity of old hearts would also spark interest in its use by large segments of a "healthy" but rapidly aging population.

Alagebrium proved its mettle quickly, doing everything that PTB could do and more. Studies showed that alagebrium in the drinking water of laboratory animals could deliver the same kind of restoration of heart and artery flexibility that PTB had elicited only via injection, and more easily. And there were things that alagebrium could do that PTB had never been able to pull off. For example, PTB had cleaved *some* of the AGEs that had accumulated in the kidneys of diabetic rodents, but not enough to restore the organ's functionality. Treat the same animals with alagebrium, and not only does their kidney collagen become more soluble, but also the fibrotic damage to their kidneys recedes, and the organs get better at filtering proteins out of the blood, preventing their spillover into the urine.

And rodents were only the first order of mammals to benefit from alagebrium. Alteon and their collaborators soon proved that alagebrium could rejuvenate the hearts and blood vessels of dogs and monkeys. These studies were much more informative about the prospects for alagebrium as a true anti-aging drug than anything that had come before, for a couple of reasons. First, they were carried out in animals that were undergoing "normal" aging, whereas the rodent alagebrium studies had used severely diabetic animals. Second, dogs and nonhuman primates enjoy longer lives, and the extra years give the forces of aging more time to induce the same kinds of pathological cardiovascular system changes that are observed in elderly humans, making them better models of human disease from a clinical and theoretical point of view.

As in elderly humans, older dogs' heart chambers stretch less to accommodate incoming blood than do those of younger animals, leading to reduced filling with blood and a simultaneous increase in the pressure within it. In other words, old dogs suffer from mild diastolic heart failure. When older animals were given a moderate dose of alagebrium for a month, their hearts became about 42 percent more flexible, as demonstrated by an increase in the volume of blood taken up in the absence of any increase in blood pressure inside the chamber. The contrast was even more marked when the volume of blood *delivered* into the heart pump chamber was increased using drugs: just weeks before, this treatment had even further widened the performance gap between old and young dogs in cardiac flexibility, but after alagebrium treatment their hearts were nearly as elastic as those of the young controls.[18]

The results seen in our fellow primates were even more striking.[19] In 2001, scientists from Alteon—working in collaboration with researchers from the National Institute on Aging (NIA) who were studying the effects of aging and calorie restriction on nonhuman primates, as well as with NIA specialists in cardiovascular medicine—published the results of a study on the effects of alagebrium in the cardiovascular systems of rhesus monkeys. Their test group was old, but "healthy" as biologically old monkeys go—and in particular, free of diabetes.

At the beginning of the study, the monkeys' arterial flexibility was assessed, as was the degree to which their heart chambers ballooned outward during their blood-filling (diastolic) phase, as a measure of the flexibility of the tissue. The monkeys then received alagebrium every other day for three weeks, after which their tissues were retested every few weeks for the next nine months.

Surprisingly, there was no measurable effect on blood pressure—systolic or diastolic. But three weeks after treatment, and even more profoundly at the six-week mark, their cardiovascular systems' tissues had clearly undergone a restoration to more youthful elasticity. Using one rough, easy-to-administer test of arterial flexibility, their arteries had become an astounding 60 percent more pliable; a more direct assay revealed a 25 percent improvement. At the same time, their hearts were also opening up more easily: they were taking in 16 percent more blood during the diastolic phase, and other measures of cardiac function at least partially dependent on improved diastolic filling likewise improved after alagebrium treatment.

Alagebrium wasn't expected to prevent *new* bonds from forming between sugars and proteins, so it was no surprise that the withdrawal of the drug was followed by the loss of these gains once the gradual molecular manacling of the monkeys' tissues was no longer being counteracted by an even more rapid breaking of those bonds. Within a few weeks of the peak of their alagebrium-induced return to more youthful suppleness, the monkeys' arteries were once again as stiff as they had been in the initial run-up to the study. Their hearts held onto their gains a little longer than the arteries, but then they too began tending to fall back into their old recalcitrance. Quitting the drug didn't leave the monkeys any *worse* than they had been to begin with—but it was clear that the AGE links being broken by alagebrium could be quickly reforged. The implication is that users of alagebrium would have to take the drug on an almost continual basis in order to keep enjoying their newfound arterial plasticity.

But that didn't much dampen anyone's spirits. The results of these studies marked a clear landmark in the development of alagebrium. Toxicity was low; no serious side effects had been observed; and the promise of a new treatment for stubborn diseases was clear.

It was time for human trials.

From Darkness, Light

The first human alagebrium trial, published in the prestigious American Heart Association journal *Circulation* in 2001,[20] looked like the tentative beginning of something big. Seventy-three older men and women with signs of vascular stiffening had their blood pressure and arterial flexibility assessed and were then placed at random into one of two groups. For two months, two-thirds of the patients took alagebrium in pill form; at the same time, the remainder received a look-alike pill with no active ingredient as a placebo control. At the one-month mark, and again at the end of the trial, their parameters were reassessed.

The results were not altogether clear, allowing for a range of interpretations, but the study was understood to be preliminary by its very nature, and most researchers were willing to give the drug the benefit of the doubt based on the remarkable results achieved in the animal models. Systolic and overall blood pressure had gone down in *both* groups, probably because of an unusually strong "placebo effect" in the group getting the dummy pill: the influence of the power of *belief* on the actual state of the body, which is a notoriously important confounder in hypertension studies. Whatever the reason, the result was that the drug conferred no clear advantage in blood pressure results. At the same time, the arterial stiffness of the people taking alagebrium seemed to have improved in comparison to the placebo group using two different measures, but there were technical objections to the method used in one of these assays and the nature of the *comparison* between the groups also made its results less than decisive.

Subsequent trials, however, have provided results that, in aggregate, allow us to draw firmer conclusions—and, unfortunately for Alteon, they suggest that alagebrium will never be approved by regulators for clinical use. Over a thousand patients with systolic hypertension, diastolic heart failure, systolic heart failure (with and without an associated, compensatory overgrowth of the heart's main pumping chamber) and even erectile dysfunction, as well as some healthy individuals, have been treated with alagebrium in early clinical trials,[21,22,23,24] and while the results provide enough evidence to suggest that the drug is safe and is breaking AGEs in these patients, the effect is clearly insufficient to have much impact on actual *function*. The results on diastolic function in the heart have not been impressive; the benefits in improved arterial flexibility have not been clear-cut; and little, if any, effect on hypertension per se has been observed. Often the main objective of the trials has not been achieved, with the benefits mostly accruing in less-important markers of the disease process that are not clearly linked to clinical outcomes (cardiovascular disease, heart attack, or stroke). Moreover, the benefits that *have* been observed have not been cleanly tied down to any particular dose of the drug. This is paradoxical because one might well expect that, in a drug that breaks AGEs, benefits would increase with the dose: more drug ought to mean more broken AGEs and therefore more youthful cardiovascular systems.

To date, the sum of the data from the animal studies clearly suggests that alagebrium can break AGEs; the question is why this benefit is not translating into improved vascular and cardiac health in human users as they do in so many other species.

Some critics hold that alagebrium is not actually an AGE *breaker* after all, but an AGE *inhibitor,* just as Ulrich and his colleagues originally designed it to be. There is a certain superficial plausibility to this view, but these arguments can't stand up against the irrefutable fact that, in animal studies, alagebrium doesn't just *slow down* the development of complications in diabetic rodents or *prevent* the AGE-related tissue stiffening of the cardiovascular system in normally aging dogs and monkeys: it *reverses* them. A drug that only *inhibited* the cross-linking of tissues would be able to reduce the rate at which *new* cross-links would form, and thereby slow down the *degeneration* of cross-linked tissues—but it would not have the kind of rapid *restorative* effects that have been elicited by alagebrium.

The fact that the tissues of alagebrium-treated animals become inflexible again so quickly after withdrawal of the drug also seems to weigh in against the suggestion that the drug is actually just reducing AGE formation, since the underlying cross-links are clearly reforming much more rapidly than they did over the many years that were required for their initial buildup. This observation suggests that the severing of the alpha-diketone bridge in these AGEs exposes a highly reactive carbonyl group, which soon sticks itself back onto an adjacent proteins. Because of the ongoing breaking of *other* cross-links, users of the drug keep ahead of this problem in "two steps forward, one step back" fashion—but take it away, and they undergo a rapid return to their old, AGEd state.

What about the inability of researchers to find alpha-diketone cross-links in the body? The reason is almost certainly that these structures are, ironically, a bit too easy to destroy. The difficult thing about designing an AGE-breaker drug is not that there's any lack of chemicals that can break apart a given cross-link; the problem is to come up with something that won't also tear normal, healthy proteins to shreds in the process. The common ways of finding AGEs in the body involve soaking a tissue sample in strong acids and examining whatever's left over. This technique catches the most extremely hard-to-destroy AGEs, such as *pentosidine,* but wipes out all trace of more delicate cross-links.

It was long suspected, and has in recent years been confirmed, that the crudeness of such assays introduced serious distortions in AGE research. In the last decade, new methodologies have been developed to uncover AGE cross-links in tissue through a painstaking process of breaking down the normal, healthy chemical bonds in a tissue almost one by one, leaving behind only *abnormal* chemical linkages such as AGEs. Using these techniques, researchers have proven that the AGEs we previously thought to be the most abundant are actually just the most resistant to the chemical carpet-bombing that had previously been used to drive them out of hiding. The most readily-assayed AGEs (like pentosidine) are in fact relatively rare in the body (and therefore make little contribution to the overall state of tissue stiffening), while other cross-links that are much more common (and therefore impose a much larger *total* protein-shackling burden on living tissues) remained invisible to our testing methods.

I believe that this is the explanation for our inability, thus far, to identify the molecular targets of alagebrium. The predicted structure of alpha-diketone cross-links is such that they would be *relatively* easy to break apart: indeed, you'll recall that this is why Peter Ulrich originally didn't think that they would even hang around for long enough to be worthwhile pursuing as a drug target.

In turn, the unfortunate difference in the functional impact of alagebrium treatment in human patients, as compared with lab rodents, dogs, and monkeys, may be the result of alpha-diketone cross-links simply being a much more common kind of AGE in those species than in our own. It's clear that there are differences in the metabolic pathways underlying AGE formation amongst the species. For instance, as we saw above, diabetic rats' bodies suffer much more oxidative stress than ours do in response to the disease. This should affect not only *how* AGEs are generated, but *which specific cross-links* are formed: structures whose formation involves free radicals will probably be a much bigger burden in rat tissues than in human ones.

Another reason to think that alpha-diketone cross-links may be less important to our own species than to others is the simple fact that we're much longer-lived than those other organisms. Long-lived, recalcitrant AGEs like pentosidine are very hard for the body to break down, so they tend to accumulate pretty linearly with age: the result is that, while longer-lived organisms accumulate them more *slowly* than shorter-lived ones, they wind up with higher *absolute levels* of such AGEs by the time they end their lives, simply because they've had many more years to accumulate them. So, if you look at **Figure 2**, you'll see that at fourteen years of age, an extremely "elderly" dog has about forty units of pentosidine in a milligram of its collagen, while a minipig of the same age—but with half of its maximum life expectancy still ahead of it—has only fifteen units. A monkey could possibly live for forty years, and at age ten it has only accumulated five units of pentosidine. A human, with a maximum life span of more than hundred, has fewer units still. Yet by age sixty, when AGE cross-links are beginning to weigh in seriously on a person's chances of surviving for another year, human skin is burdened with some fifty units of pentosidine cross-linking its proteins per milligram of collagen—more than any of the shorter-lived species had time to accumulate.

Now: remember that, in contrast to a supremely tenacious cross-link like pentosidine, alpha-diketone cross-links—the kind broken by alagebrium—are predicted to be relatively fragile as AGEs go, so the level of these cross-links is an equilibrium between rather rapid formation and breakage. Like all AGEs, the decline of metabolic control of fuel as we age would lead to an increase in the rate of its formation with age—but its relative ease of elimination should allow the body to largely keep on track of this increase, leading to a much slower rate of accumulation than the stubborn pentosidine.

The net result of this would be that, late in life when AGE-induced stiffening is becoming rapidly fatal, the contribution that alpha-diketone cross-links would make to the *total burden* of AGE (and thus, to loss of needed flexibility) in a tissue would be less in a long-lived species like ours than in a monkey or a dog (let alone a mouse), for the simple reason that we would have accumulated so much more of the more resistant types of AGE than shorter-lived creatures ever get the opportunity to do. Thus, an alpha-diketone breaker like alagebrium would, even if highly effective at its specific molecular task, leave a much greater burden of *other* cross-links behind than would be the case in model organisms, resulting in much less effective restoration of youthful tissue plasticity.

Alagebrium and Beyond

So where does this leave alagebrium in the SENS agenda? Clearly, the lack of clear-cut clinical benefits in humans indicates that this drug itself will not play a major role in the reversal of cross-link damage in our tissues. Its value, instead, is as a proof-of-principle: it shows us that AGEs *can* be cleaved in the body, and tissues regenerated, long after metabolism has done its dirty work in binding our proteins together. What will be required is a new generation of AGE-breakers that slice through more abundant AGEs, and thus free us of the structures that are really holding our tissues immobile. Such agents will yield the same kind of benefits in humans that alagebrium does in animals—benefits first demonstrable in diabetes, ISH, and diastolic heart failure, and ultimately in aging itself.

It's important to stress that no *single* drug will totally save us from tissue cross-linking. As we've seen, glycation leads to the formation of many different AGEs, each with a distinct structure. Drugs that will sunder any given AGE will probably leave most others untouched: no one molecule will be able to sever all of these distinct intermolecular linkages. Therefore, as we saw with amyloids in Chapter 8, we will need to develop a range of drugs, each of which cleaves either one specific cross-link structure or at most a small family of similar ones.

But the *eventual* need to develop drugs to break a number of distinct AGE structures doesn't mean that we can't effectively stop AGEs from contributing to the aging of our hearts, arteries, and other tissues until we have a solution for all such cross-links in hand. The insight underlying the "engineering" school of anti-aging biotech tells us that we don't have to solve all of our problems at once to intervene in the process: We can "rejuvenate as we go," taking one challenge at a time.

To see why this is so, remember that the molecular damage underlying aging begins accruing while we are still in our mothers' wombs, yet we remain youthful well into our thirties: it takes many decades of these insults before the amount of damage is sufficient to exert a functional impact on our bodies. Until this threshold level is reached, a given form of aging damage is essentially harmless to us *in itself*.

Therefore, to restore more youthful flexibility to a tissue, we do not need to sever *all* the various kinds of AGE cross-links in our bodies, but only the ones that make the greatest contribution to the stiffening of our tissues. For practical purposes, rejuvenation will be effected as soon as we can cleave a sufficient *proportion* of the AGE structures in our bodies to keep the *total* amount of cross-linking beneath the threshold level that actually impairs tissue function.

Once we have a drug that breaks a given kind of cross-link, we will be free of its baleful influence. The solution, by its nature, will not be once and for all: we will undergo a new course of treatment once every few years or decades, taking the drug for a few weeks or months—long enough to break *enough* AGEs to leave our tissues as flexible as they were in our youth. These cross-links will immediately begin to build up again, of course—but we have the luxury of *letting them do so* until that critical threshold level is approached again. (Note that this scenario assumes that the AGE-breaker leaves the broken AGE in a chemically inert state, which alagebrium evidently doesn't do to alpha-diketones; if the broken AGEs are reactive, the drug will have to be taken continuously.) However, the first effective AGE-breaker will not solve the entire problem of AGEs. It will probably reach the clinic first by virtue of targeting the most abundant AGE, but it will surely not break all AGE species. Thus, other AGE cross-links will continue forming in our bodies unabated, albeit at a slower pace. These cross-links will first reach pathological levels once our lives have been extended for long enough that they stiffen our tissues *on their own* as much as they and the first-targeted AGE *jointly* do in a currently normal lifetime.

So, yes, we will need to identify these AGEs as well, and to develop treatments that target them. But the important point from an engineering perspective is simply that this doesn't emerge as an actual *biomedical problem* until the first wave of anti-aging (including anti-AGEing) treatments has extended our lives quite a bit on its own. The first breaker of an abundant AGE will buy us the time to identify such AGEs and to develop new treatments that will free us from them in turn.

Eventually, we will develop a lifelong regimen of AGE-breakers not unlike the childhood vaccine schedule today, under which we will receive a series of specific cross-link-severing drugs, each administered on its own cycle of years, decades, or perhaps even centuries, based on the rates and sites of their targets' formation in the body. But to achieve the first great leap forward in restoring the suppleness of youth to AGEd tissues, we need only prioritize the development of cross-link breakers that will carve their way through the cross-links that cause us clinical problems within a presently normal lifespan.

The above section only discusses AGEs, but similar logic applies right across the SENS spectrum. I'll be discussing it in more detail in Chapter 14.

Know Your Enemy

Today, for the first time in history, we are in the position to design such drugs on a rational basis. Just over a decade ago, when Peter Ulrich and the Alteon chemists were doing the work that ultimately led to the development of alagebrium, they were working in the dark. They didn't even know for sure that the types of AGE that they were targeting actually existed in the body—they were simply *guessing,* from studies carried out in test tubes, that such links might form and contribute to stiffening of living tissues. But the new enzyme-based methods that I mentioned earlier now allow us to slowly take tissues apart at the molecular level, layer by layer, revealing the presence and levels of pathological cross-links in our body, exposing our previously hidden opponents to the light of science.

Researchers who have taken the time to develop and apply these new, painstaking procedures to aging human tissues have given us reliable targets on which to fix our crosshairs: a complex structure called *glucosepane,* which was only identified using these new techniques in 1999,[25] and probably another AGE called *K2P,* which is prominent in the lenses of our eyes and possibly other tissues.[26] Glucosepane is the single most important contributor to the body's AGE burden known to date, tying up as much as one out of every five molecules of the key structural protein *collagen* in old, nondiabetic humans. Glucosepane levels are around one hundred times as high as that of any other AGE that had previously been found in collagen or the lens. A drug that could free our tissues from *these* AGE shackles would have a much larger impact on total cross-linking burden than alagebrium does, and would thus bring our tissues much closer to their full youthful flexibility and functionality.

Today, while we will still use the same sort of blackboard molecular engineering that Ulrich and his copanelists did in the early '90s to design new AGE-breakers, we can fashion molecular bolt-cutters that are precisely tooled to break *specific* AGEs, whose exact molecular structures are in our possession, and whose presence in our bodies and biomedical significance as major contributors to the total cross-linking of our tissues with age are certainties.

We also have the advantage of having faster screening tools in our possession. We can use software to simulate the behavior of AGE-breakers, and to automate the generation of variations on a core molecular theme. We can put robotics to work to synthesize thousands of ampoules of candidate drugs, and use mechanized, high-throughput techniques to assay their effects against a known culprit in the loss of elasticity in our hearts and other tissues.

Being able to look at glucosepane's exact chemical structure gives us another advantage in developing its nemesis that Ulrich no doubt wishes he had had so many years ago. The hard part of developing a glucosepane-breaker for clinical use is not identifying chemicals that can destroy it: as we saw earlier, the acids we used in our old AGE-assaying techniques did a regrettably excellent job of it. The problem is to create molecules that will do it *selectively,* without damaging *healthy* biomolecules that share the same vulnerability exploited by the would-be drug. Having reconstructed glucosepane's molecular identity, biochemists can now see just how wide is the latitude they have in designing molecular shears for it.

Fortunately, this latitude may be very wide indeed. The structure of glucosepane is so different from any functional structures in ourselves or other mammals that a drug that selectively targets them should be harmless to any molecule that is supposed to be found in our bodies.

As I've explained, the resulting glucosepane-breaker should be the first in a series of AGE-cleaving drugs that we will need to unbind our proteins from their molecular shackles. Then we will have achieved for the first time in humans what was observed just a few years ago in dogs and monkeys treated with alagebrium. Old hearts will open wide again, free to fill with life-giving blood. Hardened old arteries will once again readily expand in response to the surge of the blood of life. The stiffened, inflexible tissues of the aged will move with the suppleness of youth. The absurdity and outrage of a body tying itself into molecular knots will come to an end, and we will bend and flex within and without as children in the jungle gym of life.

10

Putting the Zombies to Rest

As we age, we build up an increasing collection of "death-resistant" cells in our tissues. This is one part of our biochemical program to avoid cancer: shutting down the activity of potentially cancerous cells before they can cause trouble. Unfortunately, rather than simply remaining silent and harmless, such cells still manage to damage surrounding tissue through chemical signals gone awry. But by taking a leaf from the world of new, targeted cancer therapies, we can foresee the development of safe, effective methods to remove these senescent cells from the picture.

So far, I've mostly been talking about specific forms of damage that happen at the molecular level to our cells and their components, and how we can restore functionality to our cells and tissues by undoing, or rendering harmless, that damage. But there are a few cases where the aging body accumulates cells that are damaged in such a way that they don't just stop *contributing* to the economy of the body, but actually become *toxic* to the system that supports them.

I've already discussed one such case, back in Chapter 5: cells that have been taken over by mutant mitochondria. When mitochondria lose the ability to process fuels, as a result of mutations to their internal DNA, what ultimately causes us harm is (in my view) not the resulting failure of these organelles to carry out their job. Rather, it's the maladaptive way that their host cell alters its metabolism in order to survive that failure. This metabolic alteration keeps such cells limping along by dumping oxidative stress outside their membranes and on to far-flung areas of the body.

At first glance, one might think that the best thing for the body to do with such cells would be to kill them off, thereby saving the rest of the body from their toxic influence. But the nature of the specific cells that develop this problem makes any attempt at simple removal fraught. The most notable case, arguably, is skeletal muscle. The design of muscle means that destroying a single muscle cell-like structure will snap the entire fiber in which it's embedded. Loss of muscle cells to aging (rather than to disuse) is already a major source of age-related frailty; we can't afford to add to that problem by killing off more of them in self-defense.

So in this particular case, as I described in Chapter 6, what seems to make the most sense is to find a way to preserve and restore normal metabolic activity in affected cells in the face of their colonization by mutant mitochondria, instead of killing them.

However, there are plenty of other cases in which the costs of destroying a toxic cell are negligible, and the benefits clear and direct. Everyone is familiar with one such case—cancer—and no one disputes that destroying cancer cells is unambiguously positive. I won't be discussing cancer in this chapter, however (except for the applicability *beyond* cancer of some existing anti-cancer treatments), because that disease poses such unique challenges that I've devoted a whole separate chapter (Chapter 12) to it. Instead, I'll focus on three cell types that pose a much less catastrophic threat than cancer, but that still make a substantial collective contribution to age-related descent into illness, frailty, and death. From what I can see, there is no reason to attempt to rehabilitate these cells: it seems best that they, like cancer cells, be destroyed. I'm choosing to discuss them together because of the similarity of the threats that they pose and of the strategies that I advocate for dealing with them.

Attack of the Clones

The decline of the immune system is one of the most deadly effects of aging. Infections that young people shake off as mere inconveniences are commonly fatal in the biologically old. Influenza, for instance, puts 114,000 Americans into the hospital each year, and flu and flu-related illnesses claim the lives of about 51,000. But the disease burden is dramatically skewed

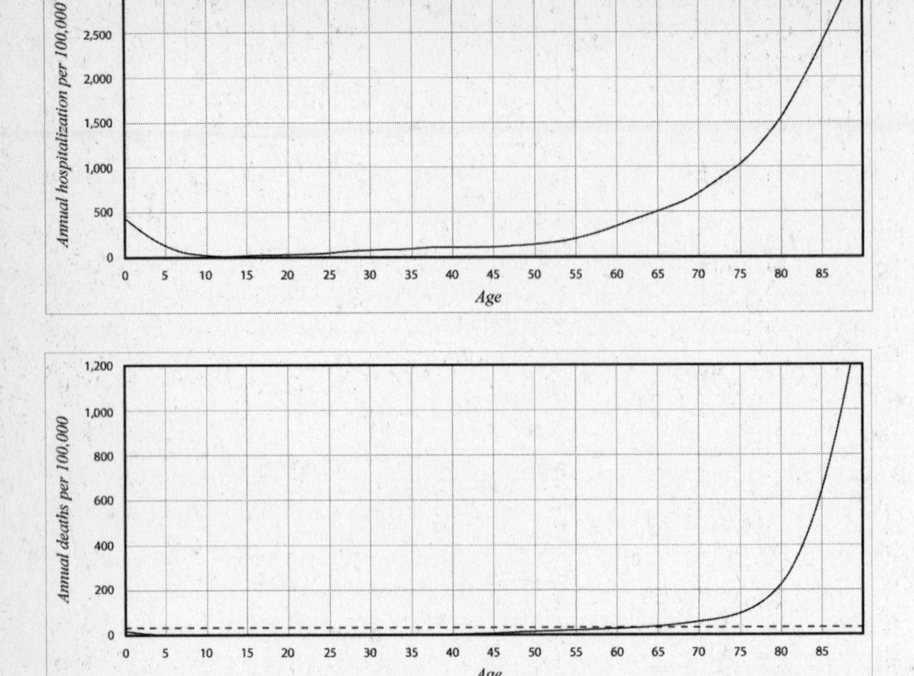

Figure 1. Aging vulnerability to pneumonia and flu. (a) Hospitalization rates by age, Connecticut, 1993–1997 (females). Redrawn.[1] (b) Death rates by age. Redrawn from CDC data.[2]

toward people who have previously suffered the ravages of aging (see **Figure 1**). Deaths from influenza and influenza-associated pneumonia are almost unheard-of in adults until the seventh decade, after which rates climb exponentially. In the United States, over 90 percent of all deaths from the two diseases are in people over sixty-five years of age.

We could, of course, do *something* about the death toll through vaccination. But not much: between 30 percent and 75 percent of older people fail to respond to flu shots, compared with just 10 percent of young adults. Add to this the facts that we sometimes vaccinate against the wrong strain of the flu, limiting the effects of even successful vaccination. There are various reasons for the weakening of the immune system that happens with aging, some of which are ultimately downstream effects of aging elsewhere in the body (like the systemic increase in free radical stress spread by mutant mitochondria). But one of the most profound—and unexpected—factors underlying our inability to mount a defense against infections that young people shrug off without ever suffering a sniffle is, believe it or not, a form of immunological *overcrowding.*

State of the Forces Report

There are two main branches to the immune system. One is the so-called *innate immune system,* whose "innateness" comes from the fact that its job is so general that it doesn't have to "learn" to identify a specific enemy. Its job is similar to that of regular soldiers on patrol in a demilitarized zone, trying to maintain order but unsure of who might be the enemy, ready to confront anything suspicious-looking that they happen upon. There seem to be very few changes in the innate immune system with aging, and those that have been reported appear to be secondary to other aspects of aging and age-related disease (and we'll be discussing a few of those later on), or to factors that are common in the elderly but not a result of *biological* aging at all, such as vitamin and mineral deficiencies.[3]

The other main branch is the *adaptive immune system.* The adaptive immune system is more like a division of highly trained special forces units, each of them expert in waging targeted, tactically sophisticated warfare against specific enemies. This branch is responsible for the ability of the immune system to *learn* about invaders—and, thus, for the effectiveness of vaccines.

Within the adaptive immune system are the *B cells* and *T cells.* B cells are mostly responsible for defending us against pathogens like bacteria and parasites that are purely foreign to the body, and that can therefore be targeted *directly* for destruction. B cells recognize specific markers (*antigens*) on the surface of such an invader that reveal it as foreign, and churn out *antibodies* to them. I talked about antibodies back in Chapter 8, in discussing vaccination as a way of clearing out amyloids: they destroy alien cells by binding to the antigens of the organisms they're intended to fight, acting like homing beacons that attract "missiles" fired by other components of the immune system, or blocking receptors and other proteins that are needed for the pathogen's survival.

By contrast, *cytotoxic T cells* (also called *CD8 cells* because of the characteristic receptor they bear) are responsible for rooting out *the enemy within*: cells that are native to the body but that have now been turned against it, such as cancer cells or cells hijacked by viruses. (There are other types of T cell in addition to CD8 cells—more on them later.) CD8 cells also use antigens to target their foes, but because the targets are hiding in—or in the case of cancer, *as*—the body's own cells, CD8s don't get the chance to catch the pathogens' calling cards on their own surfaces, and can't target the invaders directly. Instead, CD8 cells pick up antigens on the surfaces of host cells that have been infected by a pathogen, or more often on other cells in the immune system that act like reconnaissance agents, scooping up copies of the antigen left in the wreckage of cells destroyed by the enemy and reporting their findings back to CD8 cells to alert them to the threat. Having spotted enemy colors, CD8 cells seek out and destroy the infected host cells, eliminating their threat to the rest of the body.

The problem, once again, begins with the body's need to balance competing priorities in its metabolic processes—and to do so in the face of limited resources. On the one hand, it's critical that the immune system be able to identify and fight off infectious agents that it's never seen before, so it needs to have a reserve of CD8 cells that are ready to respond to new threats, "learn" about their key antigens, and then mount an attack; these are called *naïve CD8 cells*. On the other hand, the process of ferreting out an enemy that you don't recognize takes time, during which an invader could gain a life-threatening foothold in the body, so we also have a complement of *memory CD8 cells*—veterans of old immunological battles, which remember the enemy that they defeated and stand ready to identify and fight them off again.

Balanced Budgets

This would be fine if we could keep on hand as many T cells as we might like, including plenty of naïve cells and large contingents of memory cells specific to each of the many pathogens that our body has racked up in its rogues' gallery over the course of the years. But producing and maintaining these armies is a resource-intensive investment, and as with everything else, the body's "budget" for the immune system is limited. To avoid going into deficit on its "military" spending, the body maintains a strict policy of balanced budgets—a limited amount of "immunological space" (as it's been called) for the entire T cell population in aggregate. The immune system ruthlessly maintains a cap on the total number of naïve and memory cells *combined* in the body at any given time, although the *specific makeup* of that population is in constant flux, shifting dynamically as the body responds to the threat of the moment.

When this system is working well—as it does in most young people—it is the very model of the kind of flexible, low-cost, highly mobile, well-trained army that many of today's generals and world leaders dream of constructing. During an infection with a particular pathogen, there is a rapid redeployment of forces to meet the threat on the ground. Whether it's memory cells mobilizing against an enemy that they've seen before, or naïve cells uncovering and mounting an attack on a brand-new threat, CD8 cells appropriate to the enemy at hand expand their numbers, dividing rapidly in a process called *clonal expansion,* and then fan out, identifying and destroying cells bearing the foreign protein markers against which they specialize. (This use of the term "clone" is one of several in biology; it must not be confused, let me stress, with the popular nonscientific use of the word. I'll have more to say about the various meanings of "cloning" in the next chapter.)

But once an enemy has been defeated, maintaining huge numbers of CD8 cells whose only mission is to wage war on a foe that has just been driven away would be a waste of limited resources. With the body's iron discipline on its immunological budget, it can't afford to have so much of its army be specialized for combating just one opponent if that enemy isn't actually in the process of waging a campaign. So the body initiates a rapid and massive scaling back of these cells, ordering the bulk of the veterans to engage in a carefully orchestrated self-destruct program (*apoptosis*), after which it can rebalance its deployment of forces to a more generic defensive posture. But a few veterans of the recent conflict are kept on after the cessation of hostilities as memory cells, on the lookout for signs of a renewed attack from the invaders that they know so well. The small numbers required to maintain the body's vigilance against a known enemy make this expense quite tolerable, so that the cost of keeping these cells on the payroll never puts significant strain on the "budget" of the immune system. That's the plan, anyway.

Old Soldiers Never Die . . .

Unfortunately, this model of fiscal and military discipline only works well for infections that can be totally eliminated from the body. It begins to break down when the body faces enemies that it can fight to a standstill but not quite wipe out entirely. One class of such enemies is viruses of the herpes family: not just the infections commonly called "herpes" (*herpes simplex* of the mouth or genitals), but also *Epstein-Barr virus* (the one that usually causes glandular fever), *varicella zoster* (which causes chicken pox), and most especially a little-known infection called *cytomegalovirus* (CMV). All these viruses can be beaten back enough to put an end to active, symptomatic disease, but they are never completely defeated. A few copies of the virus continue to lurk hidden in some hard-to-reach corner of the body, dormant and out of sight of the immune system, waiting for the day when the tissue or the body as a whole is in such a weakened state that they can flare up again. In fact, the very name "herpes" is taken from the Greek *herpein,* "to creep," in reference to their ability to sneak about the body while they await conditions favorable to their reactivation.

You may never have heard of CMV, even though the odds are good that you are carrying it (up to 85 percent of adults over the age of forty do). That's because CMV rarely causes a recognizable illness, even briefly: about half of those undergoing CMV infection or reactivation suffer no symptoms at all, while the other half are afflicted only with hard-to-diagnose, nonspecific complaints such as general malaise, fever, and sweats.

But new research is showing us how CMV (and probably some other viruses) can also cause serious long-term harm to those of us who only suffer mild and transient activation and reactivation of the virus. Because the body can never quite consolidate its victory against these viruses, anti-CMV memory cells get called up to active duty again and again, and over successive iterations they gradually begin to ignore the apoptotic signal that is supposed to scale back their forces at the cessation of hostilities. There are various theories as to why this might happen, but I think it's most likely to be part of a complex adaptation to protect us against uncontrolled cell division (i.e., cancer) in these cells.[4] Whatever its origin, the inability to recall these veteran troops progressively weakens the immune system's ability to fight other infections, new or old. The iron limitations on the "immunological space" or "military budget" ensure that, when the body can't cull unneeded T cells specific to CMV or other infections, it has to make up the numbers with *other* immunological soldiers. As a result, the numbers of naïve cells available to keep the body ready to face *new* threats, and of memory cells for other pathogens, dwindle to dangerously low levels.[5]

. . . They Just Fade Away

It's bad enough that death-resistant anti-CMV veterans refuse to take their scheduled retirement, preventing needed redeployments and the hiring of new recruits. But the situation is actually worse than this. These problematic clonal expansions don't just refuse to make room for other soldiers to do their job: like crippled or aged fighters, these weakened (the immunologist's term is *anergic*) T cells can't even carry out their *own* duties.[6]

One of the most important factors crippling these anergic T cells appears to be the loss of a key cell-surface receptor called *CD28*—an effect that has been observed in humans[7] and animals.[8] T cells are alerted to the presence of enemy forces by *antigen-presenting cells* (APCs), the immune system's reconnaissance teams, which identify enemy combatants' antigens through direct encounters with them or by digging through the rubble of old battlegrounds (the remains of cells ravaged by them). When T cells lose CD28, APCs can't recognize them to alert them to the danger, and their intelligence report gets filed away unread. CMV-specific CD8 cells are unusually susceptible to this loss.

Another problem with anergic CD8 cells is that, having staked out a huge territory for themselves and thereby squeezed out other T-cell populations, they simultaneously lose the ability to reproduce themselves. Memory T cells normally express a receptor called *KLRG1,* which is there to keep them from proliferating when no infection is present. But healthy cells bearing KLRG1 *are* able to reproduce when a threat is actually present. Usually. Anergic CD8 cells have KLRG1 present on their surfaces,[9] but they also have another marker on their surfaces called CD57, which is lacking in normal memory cells.[10] When CD57 is present at the same time as KLRG1, the cells' ability to reproduce themselves is locked down tight, so that they *still* can't transform their reserve division into a full-fledged army when their sworn enemy is swarming over the ramparts.

Also contributing to this cellular "infertility" is the fact that the same cells have short *telomeres*—the long stretches of nonsense DNA that cap our chromosomes' ends. Anergic T cells hit this problem because, unlike most immune cells, they have lost the effective activity of the enzyme *telomerase,* which is required to renew telomeres. Most cells don't express telomerase, but it's essential to the healthy functioning of CD8 cells because they are called upon to expand their numbers quickly and frequently over the life span in response to new infections. So the lack of strong telomerase action is a further mechanism of the enfeeblement of these cells. See the sidebar, "Backgrounder on Telomeres and Telomerase," for more information, which will be amplified in Chapter 12.

BACKGROUNDER ON TELOMERES AND TELOMERASE

Every time a cell divides, it must make a new copy of its DNA. The enzyme responsible for doing this—called the *DNA polymerase*—is a bit like a molecular monorail train, zipping forward along the "guide rail" provided by the DNA strand that it is to replicate. As it travels along the "guide rail," the polymerase enzyme makes a letter-by-letter copy of the strand beneath it, spooling the new, replicated strand out to the side as it goes.

DNA polymerase has a fundamental shortcoming, however. For reasons whose details needn't concern us here, the machinery never quite manages to replicate the *entire* strand of DNA. A small amount of a chromosome's DNA material is therefore lost with every round of cell division, leaving the copied strand shorter than the original. Eventually, the end of the chromosome is eaten away.

A second problem that our cells have to solve regarding our chromosomes is that they often break, due to radiation and other stressors. The cell needs to repair such breaks. But what it must scrupulously avoid doing is stitching two intact chromosomes together end-to-end, mistaking the unjoined ends of the chromosome for the unjoined ends of a broken chromosome. Thus, it needs a way to recognize that the bona fide end of a chromosome is not just one side of a chromosome break.

Telomeres are half of Nature's solution to both these problems. Telomeres contain no genetic information—they are extremely boring DNA consisting of many copies of a short sequence—and they are present at the ends of all our chromosomes. If this repeated sequence gets a bit shorter during successive rounds of cell division and DNA copying, no harm is done until it is mostly eroded. The other half of the solution is the enzyme telomerase, which is able to add copies of that sequence to the end of a DNA strand. This solves both problems: cells expressing telomerase can compensate for the telomere shortening that occurs during cell division, and cells with or without active telomerase can avoid stitching chromosomes end-to-end because the break-repairing machinery recognizes the distinctive telomeric sequence and leaves it alone.

Humans and some other species have made ingenious use of the telomere/telomerase system to protect themselves against cancer. Cancer can only kill us by its cells dividing a lot; this is impossible without telomerase, because without a way to renew the telomere, it will slowly erode away, the chromosome ends will become indistinguishable from chromosome breaks, and the cancer cell will be stopped in its tracks by a joining-together of some of its chromosomes. Humans therefore turn off their telomerase genes as thoroughly as they dare, so that a lot of mutation is needed to turn telomerase on again and thereby allow a cancer to divide often enough to kill us.

Although there's less research on it, old CMV carriers also suffer from the expansion of defective *CD4 cells*—the "T-helper" cells that help other immune cells to ramp up their counteroffensive when pathogens first invade. Outwardly healthy older carriers of CMV infection have the same large clonal expansions of CMV-targeting but CD28-lacking CD4 cells as are seen in their CD8 populations, leading to the same crowding-out of other T-cell specialists and lack of responsiveness to activation by antigen-presenting cells.[11]

As with their CD8 cousins, CD4 cells stripped of CD28 can't respond to antigen-presenting cells by deploying CD8 and other immune cells to face the threat. Put this together with the inability of those very CD8 cells to attack their targets effectively, and CMV would be left to run rampant, generating yet more clonal expansions and wider immune dysfunction.

Clonally expanded anti-CMV CD8 cells are anergic (ineffectual) in other ways, too. When they are first infected with their species' version of CMV, young mice produce very effective CD8 cells targeting the virus, which recognize at least twenty-four proteins specific to it; but after the infection becomes chronic, their anti-CMV forces become restricted to clones that recognize an average of only five such proteins.[12] And older CMV-infected humans' anergic CD8 cells mount a weaker response to the threat than do younger infectees' cells, producing significantly lower amounts of *interferon gamma,* a key chemical messenger responsible for ramping up the T-cell response to the virus.[13,14]

When Bad Generals Lead Good Armies

The failure of anergic T cells to clear out CMV infections then probably leads to many of the other failures of immune function commonly seen in the frail elderly that can't be chalked up to any *direct* effect of aging of the cells in question. Some of these effects might be expected to flow from changes in the production of cytokines by such cells, which influence the activity of many other soldiers in the adaptive and innate immune systems, but others exert much more lasting changes than just problems in signaling.

Notably, it's now widely accepted that the aging of T cells is responsible for the age-related losses in effectiveness seen in our B cells—the immune cells that produce *antibodies* to foreign antigens, which flag pathogens for destruction by other cells. B cells rely on signals from CD4 (T-helper) cells to mature and to develop antibodies, so it was only a matter of time before someone confirmed that old T cells cause declines in the development and effectiveness of B cells, independent of the aging of the B cells themselves. [15,16,17] Unfortunately, no one has (to my knowledge) yet directly looked to see if these effects are due to the effects of CMV-induced clonal expansion that are so central to other aspects of T-cell aging, so we don't know how much the specific phenomenon of anergic T cells contributes to these declines. I'd be very interested in the results of such experiments.

Moving beyond the mechanistic studies and molecular biology, the real impact of the creeping takeover of the immune system by anergic CD8 clones on the health of people bearing them is also becoming clear as scientists begin to study its influence. Animal studies show that age-related clonal expansion of specific CD8 populations reduces the variety of T cells present in their bodies and compromises their ability to mount an effective immune defense.[18] The parallel in humans can be seen in findings such as a poorer CD8 response to flu shots[19] and a blunting of the reinforcement of T cell immunity to Epstein-Barr virus that can otherwise occur later in life,[20] in people with clonal expansions of anti-CMV memory cells.

Taking the Full Toll

If the cost of anergic T-cell clones to the body were limited to the increase in deaths and disabilities that can be directly chalked up to infectious disease, that would be plenty enough reason to want to do something about them.

But there's considerable evidence that anergic CD8 cells contribute to age-related morbidity and mortality from causes with no obvious immunological link.

For starters, when you throw an influenza or influenza-induced pneumonia attack onto an aged body, you wind up with shocking long-term consequences that can greatly accelerate other disease process and hasten a person's slide into helplessness and the grave.[21] A significant body of evidence shows that influenza in the elderly increases deaths from unexpected sources like heart attacks, strokes, and seemingly unrelated respiratory disorders; it also worsens the course of congestive heart failure.

Also, the fact that it takes biologically old people so long to recover from the flu, when overlaid on the general frailty induced by other aspects of aging, probably contributes to serious, often permanent functional decay and disability. A bout of influenza often lays an older person in a hospital bed for as much as three weeks, and studies show that for *each day* that they spend "resting" this way, elderly people lose up to 5 percent of their muscle power and 1 percent of their aerobic capacity. But no one thinks of influenza or immunological aging when they see an elderly woman struggle to open the doors at the mall, or slip on the ice and break her hip.

There are other age-related diseases in which anergic T-cell clones appear to play an important role, but where the evidence is not nearly so clear-cut. One is osteoporosis. Older women who have suffered an osteoporotic fracture have been found to carry higher levels of anergic CD8 cells than matched women with no bone disease, and there is a molecular basis for thinking that defective CD8 cells are actually a cause, rather than an effect, of the underlying thinning of the women's bones.[22]

Additionally, albeit more speculatively, even the course of atherosclerosis could be affected by the creeping "clonalisation" of the T-cell population, by leading to a state of chronic inflammation that could hasten a heart attack. In support of this hypothesis, patients with coronary artery disease have higher levels of anergic CD8 cells than otherwise matched healthy people—a fact that is independently related both to CMV infection and also to the presence of the disease itself.[23] Thus, the weakening of the immune system appears to be both *facilitating* and also the *result of* arterial infections, which may in turn be the itchy trigger finger toying with the loaded gun of atherosclerotic arteries.

As I said, the evidence for many of these downstream effects of anergic T-cell clones is still not conclusive. But a couple of remarkable studies now coordinated through the European Union *T-CIA* (T Cell Immunity and Ageing) project have gone some way towards giving us a clearer picture of the total cost, in deaths, of this driver of immunological aging, whatever may ultimately wind up written on the death certificate.

These researchers hunted through two cohorts of Sweden's "oldest old" (people in their eighties[24] and nineties,[25,26]), selecting only people who were particularly healthy compared to most people of their chronological age: free of preexisting serious diseases of the heart, brain, liver, or kidney; without diabetes or cancer or signs of existing, active infection or chemical markers of inflammation; and not taking any drugs that have significant effects on the immune system, including recent vaccination. The European team found that even amongst these relatively healthy but old people, a few are silently suffering a complex of immunological defects (the "immune risk phenotype") including several forms of aging damage that can be caused by CMV infection—not least, the clonal expansions of anergic anti-CMV CD8 cells.

The facts that the resulting study population was healthy but very calendar-old (by today's standards), and that some did and some did not harbor anergic T-cell clones, allowed the T-CIA team to study their effects "cleanly," in a population where its presence could really be said to *predict,* rather than *follow,* preexisting disease, over the course of the next two years.

It was hardly a surprise that having the immune risk phenotype increased these people's chances of death—but the *size* of the effect was a shock. The effect was especially powerful in the population in their nineties, amongst whom its presence could predict 57 percent of the deaths. This, remember, from the immunological aging damage induced by a virus whose active infection state is passed through without *any* notice in many people, and even in the rest of us usually causes only low-level malaise and fevers.

It's important to see the full implications of this finding. The impact of having the immune risk phenotype was seen at the level of *all-cause* mortality, not just in risk of death from infectious disease. While pathogens do claim many very biologically old people's lives, such deaths can't entirely account for the result.

Slash-and-Burn for New Growth

As more and more evidence has accumulated fingering anti-CMV CD8 cell clones in the age-related enfeebling of the immune system, immunologists have begun to see the hopeful side of the phenomenon. If so much of the aging of the immune system is indeed the result of this overreaching expansionism, then preventing or reversing it should (respectively) protect or restore a youthful immune system in chronologically aged people. Vaccines would again be as effective in people at presently advanced ages as they were in their youth, and the enormous burden of suffering caused in the old by infections that young people escape after an unpleasant day or two home from work or school would be lifted.

One option for prevention, advocated by many immunologists, is vaccination against CMV. Even before we understood that CMV infection is a central driver of the age-related weakening of the immune system, a 1999 report on the sluggish pace of development of new vaccines put out by the Institute of Medicine (IOM) of the National Academy of Sciences ranked the pursuit of an effective anti-CMV vaccine as the highest priority item on the list, based only on the *then-known* lifetime human and financial costs of the virus. The U.S. National Vaccine Program Office would later agree, calling for more government dollars to go into CMV vaccine research. Today, confronted with the strong evidence condemning CMV infection as a major reason for the aging of the immune system, many immunologists are sounding the call for such investments even more loudly.

Though the merits of this case seem strong, it's worth noting that this strategy is fundamentally *preventative.* While it can reduce the risk of CMV infection, and possibly improve the immune response of existing infectees to the virus, vaccination can't eliminate it—and it certainly can't reverse the accumulated effects that a lifetime of CMV infection has had on the immune system. Thus, a CMV vaccine might save a relatively small number of babies from suffering tragic birth defects, and prevent the deaths of many AIDS and transplant patients, but it would do little for the many millions of people *already* suffering with the chronic infections and continuous vulnerability of having immune systems worn down by clonally expanded anergic CD8 cells.

Other proposals, which would at least have the *potential* to undo some aspects of immunological aging, involve trying to remedy the defects of the existing anergic T cells using gene therapy. The idea is that by delivering copies of genes for proteins that are either missing or underactive in these cells (such as those for the CD28 receptor or telomerase), we could restore their effectiveness at doing their particular job, and prevent their suppressive effects on other T-cell populations. While there is *some* merit to these proposals, there are also limits to their likely effectiveness and a lot of uncertainties to their path to clinical development. And in the case of telomerase, we'd still face the big worry that has to be taken seriously when contemplating the introduction of telomerase into *any* cell, let alone one that we know is wracked with aging damage: cancer. I'll talk about this problem much more in Chapter 12, but here's a brief taster: because cells require a minimum telomere length in order to keep reproducing themselves, and because each cell division shaves off a little nub from the cell's telomeres, cells with potentially carcinogenic mutations require a way to renew their telomeres if they are going to go on to become full-blown malignancies. Nearly all cancer cells accomplish this by wrenching out the self-imposed parking brake from their telomerase genes. Do we really want to introduce this gene into defective cells—or worse, into "bystander" cells in which telomerase should *never* be turned on, should some of the gene therapy vector "infect" them too?

No: the solution here is not to try to rehabilitate these cells, but to get rid of them. Older CMV infectees appear to have no shortage of *functional* T cells targeting cells infected by the virus: it's just that these cells are suppressed by the crowding influence of huge populations of anergic ones. And remember that even if we could restore all of these defective T cells to their full immunological power, they would still cause problems so long as they continue to sprawl out over precious, limited immunological real estate, preventing the retention of both naïve and memory cells needed to protect us from other pathogens.

The solution to this is conceptually simple. Remove the anergic T-cell clones, and immunological space will be opened up for healthy cells of other types and specificities to move in—and the repressive effects of the anergic clones on their healthier anti-CMV cousins will be lifted.

The problem, of course, is how to purge anergic T cells from our systems while leaving behind all (or, at least, nearly all) the healthy memory and naïve cells that we're trying to liberate from the former's repressive domination. While oncologists can to some extent increase the effectiveness—and decrease the toxicity—of drugs or radiation by applying them as narrowly as possible to a relatively large *lump* at a fairly well-defined spot in the body, we can't do the same against anergic T cells, which are spread all over the body rather than being concentrated in one place. The same feature rules out surgery: tumors can often be removed (or at least beaten back) with the knife, with varying degrees of safety and clinical benefit, but we will not be in a position to pluck individual anergic T cells from the body one-by-one for the foreseeable future.

But even though the cancer therapies of the recent past don't offer a good model for the development of the required biotechnology, the most exciting of the currently available and imminent treatments for cancer suggest a path to therapies that would indeed selectively eliminate the burden of cells that will not die.

Smells like Gleevec

Even if no one you know has cancer, there's a good chance you've heard about *Gleevec* (a.k.a. *STI–571* or *imatinib*), *Iressa (ZD1839* or *gefitinib*), *Herceptin* (*trastuzumab*), and others less famous or still working their way through the approval process. These so-called "targeted cancer therapies" have been rightly hailed as breakthroughs; even the language of "miracles," although absurdly overused in popular books on health, seems justified to many people who have seen tumors disappear from their own bodies or from those of their loved ones, without the horrific side effects associated with radiation and chemotherapy. Even so, these drugs are not completely without side effects—no drug that "messes with metabolism" can be. Herceptin, for example, targets a growth receptor called HER-2: by tying up HER-2, it prevents the excessive growth of cancer cells that get their growth-stimulus fix by producing too much HER-2 on their surfaces. But other, healthy cells rely on a low level of HER-2 stimulation to proliferate normally. Because of this, Herceptin users can suffer deadly congestive heart failure—a side effect that recent research has also uncovered in a small number of users of Gleevec, which was thought to be an extremely clean drug precisely because it only targets an *abnormal* form of a growth-signal transducer.[27]

In the same way, interfering with anergic T cells' resistance to apoptosis might lead to their death, but this still leaves open the question of how to undo this resistance without killing needed cells elsewhere in the body.

I am confident that we can perform reverse-engineering, to adapt the new targeted cancer therapies—and even newer ones that are now in various stages of clinical development—to develop the ability to create "smart bombs" that will destroy anergic T cells (and also the other kinds of toxic cells that we'll be discussing later on) with minimal harm to healthy ones.[28] We can foresee the ability to couple carefully chosen toxins to molecules that home in selectively on the tell-tale signatures of anergic clones and thereby to directly, decisively kill them flat out instead of just interfering with their metabolism.

Light Kills Vampires

One cancer treatment that suggests ways to take out anergic T cells is *photodynamic therapy* (PDT). PDT starts with a drug that, when illuminated with laser light, either heats up greatly or produces a massive burst of free radicals. Drugs exist that have this feature and are also taken up selectively by cancer cells, allowing oncologists to cause a lot of photosensitizing drug to accumulate in the target cells while avoiding much uptake by normal cells. By themselves, PDT drugs are harmless, having no effects as long as the patient is kept away from the light. Similarly, low-energy red laser lights is harmless to people who have not received such drugs: the rays pass harmlessly through the body. But when such a laser beam penetrates cells that contain a photodynamic drug, the agent's photosensitizing properties are revealed in a searing targeted blaze of heat or a maelstrom of free radicals that destroys the tumor cells while leaving all but their very close neighbors unharmed.

The first PDT drug, *Photofrin,* was approved in industrialized countries as a treatment for advanced lung, digestive tract, and urinary tract cancers in the early 1990s, and more advanced versions are now in use clinically or are in late stages of development. The most interesting of these, *Pc-4,* accumulates more in some kinds of cancer cells than in healthy ones because it dissolves well in fats, and these particular cancers have an unusually high fat content. Once it enters the cell, certain features of Pc-4's structure allow it to insert itself into the cancer cell's energy factories, the mitochondria of which I've said so much in Chapters 5 and 6. Turn on the laser, and the free radical bombardment begins, either taking the cell out cleanly via radical-induced apoptosis, or at worst leaving some debris as the cell dies the nasty way instead, when the free radicals tear through the cell, cross-linking its proteins, turning its lipid membranes rancid, and wracking its DNA with mutations.

The Molecular Swiss Army Knife

On the frontiers of medicine, we are now seeing the emergent use of *nanotechnology*—engineering performed at the molecular level—to destroy cancer cells selectively, again providing us with a road map toward the development of a targeted therapy for toxic cells such as anergic T-cell clones. One such technology is *dendrimers:* tiny particles with exquisitely complex branching structures that extend outward like bushes, forming a spherical shape (see **Figure 2**). Dendrimers' branches are engineered in a way that allows us to bind a wide range of molecules to them. This makes them like nanotechnological Swiss Army knives: several useful tools can be united into one compact little package. Dendrimers can carry one molecule to target a given cell type, one or more deadly drugs or other poisons to kill target cells once they're located, and (if desired) a molecule that will allow researchers or doctors to track the progress of the whole package as it moves through the body.

One dendrimer under experimental development combines folic acid (yes, the vitamin) with the established anti-cancer drug *methotrexate* and a fluorescent compound called *fluorescein.* The folic acid is there to target the dendrimer to cancer cells. Many cancers suck up massive amounts of this vitamin because it is required for the production of new DNA—and because the cell needs to create a whole new copy of its DNA blueprints every time it divides, cancers have high metabolic requirements for folic acid to support the feverish pace of their proliferation. To keep their reserves of the vitamin topped up, many cancer cells "learn" to sprout veritable forests of folic acid receptors on their surface.

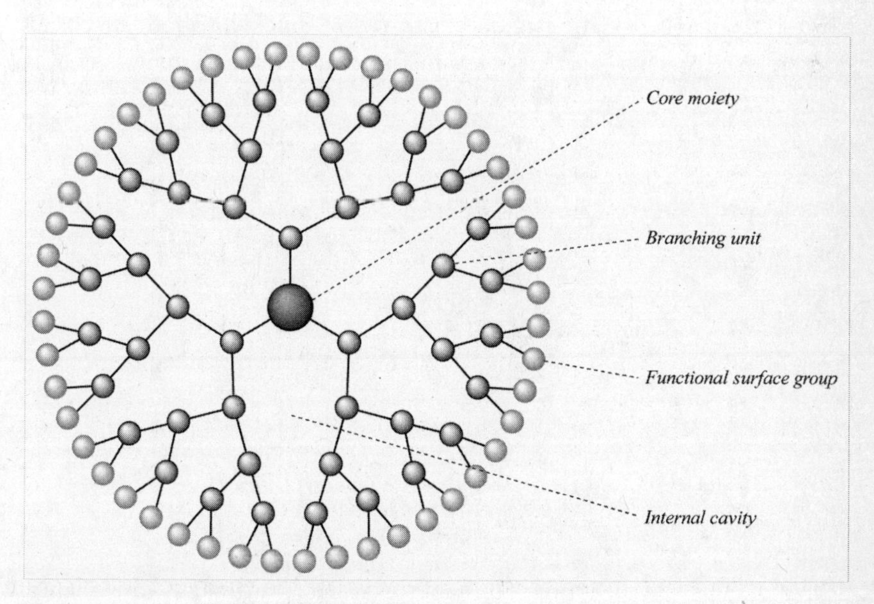

Figure 2. An "unloaded" dendrimer.

This dendrimer was tested in mice that had been injected with a human nasopharyngeal cancer line. Their tumors had grown quickly, reaching a plateau at about fifty days. Giving one group of animals a low dose of plain methotrexate had hardly any effect at all on the growth of the tumors (see **Figure 3**). A dose more than four times as high (the "medium dose" in the figure) reduced the growth rate quite significantly, but didn't actually benefit the animals much: half of them were dead of either the tumors or the side effects of the drug within thirty-nine days. Increasing the dose by a further 50 percent (the "high" dose) brought cancer growth down to almost nothing—but without doing the animals any good, because the drug's toxicity caused the rapid loss of a third of the animals' body weight, and again either allowed or caused the death of half of them just over a month into the experiment.

But look at what was achieved with the same drug targeted using a dendrimer! When targeted using this new technology, a dose of methotrexate equivalent to the *lowest* untargeted methotrexate dose was as effective at slowing tumor growth as a dose of plain methotrexate over four times higher. Moreover, the dendrimer-targeted methotrexate appeared to have very low toxicity.[30]

In a follow-up study, the same group compared the effects of the low-dose methotrexate to those of the same dose delivered using the targeted dendrimer, this time for an extended period of ninety-nine days. Left untreated, the cancer-bearing mice began dying quickly—about fifty days into the experiment—and low-dose methotrexate improved survival only

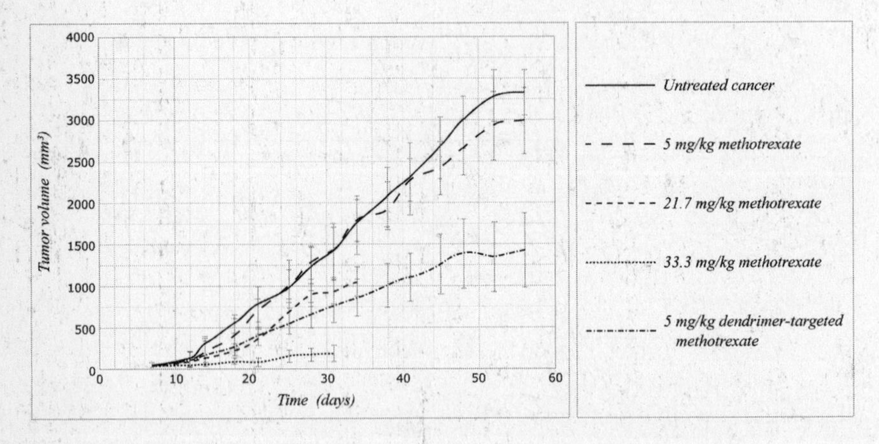

Figure 3. Targeted dendrimer technology against cancer growth in mice. Redrawn.[29]

modestly. But by the end of the ninety-nine-day study, three out of eight of the animals getting the dendrimer-targeted drug were still alive—and impressively, one of these animals was *completely cured* of its cancer by day thirty-nine. Again, the dendrimer-targeted drug was nontoxic.

Switch the Hacksaw with the Toothpick

The beauty of dendrimers, like Swiss Army knives, is that they are so readily customized. Researchers are experimenting with dendrimers bearing many different targeting molecules and cancer-killing agents. One very clever application under development is a "dendrimerized" version of a one-two anticancer punch known as *boron neutron capture therapy* (BNCT)—a neat idea that was first proposed over fifty years ago, but that until now no one has ever quite managed to make work.

The idea behind BNCT is similar to photodynamic therapy. First, you inject the patient with a form of the mineral boron. Once enough of the mineral has accumulated in cancer cells, you flood them with low-energy beams of neutrons. The boron itself is harmless, and so (more or less) is the neutron beam. But when incoming neutrons hit the form of boron used in BNCT, it absorbs them into its atomic nucleus and suddenly becomes extremely unstable, releasing radioactive *alpha particles.* Alpha particles have enough energy to nuke the cell in which the original boron is located and a few of its neighbors, but they run out of energy quickly and the low-energy leftovers are harmless. Thus, the boron-loaded cells are killed but no widespread damage ensues.

The trick, of course, is to find a way to target the boron *selectively* to cancer cells, so that when you flip on the neutron beams you don't destroy healthy brain and other tissue. Scientists have been trying to turn BNCT into a viable clinical therapy for a rare, extremely aggressive, and hard-to-treat brain cancer called *glioblastoma multiforme* since 1951, with some limited success, but they've never achieved good enough responses to justify using it as a standard treatment for the disease.

But scientists have recently reported very promising results in an animal model of glioblastoma multiforme that was treated with a BNCT using dendrimer-targeted boron. Human glioblastoma cells bearing a mutated version of the *epidermal growth factor receptor* (EGFR) called *EGFRvIII* that is implicated in the majority of these cancers were implanted into the brains of rats. The researchers then heavily loaded their dendrimers with boron, and then sent them hunting for the gliomas by attaching a monoclonal antibody to the mutated EGFR. To get a handle on just how promising the dendrimer really was, they compared its effects not only to what happens in animals given no therapy at all, but also to animals given *p-boronophenylalanine* (BPA—the most promising preparation of boron being used in BNCT clinical trials), or else the boron-loaded dendrimer in combination with (but not bound to) BPA. Within a day, about 60 percent of the injected BPA-bearing dendrimer had homed in on tumors carrying the mutant receptor, achieving concentrations that were about *triple* those of BPA alone; the uptake by normal tissues was negligible. Impressed by its homing ability, the researchers waited to see whether it could actually cure the animals of their gliomas.

The results were decisive (see **Figure 4**). Untreated animals lived an average of just twenty-six days. Animals who received BPA could expect to survive for forty days: a significant improvement, but still a grim prognosis. But animals given the dendrimer-targeted boron lived an average of seventy days, with 10 percent of them surviving for six months—considered a "cure" in the same way that five-year survival is considered a "cure" in humans, since healthy rats have a life expectancy of about thirty months. And animals that were lucky enough to get the dendrimer along with BPA survived, on average, for a remarkable 85.5 days, more than three times the life expectancy of the untreated animals, and more than double the survivorship of animals getting the best experimental therapy available. Plus, an impressive one in five of the BPA-plus-dendrimer-treated animals achieved a "cure" as just defined.

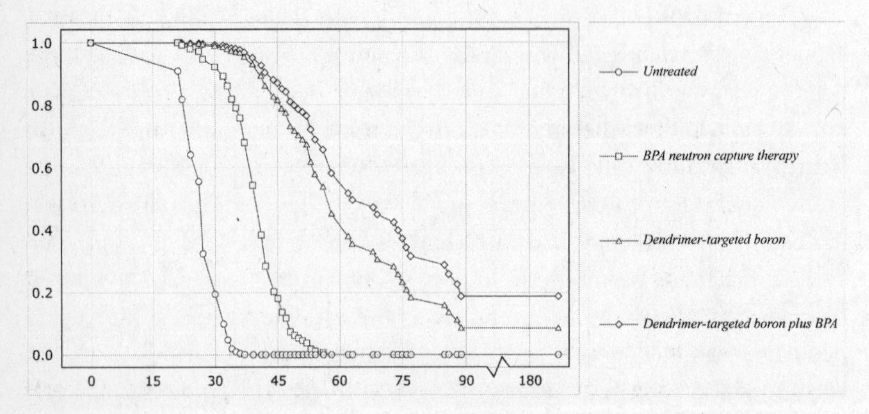

Figure 4. Targeted dendrimer technology dramatically improves the effectiveness of BNCT. Redrawn.[31]

Other targeting molecules, tumor types, and cancer-killing agents have been successfully treated by dendrimers in experimental models. These first-generation devices are turning out to be a very effective, versatile way of creating specific, lethal missiles for seeking out tumors that express known cell-surface receptors as hallmarks—and they offer promise for anergic T cells, too.

PRO-Suicide Counseling

Today, gene therapy is a routine practice in mice, used to do everything from testing experimental gene-based therapies, to investigating the effects of turning genes on and off in an organism, to creating new models of human disease by modifying animal cells to be more like human ones. Gene therapy for humans is still highly experimental, but it's clearly only a matter of time before we master it: the need for cures for congenital diseases, and the potential utility of gene therapy in medical challenges as wide-ranging as rheumatoid arthritis, trauma, dental tissue engineering and AIDS (to name just a few), is providing the impetus for the basic and clinical science needed to bring it into our therapeutic armory.[32]

One option that gene therapy will furnish us with is the ability to build a new suicide mechanism into our T cells that would cause them to self-destruct should they ever turn anergic. Scientists have for some time been able to introduce into mice (and other laboratory animals) genes that will only turn on in the presence of a particular factor, such as an antibiotic, UV light, a sugar, or even a signalling factor like calcium. This allows us to turn such genes on and off at will, simply by administering the relevant factor.

The ability to introduce a gene that is only expressed when researchers want it to be has been a powerful new tool for studying those genes' effects. But these techniques are also now being turned to medical purposes. If, instead of designing these genes to be turned on in response to *externally supplied* factors, we instead make their activation dependent on the presence of a particular protein whose *internal synthesis* is diagnostic of a cell that we want to be rid of, then we have yet another way to selectively target cells for destruction.

Just as with the other targeting technologies I've discussed, the first work in this direction has been in the cancer field. As I've noted, the one absolute requirement for a cancer cell to threaten us is that it have a way to keep renewing its telomeres: otherwise, its furious growth will grind to a halt when it reaches the end of the telomeric line, which is long before it can meaningfully threaten our health. Usually, this is accomplished by activating the repressed gene for the telomerase enzyme—a gene that all of our cells contain, but which is turned off in healthy cells most or all of the time. So by "infecting" the cells of a patient with a "suicide gene" that would be activated in the presence of high levels of telomerase, cancer cells could be killed from within. This would eliminate the need to target a drug or the immune system *to* the offending cell: every cell would hold within it the seeds of its own destruction should it ever turn to the dark side.

In principle, we could generate a *literal* "suicide gene" that would destroy the cell in the presence of the tell-tale protein. Indeed, this has already been done in animal models of cancer, using genes that regulate apoptosis;[33] anergic T cells are resistant to apoptotic signaling, but this resistance might be overcome by bombarding them with insistent messages to shrivel and die. But there's an even better alternative that's under more advanced development. This uses the somewhat more readily controlled—and therefore safer—technique of installing the gene for a protein that is largely harmless in itself, but which activates an inactive form of a deadly drug—a so-called "prodrug."

Prodrugs are substances that are inactive and harmless until they are metabolized in some way, whereupon they are chemically transformed into a pharmacologically active product. Most prodrugs are activated by enzymes in our livers and are then released in active form to the rest of the body, but others act more like molecular "sleeper agents," going about the body unobtrusively, minding their own business and blending in with their environment, until a prearranged signal is given—and their hidden purpose suddenly becomes revealed in the form of a precision strike on their target.

Several antiviral drugs, such as the herpes drug *ganciclovir* (Cytovene/Cymevene), work somewhat along these lines. Ganciclovir stops viruses from using the DNA-replication machinery of their host cells to reproduce themselves. It does this by interfering with the action of the virus's unusual version of *thymidine kinase* (TK), an enzyme that is required for the synthesis of DNA.

Thymidine kinase's job is to make *thymine* (a "letter" in the DNA code's "alphabet") available to be added onto the new DNA chain, by joining thymidine to phosphate molecules taken from the "energy currency molecule" ATP. Ganciclovir acts like a molecular impersonator on a mission to sabotage an enemy factory. It first uses its strong structural resemblance to thymidine to fool the viral TK into thinking that it *is* that molecule. Duped, TK hands over thymidine's rightful phosphate group to the drug.

Ganciclovir then uses its shiny new phosphate group to perpetuate its identity theft, presenting its phony credentials to the cell's DNA synthesis machinery, which unknowingly inserts it into the emerging DNA strand in thymine's place. At this point, the sabotage of the hijacked equipment is accomplished, because while the machinery can slide ganciclovir onto the DNA strand, it can't add any *further* genetic letters onto ganciclovir once it's in place. Without the ability to copy its DNA, the virus can't replicate itself, and its expansion campaign is brought to an abrupt end; all that's left is for the immune system to besiege and ultimately destroy the cells in which the remaining virus is holed up.

If ganciclovir were as good at tricking the version of the TK enzyme used by our *own* cells as it is against the viral version, it would potentially be a very effective cancer-killer: again, cancer can only survive by keeping up the insane pace of its growth, and turning this off by shutting down its DNA synthesis capacity quickly tames tumors. But, of course, such a drug would come with some pretty serious side effects, because it would shut down the growth of *normal* cells at the same time. This *might* make it an acceptable therapeutic bargain for cancer—the effect would wear off after the drug was withdrawn, allowing patients to recover—but it would make it totally unacceptable for its present use as a herpes treatment.

In fact, however, ganciclovir is rather poor at mimicking the human TK enzyme, and its effects are thus mostly restricted to turning off virus replication—though it does have some negative impact on the body's ability to regenerate its blood cells and on the production of sperm. But a team of Japanese and American scientists recently realized that they could in principle use the viral TK/ganciclovir combination to shut down cancers if they could introduce the enzyme into cancer victims' cells using gene therapy, but *turn it on* exclusively in cancer cells.

As I've already indicated, there is one obvious way to distinguish cancer cells from normal ones which could provide a mechanism for controlling the activation of viral TK: active telomerase. By designing a version of the viral TK gene that would be attached to a "trigger" (*promoter*) that would turn the gene on only in the presence of telomerase, the researchers realized that they could set the enzyme to work in a cancer patient's malignant cells, while leaving it dormant almost everywhere else in the body.

At this point the flow chart of what they were designing was beginning to look like the biotech equivalent of one of those exceedingly convoluted, multistep devices that players build up in the board game "Mousetrap." The scientists would set the "trap" by first seeding a copy of the gene for the viral TK enzyme, complete with its special telomerase "trigger," into every cell in the patient's body. The patient would then swallow some ganciclovir tablets, which would penetrate all of his or her cells indiscriminately.

In most cells, the drug would have no effect, because nearly all cells have their telomerase enzyme firmly turned off. But when ganciclovir entered a *cancer* cell, the trap would be sprung. The cancer's abundant telomerase enzyme would flip on the viral TK enzyme; the TK would scoop up the ganciclovir, adding on the phosphate group needed by DNA "letters" for insertion into the emerging DNA copy strand; the next time it reached for the relevant "letter," the DNA-copying machinery would grab the phosphorylated ganciclovir by mistake, jamming it into the "letter's" place on the strand. At that point, you would almost hear the cry of "*Mousetrap!*" as the DNA synthesis machines seized up, cell division came to a screeching halt, and the cancer shut down. See **Figure 5**.

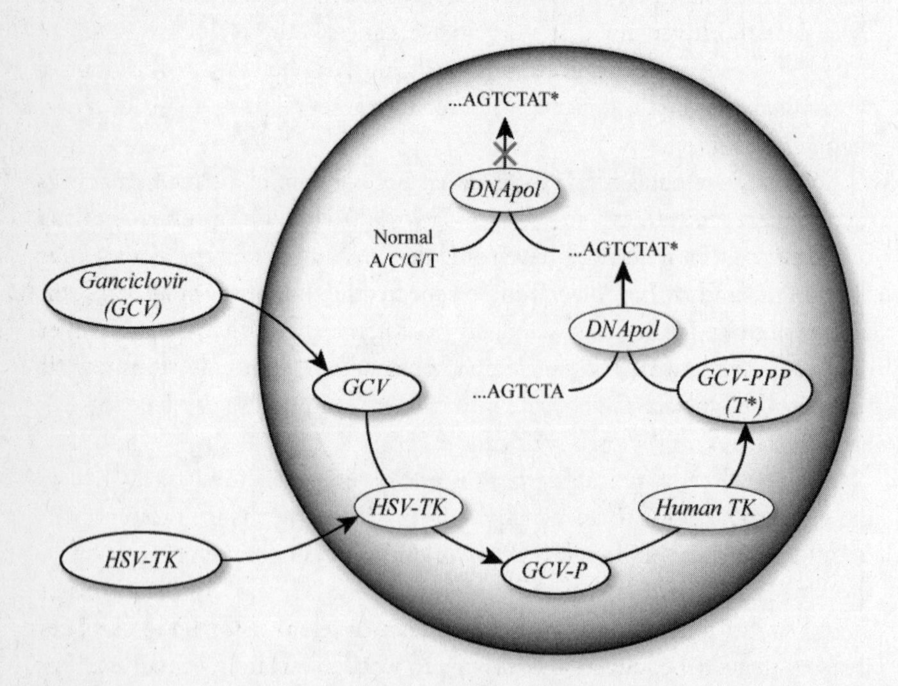

Figure 5. How ganciclovir enables viral thymidine kinase to kill mammalian cancer cells.

It was a crazy, convoluted solution—but it worked in the test-tube against liver, kidney, pancreas, and thyroid cancer cells. Moreover, the setup proved largely harmless to normal rat thyroid and human skin cells.[34] So the team took the next step in bringing something off a lab bench and into a clinic: a careful study in laboratory animals.

The researchers first cooked up two batches of customized viral TK genes: one with the telomerase-activated "on" switch, and another with a switch that could be expected to be flipped in healthy and cancerous cells alike. They then slid these genes into viruses from the same family as the common cold, allowing them to literally *infect* animals with the gene constructs. They first tried these constructs out on *healthy* animals, to see what the potential was for side effects. As expected, animals that got viral TK under the control of the nonselective promoter suffered nasty liver damage upon injection with ganciclovir, while putting in the same gene under a telomerase promoter appeared to be basically harmless, since there were no telomerase-expressing cancer cells present to activate it.

At this point, the researchers decided that their experimental therapy was ready to hit the next stage: a test in animals that had been injected with implanted human thyroid carcinoma cells. Giving such animals a copy of the viral TK driven by a promoter that did *not* rely on the presence of telomerase to activate it brought tumor growth to a complete standstill—but, as expected, it also prevented *normal* cells from reproducing themselves, leading to nasty liver damage.

But when scientists tried the *selective* targeting of viral TK to cancer cells by using the *telomerase* promoter, ganciclovir shut down tumor growth just as completely as it had when the TK was controlled by the nonselective promoter—and without the latter's toxic effects. The apparent safety of this highly selective intervention is all the more convincing when you remember that the effect can be turned on and off at will, by administering or withdrawing the ganciclovir.

Know Your Enemy

We've seen that biologists can be just as creative as weapons engineers holed up in the Skunk Works at finding new ways to target and kill cancer cells *selectively,* i.e., leaving healthy cells unharmed. It is foreseeable that the same methods in use or under development against cancer could be used to target anergic T cells. In this case, we're blessed with an enemy that is walking around with a bull's-eye painted right on its chest. The same dysfunctional receptor profile that strips anergic T cells of their ability to recognize their target antigens (absence of CD28) and to proliferate in response to infection (presence of KLRG1 and CD57), possibly along with some other markers (such as reduced levels of CD154, implicated in the failure of old T cells to support B cell development) already allows scientists to identify these cells, and could also be used to target such cells for destruction.

Unleashing the wrath of the immune system against its oppressors would be poetic justice, but vaccination (either passive or active) against these cells might be tricky. For one thing, their most prominent antigenic feature is the *lack* of a cell-surface protein (CD28), and while we could target the combination of KLRG1 and CD57, it's not yet clear whether all unwanted cells express these two proteins, nor whether other, desirable cells do. As it happens, immunologists already do identify anergic cells using a combination of immune proteins—but these could not easily be used for vaccination purposes. Also, after all, the problem that we're looking to resolve is characterized by a poor response to vaccination, so an "anergic T-cell shot" might only be effective in relatively immunologically "young" people. So while this approach is promising for many kinds of toxic cells, it may be less so for CD8s.

But that still leaves us with a lot of options. Dendrimer-based targeting approaches seem the most straightforward, because they allow for the targeting of cells via *multiple* identification criteria, and also because they can introduce any number of poisons into the cells that they select, from outright toxins to boron for BNCT.

And while using vaccines to target anergic T cells might be problematic, there could be another way to use the immune system as a proxy army, cooperating with it to restore the sovereignty of the immune system's government-in-exile. Remember that anergic CD8 cells first become a problem because they stop listening to the apoptotic order to scale back their forces that's sent out after their target pathogen has been routed from the body's frontiers. They are able to ignore these orders because they produce high levels of *bcl-2,* a protein that blocks apoptotic signalling.

This suggests the possibility of restoring normal apoptotic signaling in such cells by delivering "antisense RNA" for bcl-2's blueprints to them—strips of genetic material matched to the transcribed DNA instructions for the protein, preventing the encoded bcl-2 from actually being produced in the cell. With bcl-2 production brought down to normal or nearly nonexistent levels, anergic CD8 cells would finally hear their curtain call and bow out.

Would purging the immunological "space" of anergic T cells be enough to rejuvenate the immune system completely? I can't say for sure, because it hasn't been done—and, as I am well aware, the body is an incredibly complex machine whose parts have not yet all been identified, let alone their purposes and interactions. Existing research tells us pretty clearly that the direct and indirect immunosuppressive effects of these cells are powerful enough that a thorough spring cleaning of them will profoundly improve T cell-mediated immunity, and also very probably the functioning of other aspects of the immune system that T cells support and govern. But we'll only know just *how* profoundly once we've done it.

I can tell you right now that there is at least one aspect of immune aging that the removal of anergic T cell clones will not address: *thymic involution*. The thymus is a gland located just behind your breastbone. It's where immune cells first produced in your bone marrow go to learn to become T cells. As we age, the thymus loses cells and shrinks away, and in the process its output of naïve T cells plummets. This, of course, imposes further limits on the body's ability to respond to new threats.

In principle, there is a fairly straightforward way of dealing with this, however: stem cell therapy. This is foreseeable biotechnology, as we will see in the next chapter (see the sidebar there, titled "Rebuilding the Thymus").[35] The accomplishment of such a goal would entail significant advances in the stem cell field, including mastering the art of turning embryonic stem cells into the progenitors of the different cells of the body, and then engineering new tissue to rejuvenate the old one, renewing old tissues with pristine new cells—but these are just the same problems that are being solved quite rapidly for tissues all over the body, so there is ample reason for optimism.

That's all I have to say about the immune zombies; now it's time to look at some other types of supernumerary cells.

Deadly Combat in the Battle of the Bulge

The second kind of toxic cell that we'll want to rid ourselves of is excess fat tissue—most important, the so-called *visceral* fat that surrounds your internal organs, as opposed to the *subcutaneous* fat that lies under your skin all over the body. It's widely believed that, as people get older, they just "naturally" become more resistant to the effects of the hormone *insulin,* whose job it is to move carbohydrates and amino acids into fat and muscle cells.

This change causes a range of threatening metabolic changes, the most extreme of which manifest in people with full-blown type II ("adult onset") diabetes. It's also common wisdom that older people "naturally" enter into a more inflammatory state, with the body slowly burning up from within because of an excessive production of inflammatory signal molecules.

When you take in more calories than you expend, your body hangs on to them rather than allowing them to go to waste. This is not the result of perversity on the part of evolution, but a survival strategy: until very recently (by evolutionary standards) there was a good chance that quite soon you'd be in a period of famine, when those stored calories would be your lifeline. If your body isn't under the kinds of challenges (like weight-bearing exercise) that signal the body to build up metabolically expensive muscle or bone tissue, it will take the easy way out by storing the calories as fat. But in an environment where feast is never followed by famine, and where exercise is almost entirely a voluntary affair, we fail to shed that extra fat tissue, and it slowly accumulates as we age. Because this accumulation is a difference of aging versus healthy young bodies, it qualifies as "aging damage" under my engineering definition, even though it might not be considered as such from a purely theoretical point of view.

It's long been known that this damage—in the form of being overweight or obese—puts you at greater risk of diabetes, heart disease, and various other ailments, but it's only recently become clear why and how. You may have heard that fat stored in different parts of the body has different health implications: having an "apple shape" (fat centralized in your midsection, as in a "beer belly") puts you at great risk for diabetes and heart disease, while having a "pear shape" (fat clumped on your bottom or thighs) is unsightly but much less hazardous to your health. To the extent that there is some biomedical basis for this distinction, it lies in the difference in the locations of visceral and subcutaneous fat. Visceral fat is most visible around your middle because it clumps around major internal organs like the liver and kidneys. By contrast, subcutaneous fat just lies under your skin—and of course, there's skin all over your body, though there are more prominent depots for this fat type in some places than others.

Recent studies have found that nearly all observed age-related insulin resistance, and much of the age-related pro-inflammatory signaling shift, can be attributed to the accumulation of excess visceral fat, which precedes and predicts the development of all the elements of the metabolic storm known as *syndrome X:* insulin resistance, low HDL ("good") cholesterol, and high blood pressure, *triglycerides* (blood fats), and blood sugar.

Even more tellingly, when you compare people of different ages, you find that the difference in insulin effectiveness between young and old disappears when you account for the difference in fat, and visceral fat in particular.[36, 37]

In striking studies at the Albert Einstein College of Medicine, aging animals have had most of their resistance to insulin and other hormones reversed by highly invasive surgery that scrapes out most of their visceral fat, making their body compositions similar to much younger animals, or to animals of the same age that had been subjected to calorie restriction (among whose benefits are dramatic reductions in age-related insulin resistance and inflammatory signaling).[38] This latter result was especially striking because, when the scientists looked at the distribution of fat in the calorie-restricted animals, they found that they actually had *more* subcutaneous fat than their young counterparts, but less visceral fat—and their insulin effectiveness was almost the same.

Further pinning the blame on visceral fat, a recent study confirmed that undergoing liposuction—whose invasiveness and risks of trauma are kept low by removing only the relatively easy-to-access but cosmetically significant subcutaneous fat, while leaving the much harder-to-remove visceral fat behind—does not improve the insulin resistance associated with the original obesity.[39] On the other side of the same coin, several studies have now shown that putting overweight people on low-calorie diets or exercise programs significantly improves their insulin resistance quite early on—well before it has had a chance to impact their overall weight by much, but *after* it has had time to reduce their level of visceral fat, which (fortunately) is the first thing to go when energy needs aren't being met.

The reasons for all of this have become clear as scientists have increasingly come to understand the nature of fat itself. Fat tissue was once considered to be just inert storage space, like carrying around a spare tank of gas on your derriere. Instead, we now know that it is a metabolically active, dynamic tissue that secretes and responds to a range of hormonal and other signaling molecules. We also now appreciate that fatty tissue is not composed only of "fat cells" (*adipocytes*), but is a mixture of different cell types, including supporting connective tissue, nerves, and blood vessels, as well as immune cells—notably, macrophages. In fact, adipocytes are actually derived from the same precursors as macrophages, and secrete many of the same immune-system regulating molecules, such as the coagulation-enhancing enzyme *plasminogen activator inhibitor-1,* and pro-inflammatory signaling molecules (*cytokines*) such as *tumor necrosis factor alpha, monocyte chemoattractant protein-1,* and *interleukin-6.*

As the size of the fat depot increases, adipocytes begin pumping out more and more of these inflammatory molecules, some of which promote infiltration of the tissue by macrophages and some of which signal the precursor cells that give rise to both adipocytes and macrophages to take the path toward the latter instead of the former. Macrophages, in turn, produce even more inflammatory messenger-molecules, creating a self-reinforcing inflammatory feedback loop.

The most exciting finding in the emerging science of fat in the last ten years has been the discovery that these signaling molecules not only cause a potentially pathological increase in systemic inflammation, but also increase the body's resistance to insulin. This conclusion is supported by studies showing that isolated muscle and fat cells become insulin-resistant when bombarded with the very inflammatory mediators that adipocytes and macrophages produce, and that chubby lab rodents' insulin resistance is relieved by aspirin treatment, in part via blocking the effect of cytokines. And the relationship doesn't just hold up under the artificial conditions of the laboratory: during *sepsis* (the inflammatory storm generated in response to severe infection), human patients often exhibit very severe insulin resistance as part of the immune response.

Evidently, these and related questions will keep an army of basic and clinical researchers in the diabetes field busy for decades, resolving paradoxes, isolating deeply intertwined metabolic pathways, and double-checking their results in different models. But for *engineering* purposes, fortunately, we don't have to wait to have the results of these investigations: we just need to observe the presence of damage, and fix it.

Doing It the Old-fashioned Way

In this case, of course, there are two very simple, inexpensive solutions that don't involve advanced biotechnology: *diet* and *exercise*. But unfortunately, as decades of research and centuries of anecdotal experience have shown, most of us find it very difficult to lose weight once we put it on. As little as a 100-calorie daily energy imbalance—about what you get in a single medium-sized biscuit—accounts for the standard weight gain that creeps up on the average person in the decades between high school and middle age, and while it's easy to load it on, it's hard for most people to shed those extra pounds and keep them off. The situation is not nearly as dire as is often made out—studies show that about one overweight person in five successfully achieves long-term weight loss, and research is clarifying what it takes to get there—but the metabolic consequences of excess visceral fat are far too deadly, and the magnitude of the current obesity epidemic far too staggering, to leave the cure of this form of aging damage to self-help programs or public health measures designed to remediate our present "toxic food environment."[40] Realistically, unless we are ready to leave the fate of millions to a sudden wave of greater personal or political responsibility, we must look for biomedical solutions to visceral fat.

Fat for the Fire

One option that we might pursue is to shrink away visceral fat by causing it to burn off its excess stored energy. Scientists have of course been trying to develop drugs to do this for decades, but so far the only moderately effective such drugs are amphetamines, and their side effects and addictiveness clearly don't fit our remit.

For several years, the appetite-regulating hormone *leptin* seemed to offer the chance of shrinking fat and maintaining insulin sensitivity. An extremely rare genetic mutation that leads to a congenital lack of leptin makes both rodent and human victims monstrously obese, and injecting these mice with leptin leads to dramatic weight loss. Causing these rodents to produce more leptin *inside* their fat cells (using genetic engineering) makes them eat 30 percent to 50 percent less food, leading to more insulin sensitivity and an almost complete disappearance of their body fat. Moreover, the effects are stronger than can be accounted for by their newfound light dining habits alone.[41] Fat cells of animals with the extra leptin gene were expressing other genes that activate mitochondria, turning them into little "fat burning machines."[42] Paradoxically, however, while injections of leptin lead to rapid fat loss in both normal and obese rodents, the same level of leptin that would peel away the pounds in lighter mice circulates *naturally* in the bodies of their fat cousins, necessitating a much higher level of leptin to achieve similar weight loss. This is in part because leptin is, ironically, produced by the very fat cells whose swelling with stored energy it was supposed to inhibit, so that chubbier rodents naturally produce *more* leptin, not less. And indeed, after drugs giant Hoffmann-La Roche invested a fortune to develop a way to mass-produce human leptin in genetically modified bacteria, they found the hormone to be a miserable failure as a weight-loss treatment.[43]

This led scientists to speculate that overweight members of both species come to suffer "leptin resistance" in the same way that they suffer insulin resistance: levels of the hormone are high, but cells stop responding properly to its signals to turn off the appetite and turn on fat-burning. Recent studies by Roger Unger, the researcher who had originally raised hopes for leptin by showing its powerful effects in mice, show how this can happen at the molecular level. When you overfeed mice a high-fat, high-calorie chow, their fat cells scale back the expression of the genes that tell the cell how to build leptin receptor "doors" on their surfaces, so that the cell stops hearing the signal.[44] Similarly, unpublished studies of the fat tissue of grossly overweight people show that their expression of the leptin receptor gene is consistently so low as to be undetectable, while lean, young people may have anywhere from quite low to extremely high levels of gene expression at any given time.[45]

Unfortunately, as with leptin itself, the path to using leptin receptor gene therapy as a way to reverse the negative effects of visceral fat is not a clear one. Mice with the extra leptin gene may have stayed slim in the face of an overly rich diet, but they did not fully escape its consequences: they still suffered the same "ectopic" (mislocalized) fat infiltration of their livers, muscle, and heart as did mice lacking the extra leptin receptors eating the same chow, and their insulin resistance—the key negative effect of excess visceral fat that we need to address—was just as bad.

While we might find some way of avoiding some of this by turning the gene on and off again, thereby restoring our insulin sensitivity, it seems unclear how we would deal with the ectopic fat. We might expect that as soon as we turned the gene off the calorie imbalance that had led to the initial overgrowth of fat cells would begin again; while we would shrink fat cells back down with each round of therapy, we would have no way to eliminate the cells themselves, and over the course of a greatly extended life a visceral cavity full of large numbers of even relatively small fat cells might still lead to metabolic mayhem.

Really Trimming the Fat

No—what seems most likely to solve the problem of visceral fat is not to try to tame it, but to cull it: actually to remove a substantial number of the bloated, excess cells. The same sorts of cell-specific targeting of cancer or anergic T cells we've been discussing seem likely to be applicable to visceral fat too; and while, unlike in those other cases, no specific markers of visceral fat cells have been identified, we don't need to be nearly so specific. Unlike cancer or anergic T cells, *some* fat—including some visceral fat—is not only metabolically harmless, it's necessary for carrying on the business of life. Aside from being a spare tank of metabolic fuel that we draw upon and refill every day, those metabolic factors we've been talking about—energy-regulating hormones, inflammatory peptides, and others—also have healthful uses. As with all of metabolism, this was built into us by evolution for our benefit. As anti-aging engineers, it's not our job to interfere with it—just to prevent the damage that it causes in aging bodies.

Night of the Living Dead

The last class of toxic cells that I want to address in this chapter are so-called "senescent" cells.[46] They got this name because of the (rather dubious) analogy drawn between these cells and aging humans by their codiscoverer, Dr. Leonard Hayflick, then of the Wistar Institute in Philadelphia. These cells, like the others that we've been discussing, begin their lives as normal constituents of the skin, joints, and other tissues. They are normally quiescent, not dividing regularly, but they remain capable of reproducing themselves on demand, as part of their normal function (unlike "postmitotic" cells, which lose the ability to divide ever again once they reach their mature form and are only replaced by new cells coming from the body's stem cell pools—if they're replaced at all).

The defining feature of senescent cells is that they, like postmitotic cells, have lost the ability to divide. Hayflick observed, contrary to the dogma of the day, that cells from these tissues would not keep reproducing in the petri dish indefinitely: they would seem normal for several successive periods of replication, but would then suddenly enter into a twilight state in which they didn't die, but became in various ways abnormal. Their appearance became blotchy, their shapes irregular. They failed to form the neatly whorled colonies of mutually adhering cells that were the norm in younger cultures. And above all, they stopped reproducing themselves.

The use of the word "senescent" to describe these cells is, however, a bit misleading. When people hear about these cells, they often assume that cellular "senescence" is the ultimate fate of all of the cells in the body with age, and that the entry of "young" cells into this senescent state is the underlying cause of aging. Moreover, the term evokes an image of these cells as somnambulant old has-beens, sleepwalking through the remaining days of the rest of the body's life, not contributing anything to the organs in which they reside but also not doing us any positive harm. Their only downside, we might think, is a crime of omission: inability to replenish aging organs.

In fact, senescent cells are generally held to be extremely rare even in very aged people.[47] However, their possible role in aging has turned out to be far more complex—and far more *active*—than we at first imagined.

The most obvious characteristic of senescent cells is, as mentioned, their loss of the ability to reproduce. But, like worn-out lechers, senescent cells desperately *try* to stimulate themselves into activity—pumping out substances that, though essential to their healthy function back when they were contributing members of a healthy tissue, may promote the development of cancer when present in excess. Various mechanisms are involved.

For a start, some of the most common signal molecules overproduced by senescent cells are chemical messengers like *epidermal growth factor* that directly spur cell division in their neighbors.

As a second example, many senescent cells also overproduce protein-digesting enzymes such as *matrix metalloproteinases* (MMPs), which are the "demolition teams" of tissue remodelling. These enzymes perform the essential function of clearing away the old, damaged "scaffolding" in which cells are embedded in a tissue, making space for new growth. But, just as having the outside wall of your house knocked down during a renovation would leave you vulnerable to burglary, so an excessive or uncontrolled MMP activity can enable cancerous cells to escape from the restraints of the tissue in which they were originally embedded, and/or to work their way into new tissues far removed from the original cancer site—the *metastasis* process.

And most recently, scientists have found yet another way that senescent cells potentially roll out the welcome mat for nascent cancers: by churning out dangerous overdoses of *vascular endothelial growth factor* (VEGF)[48] and *stromal cell-derived factor 1* (SDF1),[49] which promote the growth of new blood vessels.

As you can see, the picture turns out to be a lot more complicated than we had once thought. As I'll discuss later on, the senescence phenomenon is probably an evolved response to DNA damage, helping to prevent that damage from becoming cancerous. From the point of view of that one cell, this is an extremely effective *short-term* protection against cancer, because it shuts down the cell proliferation that is the very heart of the disease. But it seems likely that the "unlifestyle" that a senescent cell leads ultimately facilitates cancer progression in the long term by destabilising its *neighbors.* So—you guessed it—they've got to go.

Putting the Zombies to Rest

One way to eliminate senescent cells is to make them no longer senescent. In cell culture experiments, this has been achieved in a variety of ways, such as by relengthening exhausted telomeres with telomerase or by depleting proteins associated with senescence. But reversing senescence would run the risk of cancer, because senescent cells typically get that way as a response to potentially carcinogenic changes to the cell, such as damaged DNA, hyperactive cancer-promoting genes, or (again) very short telomeres, which promote a mutagenic state. Restoring proliferative capacity in such cells could potentially take us out of the frying pan of senescence and into the fire of cancer.

Similarly, approaches based on turning off the dangerous metabolic abnormalities of senescent cells carry risks, because other, healthy cells depend upon these same pathways for normal function. Chronically jamming growth signals, enzymes, and inflammatory messengers might well prevent senescent cells from watering the seeds of cancer, but it would also lead to "crop failures" of cells all across the body.

As usual, the engineering approach to this dilemma is to rewrite the rulebook. We will maintain the body's evolved capacity to shut down cells in danger of going cancerous, by leaving the metabolic regulation of senescence as it is. Indeed, as we'll see in Chapter 12, we will ultimately need to transform the body in ways that will make it almost immune from cancer by ensuring that *all* cells run out of steam long before they would threaten us with uncontrolled cell growth. But we will eliminate the threat posed by those cells that *do* senesce by eliminating the cells themselves.

Silver Bullets

The first significant development in this area came in 1995, in a lab at the Lawrence Berkeley National Laboratory headed by Dr. Judith Campisi, one of my coauthors on the original SENS scientific manifesto. Campisi and coworkers found that a relatively easy, reliable test for the activity of an enzyme called *senescence-associated beta-galactosidase* (SA-beta-gal) could identify senescent cells not only in petri dishes, but in skin samples taken from older humans.

Unfortunately, SA-beta-gal is not a *perfectly* selective marker for senescence. As subsequent studies showed, the enzyme does occur in nonsenescent cells—usually at very low levels, but sometimes in high concentration. It turns out that, contrary to the simple interpretation of the Campisi lab's findings, this enzyme is actually identical to one that is normally found in all of our lysosomes—the cellular waste incinerators, whose clogging (you'll recall from Chapter 7) underlies many of the worst pathologies of aging. The change into the senescent state does not suddenly trigger the secretion of SA-beta-gal into the main body of the cell out of nowhere: instead, it appears that there is always some low level of SA-beta-gal floating around in even normal cells, as can be detected with techniques that assess the level of *the enzyme itself* in the cell—but the level is so low that its *activity* is barely (if at all) detectable by the methods that Campisi's lab initially used, which are unfavorable to the enzyme's functioning.[50,51,52]

But as a cell goes through round after round of replication—thereby drawing ever closer to senescence—its SA-beta-gal levels rise.[53] Probably this is because the cell begins to overproduce the enzyme in response to the stresses of aging—notably, the need for more lysosomes as they become less and less effective at doing their job (and also as their cell division rate, hence garbage dilution rate, slows and the job thereby becomes intrinsically more challenging). Eventually the *level* gets so high that its *activity* is noticeable even under suboptimal conditions.

SA-beta-gal activity is, notably, found at abnormally high levels in cells taken from tissues where cells are under stress, due to inflammatory diseases that fuel cell proliferation (such as chronic hepatitis C, atherosclerotic plaques, and venous ulcers). Most interesting is the finding that levels of the enzyme shoot up in cells undergoing "crisis,"[54,55] a period in which cells that have somehow escaped senescence are still undergoing cell division and the erosion of their telomeres. Such cells usually just run out of steam, but occasionally they undergo a mutation that removes the clamp from their telomerase genes, making full transformation into malignancy almost inevitable.

What's emerging, then, is a picture of SA-beta-gal as an enzyme that appears at high level in the main bodies of cells undergoing some kind of stress that may ultimately threaten their neighbors. This might mean that by using high levels of SA-beta-gal as an identifier for the destruction of senescent cells, we would simultaneously take out some useful "targets of opportunity."

However, we may be able to establish a system of double-checks, to help us weed out more genuinely senescent cells while leaving more innocent (but suspicious-looking) cells unmolested. This is because, in addition to SA-beta-gal, senescent cells also produce abnormally high levels of other molecules involved in the programmed senescence response. Senescent baboon skin cells, for example, contain an activated form of the protein *ATM kinase,* which responds to DNA damage by activating several tumor suppressor genes, including the famous *p53*. Senescent cells also exhibit high levels of p53, as well as the binding protein (53BP1) by which its gene interacts with ATM kinase, and p21, a senescence regulator that works under p53's command.[56] Some senescent cells also contain high levels of p16, the other main regulator of the process. Levels of this protein, for reasons as yet unknown, also climb slowly with age in nonsenescent cells, making it an unreliable marker for senescence when taken in isolation; but it—like these other features—could still potentially be used as part of a double-checking mechanism, with multiple proteins being used to distinguish genuinely senescent cells from those expressing only one of them for unrelated reasons.[57]

This chapter has focused on the accumulation of toxic cells with age, and the foreseeable biotechnology with which we should be able to purge ourselves of those cells as part of our platform for the rejuvenation of our bodies—restoring the immune system, alleviating metabolic mayhem, and protecting our cells from being goaded on to cancer. In the next chapter, I'll look at the opposite problem: cell *loss* with age, and the scientific—and, just as importantly, political—hurdles that we face in bringing forward the ability to renew our tissues with fresh, new replacements.

11

New Cells for Old

> Throughout our lives, we gradually lose cells vital to our continuing health. Many fatal diseases of aging—such as Parkinson's disease—are caused by the loss of populations of cells responsible for one or another crucial function in the body. Fortunately, therapies based upon stem cell research offer the possibility of recreating our missing cells, good as new; politics stands in our way as much as any remaining scientific obstacles.

After the enormous effort that had been required to organize the conference, it was an incredibly rewarding moment to see the man who was revolutionizing stem cell biology take the podium in front of a packed crowd of colleagues.

It was the second conference I'd run at Cambridge focusing on scientific progress toward the reversal of human aging, so the pressure was on for me to top the success of the first. I'm on the board of the International Association for Biomedical Gerontology (IABG)—one of the few biogerontological societies in the world with an explicit brief to pursue the development of biomedical solutions to aging—and a couple of years earlier I had volunteered to spearhead their tenth conference. I knew at the time what I was getting into. The society would provide little logistical assistance beyond networking opportunities, so I would have little help beyond the support (moral and otherwise) of my beloved wife Adelaide, and that suited me perfectly. With the formal authority of a society already on the progressive wing of the biogerontology community, I wanted to push the envelope a little further, and being left to my own devices meant that I would not have to debate my priorities with a committee.

Despite the society's mandate, previous IABG conferences had tended to be dominated by the same kind of presentations that I saw at every biogerontology conference I attended (and I try to get to most of them): basic science, geriatric medicine, and work in model organisms that the researchers hoped might someday be translated into a pill to *slow down* aging in humans. I took on the enormous and exhausting work of running this conference because it would give me the opportunity to highlight work that could contribute to a panel of interventions designed to *reverse* aging.

IABG 10—the meeting that would be, in retrospect, the first in a series of SENS conferences—was an enormous success. I am saying so myself, but I am making no boast of my own: the enthusiasm with which my colleagues thanked me for my efforts at the end of the week was ubiquitous and unmistakeably genuine. Attendees were surprised and excited by what they had heard, not only on its own merits but because most of it was completely novel to them. This was to be expected: while a typical biogerontology conference invites a roster of speakers almost entirely drawn from within the biogerontological community, I had introduced a strong interdisciplinary element, bringing in researchers working in cancer, diabetes, stem cells, and other fields, whose work would in my mind be critical to the development of effective anti-aging biomedicine but who were almost entirely unknown to researchers prone to pegging themselves in the "biogerontology" slot.

At the same time, those presenters had the opportunity to mix with researchers in whose laboratories the degenerative processes of aging were, if not being reversed, certainly being dramatically delayed in mice and other model organisms. This was work that often hardly raised an eyebrow amongst biogerontologists, who were immersed in a field in which it had been taking place since the first calorie restriction experiments nearly seven decades previously, but which amazed the experimental oncologists and tissue engineers that I had brought in to show the biogerontologists what they'd been missing.

IABG 10 was so successful in meeting my academic goals, and the requests from my colleagues that I run a sequel were so obviously sincere, that I felt sure that I could harness its momentum to make it the *de facto* inaugural meeting of an ongoing series of academic conferences on SENS science at Cambridge. From then on, however, I knew that the effort would be entirely my own: I could not rely on the support (nor brook the interference, little though it had been) of the IABG or any other society. Challenging as the job of directing such events was, I knew it would be worth it.

On the other hand, I also knew that I had set my own bar quite high with the first conference, and that some of my colleagues would be less inclined to attend a conference not run under the aegis of a recognized biogerontological society. This was all the more so when I was the organizer, because a whispering campaign against my credentials as a scientist had been initiated by some of my genuinely well-intentioned but old-school gerontological rivals shortly after the first conference. So if I wanted to get people to show up to SENS2, and to have the series continue, the quality of the conference lineup would have to be top-notch despite the opposition. I would have to meet an ambitious standard—and I wanted to overachieve.

The Master Cells: Accept No Substitutes

I knew that I would once again want to devote a whole session to *embryonic stem cells* (ESCs)—the primordial "master cells" from which our mature cells spring, and which play a critical role in our development into complex multicellular organisms from the simple ball of cells that is an early embryo. Thanks to the tragic confusion of the science of ESCs with the ethical, legal, and religious disputes around the status of the embryo in the abortion debate, ESCs are the one plank of the SENS platform with which you cannot help but be familiar. You have doubtless heard that, with the right kind of biochemical stimulation, embryonic stem cells can be coaxed into becoming any kind of cell in the body: nerve, muscle, heart, kidney, the lot.

These resulting, "differentiated" cells can then be used to repair or to replace cells and tissues that are lost to—and whose loss is a central pathological feature of—multiple debilitating, often nearly untreatable diseases, including many of the worst scourges of aging. ESCs will be needed to develop full cures for Parkinson's disease, spinal cord injuries, juvenile diabetes, amyotrophic lateral sclerosis ("Lou Gehrig's disease"), heart attack damage, some cancers, and other devastating conditions—including aging itself. Indeed, under the purely pragmatic, engineering definition of aging that clarifies so much about what needs to be done to keep our bodies ageless indefinitely, the net loss of cells is *itself* a form of aging damage. This makes it a central target of SENS.

However, because media coverage of the issue focuses on the political firestorm rather than on the real, hopeful *medical* story of ESCs' enormous potential as medicine, you may still not be clear on the key differences in basic biology and therapeutic potential between ESCs and *adult* stem cells. There are also important differences between ESCs derived from embryos being stored in fertility clinics and those that can be custom-made for each patient out of his or her own mature cells by fusing them with egg cells (a technique known as *somatic cell nuclear transfer,* SCNT, to which we shall return). For this reason, I will spend a little time disentangling these issues.

True ESCs are found only in very early-stage embryos called *blastocysts,* which are the very primitive balls of cells that are formed within just a few days after sperm meets egg. The embryo only remains in this stage of development very briefly; it has developed much further by the time the embryo is implanted in the womb. It is from the blastocyst that every cell in the mature organism must be derived, yet the blastocyst itself has none of these differentiated cells: no neurons, no heart cells, no insulin-producing *beta-cells,* and so on. So for the embryo to go on to transform itself into an organism with the complex structure of a human being, its cells need the ability to transform themselves into each and every one of those mature cells—a power called *pluripotency*.

Adult stem cells, on the other hand, are much more limited in their abilities, and also with good reason. These cells emerge in the late stages of development, and are retained in particular tissues during life as a reserve to replenish cell stores. They thus hold on to only the limited repertoire of possible fates that is relevant to their role in that particular tissue. Thus, blood stem cells can become oxygen-carrying red blood cells or any of the many blood-borne immune system cells, but (despite what has been claimed—see later on in this chapter) they cannot form either neurons or heart muscle cells: if asked, a blood stem cell would doubtless indignantly reply, "That's not my job." They're there to fulfill a specific role in the body, and to do it well, but not to be on reserve to heal all damage everywhere. This more limited range of developmental flexibility (or "plasticity") is called *multipotency*.[1]

Indeed, there are many areas of the body for which there are no adult stem cells dedicated for use in repair—and, as you might expect, these include the areas that suffer the worst cell loss during aging. This is the situation, for instance, in much of the brain. For many years, it was believed that the entire brain loses cells over the course of normal aging, and that there was no way for the body to replace these losses. This dogma was overturned a few years ago, largely due to the work of Fred Gage and his coworkers at the Salk Institute, who showed that the brain does indeed harbor stem cells capable of renewing *some* parts of the brain. This has led to a swing, in the popular imagination, to the impression that the entire brain has the inbuilt capacity, through its adult stem cells, to keep the *entire* brain young and functional.

In fact, however, that impression is also wrong. Only a small number of areas in the brain produce stem cells capable of developing into new neurons: a sub-subsection of the *hippocampus* called the *subgranular zone* of the *dentate gyrus,* and a part of the *subventricular zone,* where neurons are created to supply the *olfactory bulb* (the area of the brain that processes the sense of smell). There's evidence that some of these cells do attempt to repair areas of the brain damaged by age-related disease, but there's little evidence that they're much help. After a stroke, for instance, a few of the stem cells formed in the subgranular zone do change their normal habits and migrate toward the site of damage, but over 80 percent of them die within a few weeks, and the remaining cells replace only about *0.2 percent* of the cells destroyed by the incident.[2]

Why do we maintain the capacity to replace neurons in some areas of the brain and not others, like the cerebral cortex where our long-term memories are stored, or the frontal lobe where our ability to make and stick to plans for our future is centered? Most likely, it's because the olfactory bulb and the dentate gyrus are the only places where evolution has encountered the need for a regular influx of new cells *within the brain's "biological warranty period."* Both those areas have short-term functions that require the regular renewal of their cell populations. There is no built-in population of adult stem cells to deal with cell loss induced by the ravages of aging and age-related neurological diseases such as Alzheimer's and Parkinson's. As you'll have realized if you still remember Chapter 3, this is because, while these disorders have their seeds in molecular damage that occurs throughout life, that damage does not reach a threshold where function is impaired sufficiently to affect Darwinian fitness in a short, Paleolithic human lifespan.

Another example of a tissue in which cells die but are not naturally replaced is the thymus, a key organ in the immune system which acts to "mature" precursor cells into T cells. Its regeneration using stem cells is at an early stage of development, so there's not much to tell you yet, but a proof of concept exists in a rare but very serious congenital disease—see the sidebar "Rebuilding the Thymus." I've described the immune system generally, and T cells in particular, at some length in Chapter 10, so you might want to refer back to that chapter while reading the sidebar.

REBUILDING THE THYMUS

The promise of using stem cells to treat thymic involution can be seen in recent advances in treating babies with *DiGeorge syndrome*—a genetic disorder whose victims are born with a variety of defects, including having a thymus gland that is underdeveloped, or in some cases completely absent (the latter being called "complete DiGeorge"). Complete DiGeorge has, until recently, often been a very near-term death sentence: with no ability to produce T cells, these babies would die of infections that are trivial to the rest of us, within a few months of leaving their mothers' wombs.

The obvious way to solve the problem of a missing thymus is transplantation, but that's a tricky business: to do its job, the tissue needs a very good blood supply and plenty of oxygen saturation, which is difficult to achieve without the natural penetration of tiny blood vessels. There have also long been problems with rejection and graft-versus-host disease: perversely, sometimes a few of the child's bone marrow cells will "spontaneously" transform into dysregulated T cells that don't recognize either the child's own antigens or the thymus tissue donor's. This leads to a ferocious attack on both, usually killing the child; moreover, often the *donor's* T cells would turn on the transplant *recipient's* foreign tissues in an equally deadly, reciprocal attack.

Recently, surgeons and immunologists at Duke University developed a protocol using very thin slices of tissue to ensure maximum transfer of oxygen, which are engrafted into the child's thigh to give it a generous, readily accessed supply of blood, along with a novel immune-suppressing drug that targets T cells specifically. The intervention is still experimental, but it's become progressively better through new innovations and now seems relatively successful. In a 2004 report, the Duke team found that five of the six patients receiving the new therapy were still alive fifteen to thirty months later, a greatly improved survival rate.

If, instead of using transplants of foreign tissue, we could take the child's own stem cells, coax them into becoming thymus cells, and engraft them, we would eliminate the need for risky immune suppression. Then, if we could encourage these cells to grow in a scaffold in which we could build up a complex organ structure, including a proper blood supply, we could abandon the highly unsatisfactory replacement of an organ with a wafer-thin tissue slice in favor of a real organ "transplant." We may never actually be able to do this in DiGeorge syndrome, for the simple reason that we don't have enough time—but if a foreign tissue implant can generate viable T cells and increase survival in babies born with no thymus, then I can only see promise in delivering a person's *own cells,* taught to become T cells and if necessary coaxed and structured into a more complex tissue, to an existing but atrophied organ, to restore it to youthful functionality.

Similarly, in the heart, cells exist, which some researchers have called "cardiac progenitor cells" or similar names; but, while these cells can be nudged into showing *some* stem-cell-like molecular signatures in a test tube, they have not been shown to form heart cells in the body. Indeed, some closely related stem cells found elsewhere in the body (*mesenchymal stem cells*) have the same hallmarks but definitely cannot become heart cells. Whatever the ultimate truth of the matter, what we do know is that neither these nor any other cells in the body step in to heal the massive damage wrought to the heart muscle by being starved of oxygen during a heart attack—as any cardiologist or heart-attack survivor can sadly attest. Again, the reason for this lies in the cold statistical analyzes effectively performed by natural selection after generations of genetic dice-rolling in a premodern environment: heart attacks don't kill twentysomethings, so by evolution's calculus it's not worth investing in a repair system that will almost never be used before its owner is killed by something else.

In the first days of the political debate around embryonic stem cells, some very respected laboratories issued reports of ESC-like flexibility in adult stem cells—of blood-forming cells spontaneously transmuting themselves into liver and brain cells, and perhaps most promisingly of such cells being injected into the hearts of rats given simulated heart attacks, forming new heart muscle tissue, and restoring functionality to the organ. These reports were taken so seriously that several groups began early clinical trials in humans, in which stem cells have been derived from the bone marrow of heart attack victims and then injected into their ravaged cardiac tissue.

But independent laboratories have been unable to confirm these claims. Instead, what may be happening is that the cells are indeed being incorporated into the tissues in question, but are doing so by *fusing* with the existing cells.[3,4,5,6,7,8,9,10] There may be some limited benefit to this: the process of fusion may support the surviving cells in damaged tissues, either by secreting growth factors needed during repair, or by helping new blood vessels to grow into the tissue.[11] But, while such effects may help to keep a disintegrating ticker beating for a *short* while more, it cannot substitute for actually *rebuilding* heart tissue, either for heart attack victims or for the aged humans whose hearts we wish to rejuvenate.

Indeed, recently the *New England Journal of Medicine* published the results of the first trials of bone marrow stem cells as a treatment for human heart attack victims that were large enough to give meaningful information about actual clinical outcomes in the patients (as opposed to just collecting safety data and early reports of physician and patient experience). One of these trials[12] found no benefit, and the other two[13,14] reported what the *Journal*'s summarizing editorial described as "small, [statistically] significant, but clinically uncertain improvements"[15] in treated patients compared to those receiving dummy injections. They reported no evidence either way on the subject of the cells actually transforming into heart muscle cells, but the animal studies mentioned above have at this point dashed previous hopes of such an effect.

Contrast these weak effects with the results of an animal study using embryonic stem cells to treat an induced heart attack. Eighteen sheep were subjected to such an assault, and then allowed to decline for two weeks. During this time, scientists harvested ESCs and nudged them to begin making the transition into becoming heart muscle stem cells. Before the embryonic stem cells had completed their developmental journey, the researchers seeded these cells onto the hearts of half of the group, while for comparison the remaining nine animals were left to slide further down the road to disability.

Where the benefits of adult stem cells had been dubious, the healing influence of ESCs was undeniable (see **Figure 1a**). The cells took hold in the damaged hearts and were shown to transform into mature heart cells, and the animals experienced a dramatic recovery. In the two weeks since their matched relations had been given the ESC treatment, the control group's hearts had lost an additional tenth of their blood-pumping ability. By contrast, animals who had received the cardiac-committed stem cells enjoyed a 6.6 percent *improvement* in pumping capacity.

And if you dig into the details of the study, you find even more reason to be optimistic about the potential for ESCs as a therapy for the heart. For

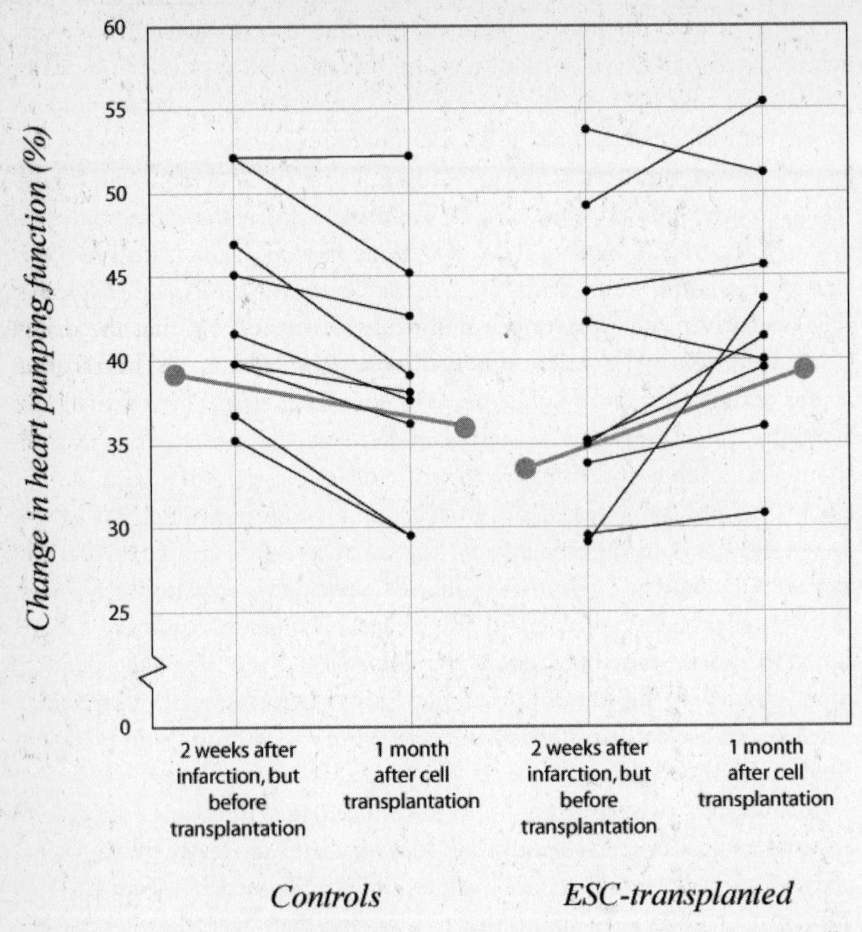

Figure 1a. Restoration of the heart's pumping ability by embryonic stem cells. (a) Controls vs. ESC recipients. (b) Controls, ESCs plus immunosuppressive drugs, and ESCs alone. Redrawn.[16]

one thing, the scientists in this study waited until two weeks *after* the animals suffered their heart attack to do anything about the damage to their hearts, and it was during this period that the bulk of the degeneration of the animals' hearts' pumping capacity occurred. Early intervention, whether with stem cells or even with more conventional medical duty-of-care, might have prevented a lot of this decline, potentially leading to much better outcomes after ESC treatment.

Second, the ESCs that were used in this study weren't even derived from sheep, but from *mice*—an important point to which we will return later. While the cells clearly did their job—maturing into heart cells, uniting with the native tissue, and restoring significant functionality to the animals' hearts—it still seems reasonable to think that using cells that were actually from their own species would have yielded a better metabolic and functional match, and, therefore, better outcomes.

And third, the average improvement in the ESC-treated group actually conceals a very positive variation in response to ESCs *within* the group. Because of the possibility that their immune systems might reject the mouse-derived ESCs and spoil the experiment, five out of the nine treated animals had been given immune-suppressing drugs. It turned out that the drugs were unnecessary: the researchers took slices from all animals' hearts after the study was over, and there was no evidence of inflammation or attack by immune cells in the hearts of the animals given ESCs, no matter whether they were dosed with immunosuppressive drugs or not.

This is positive news in and of itself, but there was even better news to follow. The reported 6.6 percent recovery of heart pumping capacity in ESC-treated animals was a pooled result, including animals that did and did not receive immunosuppressive drugs. When the researchers broke down the results according to whether animals received these drugs or not, they found that the immunosuppressed animals had actually responded *more weakly* to ESC treatment than ones whose immune systems were left to carry out their business. The sheep in the ESC-only group healed 25 percent more scar tissue from their original heart attacks than the drug-treated animals, and their hearts recovered over twice as much pumping capacity: about a 9 percent versus roughly a 4 percent gain (compared, again, to a 9.9 percent *further loss* of functionality in animals not receiving ESCs—see **Figure 1b**). So, in evaluating the prospects for human use of ESCs, we should look at the stronger results available from an ESC-only approach, rather than the weaker results from pooling these animals together with those given immunosuppressants.

After this study was published, the first head-to-head comparison of ESC versus adult stem cell therapy for heart damage similar to that endured during a heart attack were reported; the results showed clearly the superiority of the ESCs, which transformed into heart muscle cells, achieved long-term incorporation into the animals' heart tissue, and improved the animals' heart function, while the bone marrow stem cells had no significant effect.[17]

And this is only the beginning of the biomedical promise of these amazingly versatile cells. Embryonic stem cells have been used to cure animal models of some of the most fearsome diseases human beings suffer,

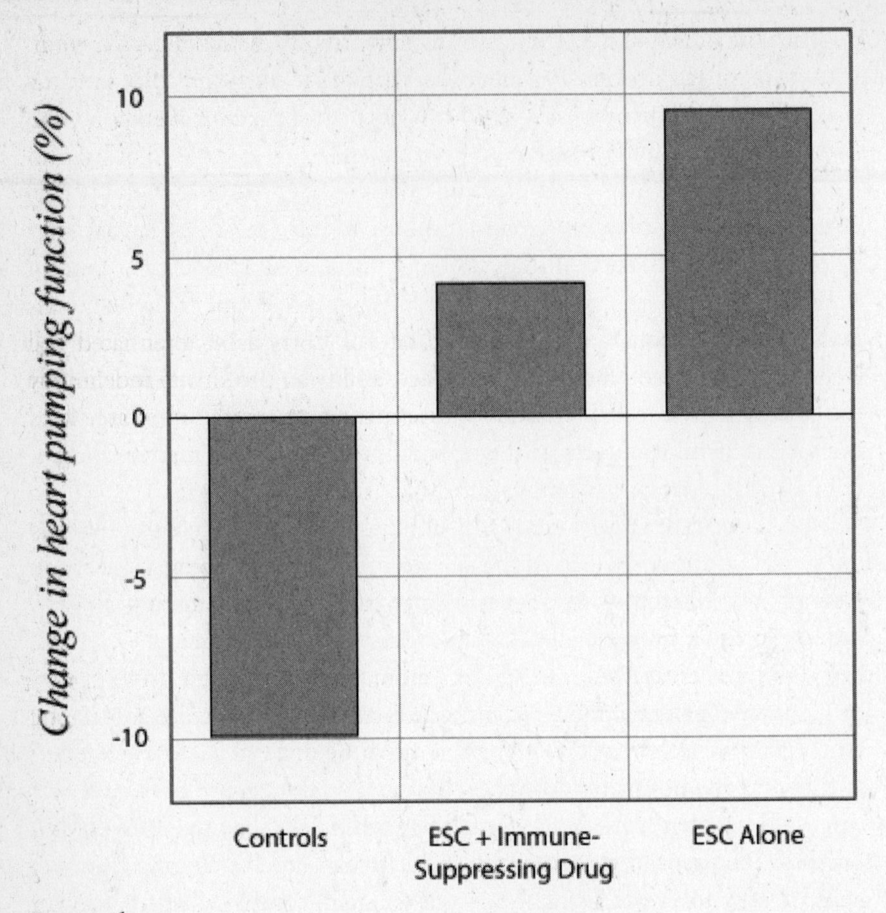

Figure 1b.

such as juvenile diabetes,[18] spinal cord injuries,[19,20] multiple sclerosis (MS),[21] cerebral palsy,[22] stroke,[23,24] Parkinson's disease,[25] a form of paralysis caused by a virus that induces a standard mouse model of ALS,[26] and—very recently—macular degeneration (the form of blindness caused by the loss of light-sensing cells in the center of the eye's retina).[27] All of these are diseases where a person's native, *adult* stem cell supply fails even to begin to replace the cell loss caused by the disease.

Of course, none of these therapies has made its way into the clinic—yet. But there's every reason to think that they will lead to dramatic improvements in our ability to treat these patients. The balance of preliminary evidence from human trials using *fetal* cells or cells derived from stem-cell *tumors* (not true ESCs) in Parkinson's disease and stroke victims, for instance, already shows a lot of promise that can only be expected to improve with the use of actual stem cells, and recently a study using ESCs in a monkey model of Parkinson's has confirmed their ability to transform into the required type of neurons, engraft into the appropriate area of the brain, and relieve many of the symptoms of the disease.[28] These are exciting times.

Why We Need Them

Because the horizons for the ultimate fate of ESCs as differentiated cell types are wide open, and because of their ability to proliferate indefinitely (unlike adult stem cells, whose replication capacity tends to be more limited), the scientific consensus acknowledges the greater therapeutic potential of ESCs over that of adult stem cells. There are certainly therapeutic uses for adult stem cells; indeed, the only stem cell-based therapies *currently in clinical practice* are things like bone marrow transplants, which use adult stem cells taken from a donor or from the patient's own body. But the oft-repeated claims by social-conservative lobby groups that adult stem cells can effectively treat "70 diseases" or "more than 65 diseases" have rightly been called "patently false" and the accompanying information on one prominent such group's Web site "pure hokum" in the editorial mentioned earlier from the normally diplomatic *New England Journal of Medicine.*

As things stand, only embryonic stem cells hold the potential—both in terms of the *range* of cells required, and in terms of the sheer *quantity* of cells needed to create large tissue grafts and in some cases even whole organs—that will be needed to make young bodies from old ones. And need them we will. In addition to cells lost to heart attacks and neurodegenerative diseases, the truth is that we are losing cells—and the functionality that those cells provide—from our tissues on a continuous basis. Parkinson's disease, for example, is the result of the loss of neurons in the brain that produce *dopamine,* a chemical messenger involved in fine control of the muscles. You get a clinical diagnosis when you have lost about half of these neurons, impairing this control enough that parts of your body begin an involuntary rhythmic shaking and your face turns into a staring mask with a fixed blank or even hostile expression. But *all of us* are losing dopamine-producing neurons every day to aging; people with Parkinson's just lose them more rapidly, reaching the clinical threshold earlier. Without the ability to replace these cells, we'll all develop the disease eventually (if, as the refrain goes, something else doesn't kill us first).

And it's happening all over your body, and not just for the kind of intrinsic metabolic reasons that are most precisely termed "aging." You are permanently losing cells every day to molecular damage caused by the reactive by-products of normal metabolism, and even after we undo such damage using the foreseeable biotechnologies of the SENS platform, we will still need to reverse these losses if we are to build ageless humans. Plus, we also lose cells to other causes. We all regularly destroy some naturally irreplaceable cells to minor bumps on the head, moments of oxygen deprivation, and the *apoptosis* ("programmed cell death") imposed on cells by the body when it senses that they are doing more harm than good.

Whether these latter cell losses are a part of "aging" is debatable, but fortunately it's not an issue that we need to resolve in order to get moving on the restoration of old and dysfunctional bodies to the full health and functionality of youth. Replacing these missing cells will play an essential role in anti-aging biomedicine, no matter what the causes of their attrition or their relationship to "aging" in the abstract. Progressive cell loss represents a change away from the healthy ideal of youth, and therefore an anti-aging engineer should work toward fixing it, just as any engineer will work to restore machinery back to the state in which it functions best.

Throwing Away the Key to the Medicine Chest

Adult humans have adult stem cells, not embryonic ones: again, true ESCs only exist in blastocysts. Thus, getting a supply of ESCs for use as cellular medicine involves somehow deriving such cells from early-stage embryos. Fortunately, there is a quite generous—and heretofore almost untapped—supply of such cells that is already being produced by an existing industry: *in vitro* fertilization (IVF) in fertility clinics.

The chances of any given IVF embryo being successfully implanted and then carried to term as a result of the procedure are still relatively low, so fertility clinics routinely create several embryos from the sperm and eggs supplied by either would-be parents or their donors. That way, they have a supply of embryos available for multiple attempts, without requiring women to undergo multiple rounds of the expensive, very unpleasant, and modestly dangerous hormonal treatments required to extract eggs from them. Typically eight such embryos are left over after every round of IVF, with the result that there were 400,000 surplus embryos frozen in storage in American fertility clinics alone as of 2002. At least 16,000 of these are *unclaimed* by any donor, an additional 45,000 have a similarly murky status,[29] and nearly none of the others will ever actually be used in fertility procedures. These embryos are ultimately discarded, or become sufficiently decayed that they cease to have any potential to form a baby.

This is what makes the debate around the use of embryos from fertility clinics such a frustration to doctors and scientists. These embryos are slated for destruction *no matter what we do with them:* there is *no* chance that the vast majority of them will ever be implanted in a womb and undergo the additional development needed to make a baby. The opponents of ESC research and therapy have proposed preventing their disposal by implantation into volunteers who would carry them to term for adoption, but even in that scenario there is no realistic prospect that even one percent of such embryos would be diverted from the rubbish tip. Once created, the fate of those blastocysts that are not actually implanted into a woman is sealed; the only question is whether scientists will be allowed to use their cells for research and as cures.

Actually, the insertion of these cells into the midst of the abortion debate is even more artificial than this makes it sound. Blastocysts are so primitive a stage in embryonic development that they have not yet made the biochemical "decision" to become a distinct human being. This is part of why they have the full flexibility to become any type of cell in the human body—and also why the confusion of stem cell technology with the abortion debate is so ethically misguided. At this early stage, for instance, an embryo could still divide into two separate cell populations, *each* of which can go on to become a *separate, unique person.* Indeed, this is exactly what happens when identical twins are formed. Since this ball of cells can still go on to become either one, or two, or even more *different people,* clearly the unified cell mass that precedes this separation does not embody the identity, the essence, or the soul of any single, personal human being. And while we can stand in justified awe of the *potential* for life (or lives) locked up in these cells, that should not cloud our ethical vision into thinking of this potential as *morally* being even in the same ballpark as the *actual* lives of patients that need its cells for medicine, when it is closest to that of skin cells in a petri dish.

The Nicodemus Solution

Powerful though embryonic stem cells derived from embryos left over from IVF may be, however, they do have one *potential* disadvantage hanging over their medical use. Cells derived from such embryos will, by definition, be immunologically alien to the patient's own cells, making them a target for attack by the immune system. Thus, the same kinds of problems that currently plague conventional organ transplantation—the horrors of rejection, graft-versus-host disease, and the dangers of living with an immune system turned off artificially with drugs to preserve the transplant—might possibly be an issue in embryonic stem cell transplants, too.

So far, the evidence suggests that we will be able to manage this issue with little hassle in many cases. Much of our confidence on this front derives from recent experience in actually using ESCs in experimental treatments for various diseases. Most such studies have just *assumed* that rejection would be a problem, and have preemptively taken steps to prevent it, either by using animals with defective immune systems, or by administering immunosuppressive drugs. But more recently, some studies have been performed using ESCs *without* taking such steps, and the results suggest that there may have been nothing to worry about in at least some cases. In the sheep heart-attack study I mentioned earlier and in several rodent studies,[30,31] ESCs taken even from another *species* have incorporated themselves into the "patient's" native tissues and provided substantial regenerative benefits, with no rejection issues.

Such results may mean that ESCs' state of development is *so* early and tentative that they may not even distinguish themselves with enough antigens to create a problem across the species barrier—let alone the barrier between individual humans. Additionally, it now appears that ESCs produce their own, very localized immunosuppressive signalling molecules that selectively protect them from immune attack, and even trigger any attacking killer T cells to undergo self-destruction (apoptosis).[32] Because these mechanisms involve either direct cell-to-cell contact or factors secreted and used very close to the stem cells themselves, this local immune shielding system is free from the systemwide side effects of taking immunosuppressive drugs.

Moreover, in some specific applications the risk of rejection will be low to begin with, because the tissues where we'll be delivering the cells are substantially shielded from the immune system. A lot of the nervous system, for instance, is largely inaccessible to immune attack (which is how the virus that causes shingles can hide out there for years after being purged from the rest of the body).

We can also lower the risk of rejection by providing patients with ESCs from isolates ("lines") that are a match for all of the major antigens involved, which we could readily do in many cases if we are allowed to pick and choose our stem cell lines from among the embryos currently slated for destruction. It has been calculated that a bank of just 150 donor embryos randomly selected from the existing stockpile could do this perfectly for one patient in five, provide a probably usable match for almost two in five, and allow for a long-shot match for almost 85 percent of potential patients—and if we were able to choose specific immunological combinations out of the surplus instead of choosing embryos at random, just *ten* such donations could give grade-A matches for nearly 40 percent of patients and good matches for over two-thirds.[33]

But we can't yet rule out the possibility that rejection may present a barrier to our effective use of ESCs in human medicine for aging and disease. In that case, the good news is that technology exists that *already* allows us to generate embryonic stem cells that are a *perfect immunological match* for animals as complex as cattle and monkeys, and several scientific teams say they're on the verge of being able to do the same thing for humans. I've already mentioned it: *somatic cell nuclear transfer* (SCNT). In SCNT, doctors begin by taking a mature cell from the patient's body (a "somatic cell"), by for example swabbing the inside of the cheek, and then *turn back its clock,* releasing it from the strictures of a mature, differentiated complexity and transforming it into a patient-specific embryonic stem cell.

This biological miracle is accomplished by a technique that is incredibly simple. The metamorphosis occurs in an egg cell, provided by a donor. This cell's nucleus is removed to make way for the one from the patient's cell. With a biochemical boost or a zap of electricity, the two become one, and the egg begins dividing just as it would if it had been fertilized, kick-starting the production of embryonic stem cells created from a patient's own genetic instructions, creating a perfect immunological match (see **Figure 2**). The cells can then be used for medicine just as any ESC would be, but with absolutely no fear of rejection.

Actually, you may well already have heard of this advanced biomedical research under a name more popular with the media: *therapeutic cloning.* While this term is perfectly scientifically accurate, it has generated an enormous amount of confusion about the nature and purpose of SCNT, splashing political napalm onto the heated fires burning in legislatures and online chat rooms surrounding stem cells. Let me try to extinguish those flames.

To a scientist, the word "clone" means simply a set of genes, cells, or organisms that are identical to one another at the DNA level because they are derived from a single ancestor. We've used the word in this strict scientific sense in the "*clonal* expansion" of T cells, and the "mono*clonal* antibodies"

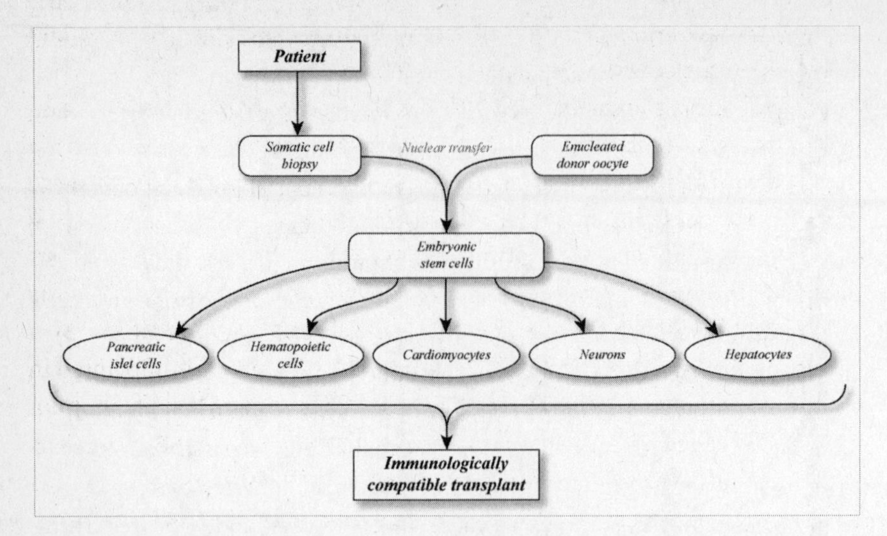

Figure 2. How SCNT ("therapeutic cloning") works.

that are currently used to treat some cancers and will probably be used as part of our panel of engineering solutions to aging. Similar uses of the word occur when scientists speak of a "clone" of common bacteria bearing a gene that turns them into tiny biological factories for the production of insulin for diabetics, or even when gardeners talk about a "clone" of strawberry plants.

But say "clones" to even highly educated people who don't work in a few disciplines of biology and biomedicine, and you evoke images of a sea of indistinguishable, zombielike drones, enslaved to technocrats or created for other sinister purposes. That this confusion is corrupting the debate about this potentially essential life-saving technique can be seen starkly in a speech delivered to the Canadian Parliament on February 27, 2003, during debates surrounding Canadian legislation to regulate stem cell research. Mr. James Lunney, a Conservative party member of Parliament for the Nanaimo-Alberni riding on Vancouver Island, began by saying that "[I]f we took one of [the speaker of Parliament's] cells, extracted the nucleus and put it into an ovum, one could stimulate it electrically and allow it to grow." So far, so good. But then Mr. Lunney rocketed off into a grotesque but all-too-common flight of misunderstanding: "The so-called therapeutic clone would be to take the immature model of Mr. Speaker and extract an organ, if he needed one, killing the clone in the process. That is so-called somatic nuclear cell transfer or therapeutic cloning." Similar outrageous confusions have been perpetrated on the floors of the U.S. Congress and elsewhere in the course of the stem cell debate.

SCNT doesn't involve making clones of people at all. It involves making blastocysts—balls of cells that, as we've seen, have not yet even made the necessary steps to decide whether they will become one, two, or more people. True, these blastocysts *could* in principle be used to make babies if they were implanted into a woman the same way as is done with blastocysts produced through IVF, but this is a *potential,* not a fact. When blastocysts are created by SCNT for therapeutic purposes, no egg is fertilized by a sperm; no new, unique DNA identity is created; no embryo is implanted in an uterus; no pregnancy results. Biomedical SCNT creates cell life, but not human life: renewed cells, not new people. They certainly have no organs that we could harvest—including, importantly, no brain, nor even the beginnings of nerve cells. We no more "kill" a blastocyst produced by SCNT when we derive stem cells from them than we "kill" a vat full of replicating skin cells when we throw it out at the end of an experiment. Fundamentally, SCNT would be the basis for therapies that cure you with *your own cells,* restored to the potential they had in their first moments of existence by the power of the stimulated human egg.

Because they derive from the patient's own DNA, SCNT cells are an exact genetic match to those in your own body, and are treated as "self" by your immune system.[34] Whatever may emerge from further research with ESCs derived from surplus embryos left over from IVF, SCNT cells offer a virtual guarantee of freedom from the specter of rejection, graft-versus-host disease, and a lifetime spent on toxic immunosuppressive drugs.

In preliminary, preclinical research, the new regenerative powers of cells derived from SCNT have already shown their promise. In animal models, SCNT medicine has already been used to cure many of the devastating conditions for which human treatments must still be found, such as Parkinson's disease,[35] heart attack damage,[36] and the animal equivalent of the "bubble baby" syndrome (SCID)—rescuing not mere weanlings, but fully developed, adult organisms that had suffered with the disease for their entire lives.[37] As we've seen, the ESCs taken from more conventionally generated blastocysts have worked some of the same marvels—but some of these studies suggest that, even where rejection doesn't happen, SCNT may still provide some advantages. And indeed, the results tend to *downplay* the therapeutic potential of SCNT, because in these studies the scientists have not actually derived the cells from each animal individually so as to provide a perfect match (as we would do for human patients), but have used one line of cells to treat an entire colony of close cousins.

In the Parkinson's study, for instance, the researchers coaxed SCNT-derived cells to produce neurons suited for use in several areas of the central nervous system (forebrain, midbrain, hindbrain, and spinal cord) and responsible for a broad range of functions. Some of them were the kind that produce the neurotransmitter *dopamine,* which is involved in fine motor control; as I noted, it's the loss of these cells that causes Parkinson's disease. Others were cells whose central functions involve another neurotransmitter called *choline,* and whose loss is characteristic of Alzheimer's disease. They were also able to derive cells that secrete other neurotransmitters in the brain (*serotonin* and *GABA*); that carry movement-control signals from the spinal cord to the muscles (and whose degeneration is central to motor neuron diseases); and that act as "support" cells to neurons proper, nourishing and protecting them. This was a much wider range of mature cells than had been successfully derived from previous protocols using conventional ESCs.

The team then put these cells to the test in the Parkinsonian rodents (whose dopamine-producing cells had been knocked down to less than a third of their healthy numbers by a toxin), comparing their effects to ESCs derived by the conventional route. Dopamine-producing neurons derived from either protocol formed solid, stable grafts, and improved behavior in their recipients, and there was no sign of rejection of either type of ESC. But even though the SCNT-derived cells came from recipient animals' cousins rather than their own, individual bodies, these cells performed better than conventional ESC cells, forming larger grafts in their brains, with *double* the number of transplanted nerve cells surviving eight weeks after transplantation, and the final graft sites containing about 50 percent more dopamine-producing neurons.

And it appeared that they might have been somewhat more effective at restoring function, too. Because all the damage had been inflicted on one side of the brain, chemically stimulating the remaining dopamine-producing cells in the brain caused an imbalance in their motion, with the larger number of intact cells on one side of the brain sending out stronger signals to the legs they control than the more damaged side does. The result was that the animals began to veer to one side, rather like what happens when you push a shopping cart that has a damaged wheel on one side of it. This "rotational behavior" is a key test of the function of the damaged part of the brain. Treating these animals with dopamine-producing cells derived either from conventional ESCs or ESCs created using SCNT reduced this aberrant motion by more than 70 percent, with a hint that the SCNT-derived cells were more effective (see **Figure 3**).

Because the range of neurons and supporting cells produced using these protocols was so broad, the researchers who performed the study believe that their technique could also be used to treat multiple sclerosis and other "demyelinating" disorders (in which the myelin sheath essential to the correct function of neurons is damaged or destroyed), Huntington's

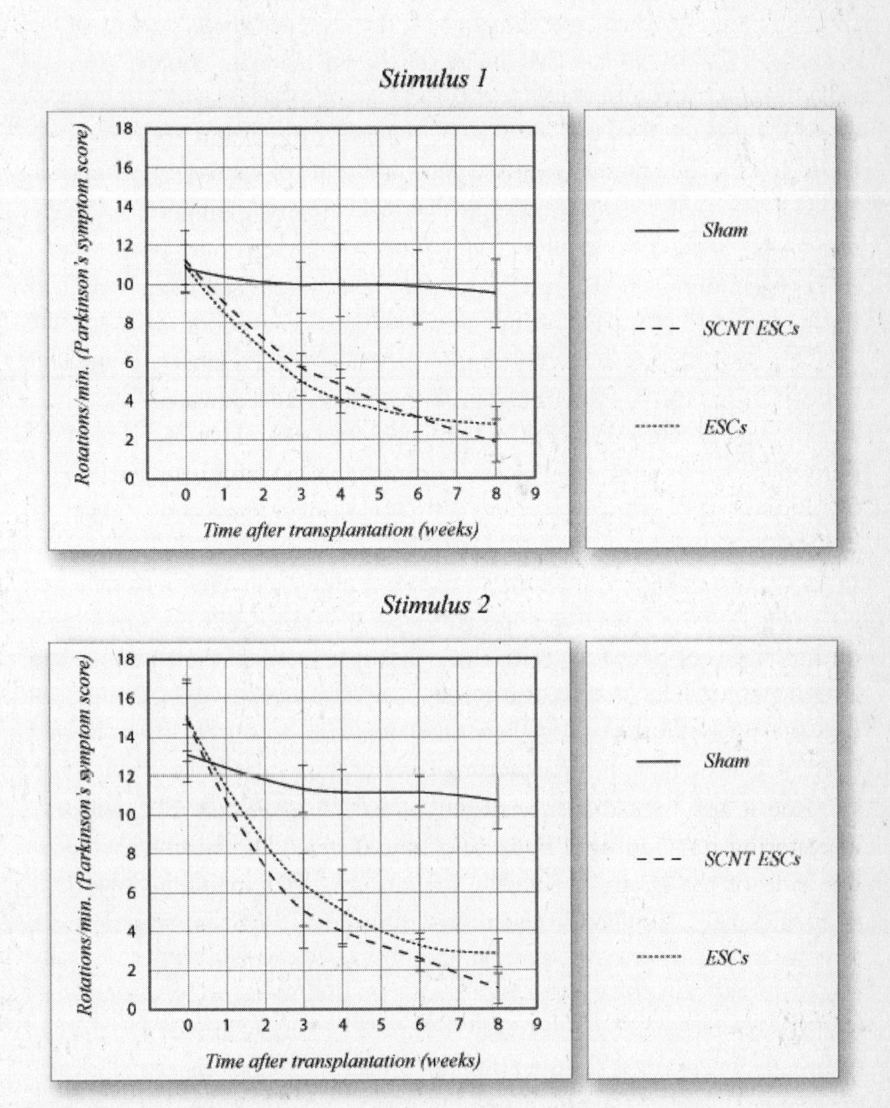

Figure 3. Embryonic stem cells, and especially SCNT-derived cells, restore normal motion in a Parkinson's disease model.

disease, amyotrophic lateral sclerosis (Lou Gehrig's disease), and other motor neuron diseases.

There remain technical hurdles to overcome in developing SCNT, but theoretical objections to their ultimate use in human medicine continue to fall. There have been concerns about the mitochondria in these cells, for one thing: SCNT is achieved by replacing the DNA in an egg cell with a patient's DNA, but this still leaves the egg's own energy factories providing the juice to keep things going. Many researchers have therefore been concerned that the resulting cells would be dysfunctional due to a mismatch between these mitochondria (created originally from the *egg donor*'s nuclear and mitochondrial DNA) and the final cell, or that the patient's body might reject the cells based just on the immune-sensitive parts of the transplanted mitochondria. So far, however, it doesn't seem likely that this mitochondrial mismatch is going to trouble us. Aside from the fact that these cells have been incorporated successfully into the patient-animals' bodies without any signs of rejection in studies so far, a very careful study that looked all the way down at specific proteins that are used to monitor for mitochondrial "foreignness" found that the cells were accepted as completely native by all available measures.

Similarly, there are concerns that the "epigenetics" of SCNT-derived cells—the "scaffolding" around the genes in the DNA code, which regulates the expression of those genes—would be abnormal, leading to cancer or dysfunction. Again, however, while this has been a problem with using embryos created using the nuclear transfer technique for making cloned *animals,* it has not yet appeared to prevent *stem cells* derived from SCNT from functioning properly when transplanted as a treatment into animals. And indeed, early in 2006, scientists from the Whitehead Institute for Biomedical Research in Cambridge, Massachusetts, reported that stem cells derived from SCNT have *identical* patterns of being transcribed and turned into proteins as do ESCs created by conventional IVF-type fertilization, with any differences attributable to the genetic differences among the animals donating the cells rather than the kind of cell involved.[38] Moreover, it appears that the very process of generating stem cells from SCNT blastocysts of necessity imposes a kind of "survival of the fittest" of its own, with any epigenetically inappropriate cells collapsing under their own dysfunctionality; this may explain a large part of why it's been so difficult to get a high yield of such cells from a given blastocyst. This might pose problems for anyone actually looking to use nuclear transfer to clone a person (a point that should itself relieve those concerned about such a use of the technique), but it appears that epigenetic problems will only make the use of SCNT for medicine more difficult, not prevent its safe and effective use.

Another open question is where we'll get all the eggs we'll need to use SCNT widely as an anti-aging therapy. The supply is limited by the number of women prepared to donate their eggs, and the hormone treatments and moderately invasive surgery needed are likely to continue to keep the numbers of such women down for some time. Many people also raise ethical concerns about offering money or other inducements to solicit more donors, especially for a procedure which is not without risk.

Even this, however, may yet be overcome on a technical basis instead of a sociopolitical one. One option would be to take the eggs from other species. Such an approach would not be without *technical* hurdles: notably, the presence of mitochondria in such eggs that are not just from a different *person,* but from a different *species,* might make the cells unable to create and sustain an energy supply. With my background in mitochondrial biology, I recently proposed a solution to this problem should it arise.[39] My solution results in the mitochondria of the eventual ESCs being derived from the eventual recipient of the cells, just like the nucleus, so it also avoids the "intra-species" mitochondrial problem I mentioned above.

Another alternative may be to mass-produce egg cells bioengineered from more common cell types, such as skin. Canadian researchers recently reported[40] having used skin cells from fetal pigs to produce cells which look—based on gene expression patterns, cellular structure, and some functional abilities—an awful lot like egg cells. Whether these cells have the full range of functions of egg cells remains to be seen, but they—or a more developed version of them—might have the same power to reset the clock in mature somatic cells that conventional eggs do. This would mean that we can bioengineer an almost unlimited source of eggs: human fetal skin tissues, which contain nineteen billion such cells per square inch. Such huge numbers would allow us to avoid entrapment in the battlefields of the culture wars, if we can simply reach agreement on the use of tissues from stillbirths rather than aborted babies.

Frozen Embryos, Frozen Science

This brings me back to my second SENS scientific conference. At the time, you could still smell the ozone in the air from the second in a pair of scientific lightning strikes from a previously obscure group of Korean scientists from Seoul National University, headed by veterinarian Hwang Woo-Suk. A few years after Dolly the sheep, Hwang had claimed to have cloned a cow, and more recently a dog, but his fame came when he announced in the winter of 2004 that he had achieved the world's first derivation of fully fledged human ESCs using SCNT. This proclamation rocketed him to international fame, but it was just the beginning: a little more than a year later, in the months leading up to SENS2, he reported a dramatic improvement on the technique. In his first report, Hwang had only been able to derive a single stem cell line from the 242 eggs that had been donated—and this line had been taken from an egg fused with DNA taken from the egg donor herself, which was of very limited biomedical use. Now, Hwang was saying that he had created eleven human lines using only 185 eggs, and using DNA taken from completely different people, including potential patients of both genders and many age groups.

Everyone in the field, as well as the popular press, hailed this result as a phenomenal breakthrough, and I was far from alone in seeing its potential for treating not only age-related disease, but aging damage more broadly. I knew that I would want someone to present not only Hwang's results—with which most attendees would be at least peripherally familiar, due to the enormous press coverage—but what they would mean for scientists working in the field.

Hwang's result was clearly going to have an enormous galvanizing effect on stem cell research. Because of the political climate in which it had taken place, the impact of the announcement was far greater than could be accounted for by the purely technical breakthrough (great as it might have been) of actually being able to make customized stem cells for healing patients. Stem cell research had been stymied for years by President George W. Bush's notorious decision, in the summer of 2001, to limit federal government funding of stem cell research to work done using lines created prior to the morning when he announced the policy.

That decision reversed a policy accepted under the Clinton administration, but not yet implemented, that would have plowed NIH funds into ESC research using lines derived either from IVF clinics or from work originally performed with private funds. It came not out of science but from the political maelstrom of the abortion debate, and the antiabortion stance held by President Bush and by his key constituency of the Christian Right. And while it is not accurate to call that executive fiat a "ban" on ESC research, it created an enormous chill over the entire field—and not just because of the direct effects of cutting off funding for research performed on nearly all available ESC lines.

The most obvious problem was the stranglehold it put on direct U.S. federal government funding for embryonic stem cell work. The administration holds the purse strings of a remarkably large share of U.S., and even global, basic research in science, with US$20 billion in research-related funding coming out of the National Institutes of Health alone each year. Bush and his political allies would argue that their policy provides scientists with plenty of opportunity to work with stem cells because of the availability of the approved lines, but that claim ignores the actual state of the lines in question.

The White House originally announced that their policy would allow scientists to work with as many as seventy-eight robust stem cell lines, but when senators put the question to NIH Director Elias Zerhouni, he admitted that only nineteen of these lines were actually viable, available in practice (as opposed to locked up by intellectual property restrictions and similar constraints), and ready for use in stem cell work. By 2004, this number was still no higher than twenty-one. In preliminary research presented at the National Academy of Science on October 12, 2004, fourteen of the lines tested by Carol Ware at the University of Washington were found no longer to grow well and to be hard to separate because of the outmoded way in which ESC lines were derived and cultured at the time. One such line was actually withdrawn from scientific use because of this finding.[41]

A supply of just twenty-odd lines also fails to represent the genetic diversity of humanity well, so it's hard to verify whether a given finding is a quirk of that line or, say, of people of a given race. A supply of hundreds of viable ESC lines is the likely minimum for healthy progress in this field of science. Indeed, the present situation is worse than this: because of their age, these lines are accumulating mutations that could skew the results of research performed with them because they no longer even represent the stem cells of those original donors. It also means that we can't study the stem cells of people with particular diseases, or how experimental drugs affect those processes—studies that would best be done with SCNT, which would let us take the cells of people already known to suffer with a given disease and wind back their clocks to the first moment of their primitive existence as blastocysts. Researchers could then watch the cells as they underwent differentiation into the cells most affected by the disease and then the late-phase changes that happened as their abnormal metabolism intersected with aging processes that happen even in healthy people's cells.

And not only are the cells of very limited use for basic research: everyone working in the field recognises that these cells will never be usable for actual therapies either. All the lines that are both approved and available are useless for clinical purposes because they were originally cultured using *feeder cells* taken from mice—supporting cells needed to secrete factors and provide structural support that is essential to keeping them in their primal, unspecialized state. Contact with these cells has tainted them in various ways: one study[42] found that their cell surfaces contained a sugar that the body's immune system recognises as foreign and attacks, and it's widely expected (though not yet proven) that they may also contain mouse cell proteins and even viruses.

It's only been in the last couple of years that scientists have developed new techniques that first allowed for the propagation of human ESC lines from *human* feeder cells, and most recently ESCs generated using no feeder cells at all.[43] And, again, it's at least possible that nothing short of custom-made ESCs produced specially for the patient with SCNT will fully address the potential problem of rejection. Thus, only ESC lines derived well after the 2001 line-in-the-sand can actually be used as medicine for disease and for the full repair of aging damage in the future.

The policy also ripples out well beyond the labs of people actually working with the approved lines. For one thing, the restriction on providing money to unapproved stem-cell work is so aggressively enforced that the NIH has to snatch away grant money awarded even for research *completely unrelated* to ESCs, if any of that money goes into facilities or equipment also used for work on "banned" ESC lines. You could be testing a cancer drug on rats and have your funding pulled if someone else on campus were sharing the use of gene-expression array equipment with you and were using it for ESC work using lines not approved under the August 2001 decision. That makes it enormously difficult for anyone to carry out work on ESCs other than the sanctioned lines on most university campuses, or in essentially all government research centers. Laboratories in which sanctioned work does take place must expensively duplicate and track equipment, all the way down to sticking colored dots on petri dishes and other glassware, and must generally work in ways that degrade effectiveness.

The policy also erects enormous roadblocks even to privately funded research—research that, in theory, is not the target of the Bush policy. Scientists in industry are first trained in universities, and when those universities can't carry out ESC work, because of a mixture of lack of federal dollars and the handcuffing of non-ESC work carried out using shared equipment with work on "forbidden" lines, young researchers don't get trained in the techniques of working with stem cells, let alone get the opportunity to perform original research that would advance the field. This, of course, means that such researchers aren't available for hire by private firms even if they had the money to bring them on.

And naturally, the political uncertainty swarming around stem cell work makes potential investors reluctant to pour money into companies focused on developing ESC-based cures. At the time, it seemed possible that the United States would follow other countries in making aspects of this research—such as SCNT—not merely ineligible for government funding but actually a criminal act. Investors will stomach most risks, but not political risk, and so they abandoned private-funded ESC research for a number of years.

It briefly seemed likely that the Bush administration's policy would be quickly brought down by political pressure. Even a very conservative selection of polls shows that the majority of people in the United States and elsewhere are in favor of a fairly open policy on scientific access to ESCs. Even in surveys in which the question is posed flat-out with no mention of potential human benefits, the majority of people say that they support deriving stem cells from surplus embryos from fertility clinics for scientific research.[44] When the question mentions the potential for human treatments, this proportion climbs into the 70 percent range. And most people even support SCNT research—a fact that fills me with optimism about their future acceptance of other anti-aging therapies.

So, in the August heat, with the controversy raging and President Bush's popularity resting on unsteady ground, it seemed possible that public opinion would mobilize against the restrictive policy, and scientists would within reasonably short order be enabled to work with ESCs from a wide range of sources.

Then, the planes hit the World Trade Center.

In a month, everything had changed. Where ESC research had been front-and-center on the national stage in August, it was off the radar screen for almost everyone in late 2001, replaced by the immediate fear of terrorism. As pressure and scrutiny from the wider public melted away, those whose organization, resources, and ideological investment were strong enough to continue to push their agenda on the subject even in the shadow of the ruins of the Twin Towers suddenly became the only voices pushing legislators—and in this case, because of the conflation of the science with the abortion debate, that meant almost entirely forces *opposed* to ESC research. Anti-abortion groups, who are well organized and well funded, made their case to Washington as strongly as ever, without the usual balancing force of either the public at large or of their usual opponents: prochoice and civil liberties groups had no particular stake in stem cell science, and the latter had their hands full with presenting the case for preserving Constitutional rights in the face of the threat of terrorism. And while patient advocacy groups might have stood in opposition to blockades on research, such groups were nascent at the time, and lacked the support from pharmaceutical companies that often sustains them, since in this case the companies had no vested interest in promoting the groups' cause.

Feeling deferential to a newly popular wartime president, handed plenty of unbalanced misinformation by the anti-abortion activists on the religious right that made up that president's base of support (and had played a significant role in their own sweep into power), and with a leadership dominated by representatives with a social-conservative worldview of their own, the Republican-dominated Congress substantially raised the threat against stem cell science. Parallel bills introduced by Sam Brownback in the Senate (S 245) and Dave Weldon and Bart Stupak in the House (HR 234) sought to ban all forms of "human cloning"—including SCNT performed entirely for scientific or medical purposes.

These measures would not only deny federal funding to, but *criminalize,* the creation of blastocysts using SCNT—imposing actual jail sentences on scientists performing the work. They would also have imprisoned scientists doing any scientific research using ESC lines derived from SCNT; in the original texts of these bills, this went so far as to threaten both doctors and patients with prison time if they administered or accepted cures using SCNT-derived stem cells. Some language even suggested that people who went abroad to receive treatment with SCNT stem cells could be penalized for it on their return to the United States.

But in the coming months, as the public slowly began to raise their heads out of their foxholes, proresearch and patient activist forces began to countermobilize. They were greatly helped by the voices of prominent patients suffering with diseases likely to benefit from SCNT and those close to them, including Michael J. Fox (Parkinson's), Kevin Kline (whose son has juvenile diabetes), Christopher Reeve (spinal cord injury), and, most powerfully, Nancy Reagan (whose husband, the former president, died of Alzheimer's disease). A bipartisan coalition favoring expanded embryonic stem cell access—and in many cases the full legalisation of biomedical

SCNT—began to form, including such prominent anti-abortion Republicans as Orrin Hatch, Strom Thurmond, Arlen Specter, John McCain, and ultimately then senate majority leader Bill Frist, and apparently extending even to Bush's own secretary for Health and Human Services, Tommy Thompson. Meanwhile, prominent scientific organizations (including the National Academy of Sciences, the American Medical Association, the Association of American Medical Colleges, and even the National Institutes of Health itself), as well as multiple disease-specific charities (such as the Juvenile Diabetes Research Foundation, the American Association for Cancer Research, the Lance Armstrong Foundation, and the American Diabetes Association) endorsed research using new ESC lines and the advancement of work on SCNT.

Hatch's coalition introduced legislation to legalise SCNT for scientific and medical research, while banning use of the technique as a means of cloning *people.* They also introduced legislation to allow access to surplus embryos from fertility clinics as a source of stem cells. Slowly, more and more legislators from both sides of the aisle signed on to the pro-research side of the debate. For the next several years, the two forces fought to a standstill, with both bills repeatedly introduced and defeated. This created a legal and scientific limbo that ultimately served the anti-research camp's agenda: the few scientists working on SCNT remained out of prison but without access to funding, potential private investors continued to wait out the political uncertainty, and the President's restrictions on ESC research remained in place.

False Dawn

Then suddenly, in 2005, came Hwang's announcement of relatively high-yielding techniques for creating individually tailored ESCs. The news acted like a juggernaut, smashing through barriers both scientific and political. Technically, the ability to make viable, customized embryonic stem cells tailored to individual patients was a massive breakthrough. Politically, it not only reenergized pro-research forces, but applied a new source of pressure on politicians. Stem cell advocates had long argued that, if the government continued to keep a tight lid on ESC research, the science would be done elsewhere: the United States would simply suffer a brain drain, as American scientists moved to more hospitable climes to pursue their vital work and as foreign graduate students (already chafing under new security restrictions) refused offers from American universities. Now, the prophesy began to come true. The Korean government was ready to back their new scientific star's work with significant resources; countries as far-flung as the United Kingdom, Israel, Sweden, and Singapore began establishing themselves as well-funded hubs for ESC research; and reports of prominent scientists packing their bags started appearing in the media.

The forces of competition began to work their usual magic. Individual U.S. states, fearful of being left behind, began putting up bills to fund stem cell research within their own borders. Federal politicians not strongly ideologically committed against ESCs—including many free-market-oriented Republicans—became increasingly willing to challenge the agenda of the antiresearch ideologues. A couple of years earlier, fifty-eight senators—most of them Democrats, but with substantial support from key Republicans—had signed a letter asking Bush to rescind his policy; a little over a month after Hwang's announcement, 206 House members followed suit.

I knew that highlighting these advances, and the opening that they afforded to researchers, would be a great way for me to further my conference's mission to promote anti-aging biomedical research. Short of Hwang himself, the best person to present the opportunities was Gerald Schatten, a stem cell researcher at the University of Pittsburgh who had been working with Hwang for the last two years, had used his veterinary techniques to clone a monkey, and had signed off on the paper announcing the new SCNT lines in *Science*. I asked him to present their results and outline the royal road to accessing patient-specific ESCs through Hwang's team at Seoul National University: a "World Stem Cell Hub" that would generate SCNT cells to order using Hwang's established facilities and experienced technicians.

I was delighted when Schatten accepted my invitation—but I was positively overjoyed when, not long after, he came back with another e-mail saying that he'd like to bring along a friend. Hwang himself had expressed an interest in presenting at SENS2, he said; he realized that it was short notice, but would I be willing to let him share Schatten's own half-hour slot in the conference? I, of course, offered instead to give Hwang his own half-hour talk as a featured presenter in the session on stem cells and regenerative medicine. I'd have been happy to do this even if it had required throwing out my original schedule and starting from scratch, begging forgiveness from presenters as I shuffled them around at so late a stage in the planning of a very packed conference schedule; but fortunately I didn't have to do this, as another presenter had recently been forced to pull out. With hardly any shuffling, Hwang was confirmed.

So it was that, with great pleasure on my part and keen attention from hundreds of my colleagues, Hwang mounted the lecture podium of Cambridge's Fitzpatrick Lecture Hall.

Of course, as you know full well, unless you spent much of the winter of 2005–2006 in a cave in Nepal, it was all a sham. Within months of electrifying my scientific audience in September, Hwang had been exposed as a fraud.

Bad Wizards *and* Bad Men

First there were ethical questions about the sources of Hwang's eggs; then, questions about the viability of four of the eleven stem cell lines that he had submitted to *Science*. And then, reporters looking at photographs presented with Hwang's data began to notice some suspicious resemblances between allegedly unique stem cell lines. Hwang brushed these off as the result of a confusion with the *Science* production staff about which of the numerous photos that he had submitted were to be used as figures in the article.

The case against Hwang quickly picked up momentum. Scientists reviewing the paper's data noticed suspect similarities in the genetic profiles of the various lines' cells. Then Schatten requested that his name be retrospectively removed from the paper's authorship because of "allegations from someone involved with the experiments that certain elements of the report may be fabricated." And on December 15, one such collaborator came forward with the flat statement that *nine* of Hwang's eleven lines were flat-out fakes, sharing *identical* DNA with one another, and claiming that Hwang himself had admitted the fraud.

As each doubt was raised, Hwang would protest his innocence, variously blaming errors, contamination, and the incompetence of others for each of them—even going so far as to claim that one former collaborator had "switched" some of his lines. But, eight days after his former collaborator claimed fraud, he proffered his resignation to Seoul National University—which was refused, on the grounds that he was now the subject of an internal investigation. He was suspended in February, dismissed in March, and indicted in May for fraud, embezzlement, and violation of bioethics legislation.

The fallout of these revelations was felt at many levels. There was, of course, enormous outrage at the fraud, and great disappointment that Hwang's breakthrough turned out to be flimflam. And it was a political fiasco, exploited by anti-ESC campaigners to cast aspersions on the entire field.

But Hwang's fraud had also set the progress of the entire field back by at least a year—an eternity in science. The Bush restrictions on ESC research, amplified by the looming threat of criminalization for SCNT from the American Congress, had kept all but a few research teams from working to perfect nuclear transfer for human patients. Hwang's claims had further diminished the incentive to put resources into the goal: No one wanted to reinvent the wheel already spinning in Korea, and private firms would no longer have a competitive edge as the first creators of patient-tailored ESCs, using in-house methods that could be kept exclusive as trade secrets or through the patent process.

One private firm, Advanced Cell Technology (ACT), had been courageously soldiering on with the work, producing a great deal of quality ESC and SCNT science (much of it, admittedly, overplayed in the media)—this despite constantly lurching from one financial crisis to the next because of investors skittish over the legal climate of their research. In late 2001, they famously announced the first "cloned" human blastocyst,[45] although the DNA that was transferred into the egg cell came from the donor's own body—in fact, from cells that normally enshroud the egg itself—and the resulting blastocysts couldn't develop beyond a six-cell ball. They spent much of the next two years perfecting that technique, issuing many publications (most of them on research done in cattle) charting their progress toward teasing out the reasons for the low yield of viable blastocysts in SCNT techniques, and working steadily to perfect the technique for human biomedical use.

In late 2003, insists ACT scientific director Robert Lanza, they were very close to resolving the sticking points in their technique, and could shortly have generated the first viable stem cells tailor-made for patients or for research on specific diseases. With Hwang's eleven-cell-line announcement, however, investors began pulling their bets out of what was perceived to be an "also-ran" in the race—a blow turned into a haymaker by ACT's simultaneous setback of losing their main source of human eggs. Cells were put into deep-freeze, and ACT's human SCNT work shut down.

Equally infuriating is the case of Professor Alison Murdoch and Dr. Miodrag Stojkovic, of the Newcastle Centre for Life, a fertility clinic and research center in Newcastle-upon-Tyne in Britain. These researchers managed to create the first human SCNT blastocysts using DNA taken from the cells of a person *other than* the egg donor.[46] Like the ACT cells, these blastocysts were not fully viable, but in the United Kingdom's more research-friendly environment, the group had received formal approval from regulatory bodies to pursue work using SCNT-derived stem cells, giving them the green light to perfect their technique. But their publication came on the heels of Hwang's, and compared to his explosive success, the creation of only three fused cells that actually began dividing, and no actual stem cell lines, seemed to be of no relevance to the progress of the field. They, too, promptly shut down their research on the technique—a decision that Murdoch says cost them at least a year.

It was the same story elsewhere. SCNT research teams in Sweden, and also at three American universities that had raised enough private or state money to set up stem-cell research centers with elaborate financial firewalls separating them from federally funded research elsewhere on the same campuses, either dropped their efforts altogether or put them on hold while waiting to see whether the Korean team's plans would make their efforts redundant.

But if Hwang's shot, heard 'round the world, turned out to be a backfire at best, it still served to wake up a lot of people. All over the globe, and especially in the United States, researchers began thinking seriously again about what they could do once they had access to the Korean team's expertise. It promised stem cells with the miraculous flexibility of the blastocyst, but perfectly matched to patients suffering with the worst of the nightmares of aging: Parkinson's disease, stroke, scarred and weakened hearts, eyes blinded by the death of light-sensing cells choked in their own waste, limbs withering away as the electrical sparks stopped flowing through nerve cells or muscle cells snapped one by one under the force of their own molecular decay. Labs began contemplating grant proposals. Young science students turned back to stem cell research as an exciting career prospect. The dawn was false—but its rays woke the slumbering forces of science all the same.

Today, after the collapse of Hwang's house of cards, and in defiance of the Bush administration's politically driven, morally misdirected obstinacy, the field is undergoing a renaissance. Murdoch and ACT have fired up their programs again. Teams are in hot pursuit of successful SCNT techniques, and the research and cures that they will enable, all over the world. Cutting-edge work is occurring at the Center for Regenerative Medicine at Edinburgh University, Scotland (taking over clinical research from the veterinary work that created Dolly the sheep); at the Karolinska Institute in Sweden; at Shanghai Second Medical University, China; and at several privately funded centers in the United States, including the Harvard Stem Cell Institute, the University of California at San Francisco, and UCLA's Institute for Stem Cell Biology and Medicine.

The legal climate is shifting, too. In addition to China, Great Britain, and Sweden, SCNT is already explicitly legal in Singapore, Belgium, Japan, Spain, and Israel. And it remains legal in the United States, despite the efforts of Senator Brownback and his allies: the Brownback-Weldon bill has twice failed to make it through Congress, though it was reintroduced in 2005 as S 658/HR1357, the Human Cloning Prohibition Act of 2005. More excitingly, Orrin Hatch's bipartisan Stem Cell Research Enhancement Act, which would have opened up ESC research using lines derived from IVF surplus embryos, passed both the House and the Senate, and was only blocked from becoming law by a Presidential veto—the first of Mr. Bush's six-year administration. Hatch's pro-SCNT bill is also back in play, although no vote is imminent.

Meanwhile, individual U.S. states are moving ahead, doing their best to sidestep the funding and regulatory vacuums at the federal level. SCNT research has been legalized in California, Connecticut, New Jersey, Rhode Island, Illinois, and Massachusetts, and although several other states have also specifically prohibited all SCNT work, many more are permitting work done using surplus IVF blastocysts. A ballot initiative in Missouri that would constitutionally protect scientists' ability to conduct ESC and SCNT research, and patients' ability to access cures based on these techniques that are available anywhere else in the nation, narrowly passed in November 2006.

The states are also digging up the funds needed to get work going inside their own biotech sectors. The most famous of these is California's Proposition 71, a ballot initiative that established the California Institute for Regenerative Medicine and gave it a bond-funded budget of $3 billion to fund ESC work including (but not exclusive to) SCNT research. The actual disbursement of these funds has so far been bogged down by anti-research legal actions and some legitimate questions about oversight and ethics, but legal rulings have been almost uniformly favorable, and Governor Arnold Schwarzenegger has recently stepped in with a bridging loan, to start the Institute's mandated dollars flowing toward development of SCNT- and other ESC-based cures.

And if California is the most famous case, it is far from the only one—or even the first. That honor goes to New Jersey, which in early 2004 became the first state to earmark state funds for ESC research. Starting in December 2005, New Jersey has put out a total of $5 million in grants awarded to seventeen research institutions for research on stem cells from embryos and other sources, and has set up the $23-million New Jersey Stem Cell Institute. Similar initiatives are starting up in Connecticut, Illinois,

Maryland, Massachusetts (despite a failed veto attempt by the state's governor), and the state of Washington.

Meanwhile, efforts *have* been under way in the private sector, despite the lack of support, adverse regulation, and a climate of uncertainty that sends all but the steeliest of investors off in search of other opportunities. ACT is one prominent example; another is Geron Corporation, a biotech company most famous for its work on the "youth enzyme," telomerase (which it is now seeking to manipulate in order to shut down cancers by turning *off* the telomerase taps—see Chapter 12 for much more on this). Geron has perfected methods of raising human ESCs without feeder cells, and is testing six different lines in animals; excitingly, it expected to be ready to begin the earliest stages of human trials using neural stem cells for spinal cord injuries in the spring of 2007.

Some scientists are also seeking purely *technical* ways to liberate science from the phony moral dilemma surrounding the use of blastocysts for research into, and as cures for, human disease. There are multiple proposed ways to create ESCs from blastocysts without eliminating the potential for these cell balls to ultimately become human lives. One is *parthenogenesis,* a label taken from the technical term for "virgin birth." In this technique, the genes in an egg cell (which naturally contains only *half* of a full complement, as it is designed to be augmented by those in the sperm cell at fertilization) are doubled up on themselves, thereby generating a complete set of DNA instructions; this allows the egg cell to behave enough like a blastocyst to produce ESCs for the egg donor, without actually fertilizing the egg. Another approach is to take stem cells from IVF clinics' embryos that have defects preventing them from going forward to make a fetus, or to actually *induce* such defects into a patient's DNA before making a blastocyst from it using SCNT, in order to eliminate even its *potential* to form a human life. Recently, ACT introduced yet another option: using stem cells derived from a *single cell* plucked from the blastocyst, while leaving the rest of them in place as a potentially viable embryo (as is already sometimes done for genetic testing of IVF embryos before implantation).[47] Yet a fourth possible option is to coax adult stem cells into behaving more like embryonic stem cells, using growth factors and other chemical messengers, instead of using the inherent renewing power of the egg to do the same thing.[48]

I have no doubt that such work is valuable—but primarily because it will tell us more about stem cell biology. The lessons learned will allow us to manipulate ESCs and SCNTs more capably when the legal environment finally unleashes the scientific racehorses, champing at the bit to bring the promise of these cells to fruition. The specific techniques involved will probably not, and most certainly *should* not, be necessary to bring cures to patients with "official" diseases or to regenerate human bodies deprived by the aging process of their capacity to self-heal. Their perceived necessity is a purely political construct, unrelated to scientific reality or underlying humanitarian need. The *real* need is to free scientists from misguided interference with the quest to turn the enormous potential of embryonic stem cells, including patient-specific cells created by fusing a patient's cells with an egg, into therapies for the sick and the old.

Fortunately, this is one area where nearly all my colleagues already essentially agree with me—and not just in biogerontology, but across the medical and basic biological science world. The grant-review scientists at the National Institutes of Health would be delighted to disburse funds to promising ESC and SCNT research around the nation, if their hands were not bound by their president's executive orders. *Everyone* not under the blinders of a misplaced sense of moral responsibility to a ball of cells recognises the need for ESC research guided by responsible ethics and regulation, not artificial restrictions created out of confusion, fear, and the grasp of political opportunity.

Certainly, there are scientific hurdles to be overcome before ESCs can be used directly as medicine. We have to develop much more reliable techniques for deriving stem cells, transforming them into the kinds of cells that we need, and making them play the same full range of roles that the corresponding cells growing under the guidance of sophisticated developmental programs within our bodies already do. The nascent field of regenerative medicine is already making surprising progress even using cells and tissues grown from patients' adult cells: tissue engineers are moving from cells to organs, seeding cells into a biodegradable scaffold that guides them to form into a structurally appropriate engineered tissue and then melts away, leaving functioning tissue behind.[49] Human patients have been given functioning urethras, starting with cell-free structural tissue taken from cadavers and seeded with the patient's cells, which have lasted seven years. Functioning bladders have been engineered and transplanted into Beagle dogs, and rabbits have received, and successfully used, engineered erectile penis tissue. And in the most ambitious work to date, cattle have been given simple kidneys created using SCNT, with DNA taken from an ear clipping. The rejuvenated cells were expanded, growing to fill in every cranny of a complex biodegradable kidney scaffold, and the resulting organ implanted. The artificial kidneys were functional, filtering the blood and producing a fluid with close chemical similarity to normal urine.

But the fundamental impediment to the dream of new cells to replenish bodies worn by the years or by disease is a political one—and so must its solution be.

Action Now for Science and Medicine

Nearly all the anticipated readers of this book are citizens of democratic states. A few live in countries that have already given their scientists the green light to pursue cures with ESCs within careful ethical and regulatory frameworks. If so, congratulations. You can help to lead the race for cures forward further by lobbying your politicians to increase funding for such research.

But probably the single largest share of my readership will be in the United States—the country that still makes the greatest contributions to world scientific progress, where young scientists still flock to chase the advancement of human capacity, and where government and industry funding could, if unleashed, have the greatest impact on the field. The National Institutes of Health need to have the brakes taken off their funding power, and to be allowed to step down on the accelerator—hard.

Your voice can help the scientific process along in a way that the scientists themselves can't match. Write letters, join lobby groups, educate yourself about local issues and your Congressional representatives' positions; then *vote* for pro-stem-cell ballot initiatives and research-friendly politicians. Excellent background information and tools to help you support favorable legislation are available from the *Coalition for the Advancement of Medical Research* (CAMR) at http://www.CAMRadvocacy.org. You can bring forward the date when animal studies become clinical cures—and when, eventually, the old grow young, their bodies renewed by their own rejuvenated cells and tissues. Scientists need your help now to bring medicine out of the lab and into the lives of suffering patients. When you and your loved ones need *their* help, you will want to know that you have done everything you could to support their lifesaving work.

12

Nuclear Mutations and the Total Defeat of Cancer

Apart from the small amount in our mitochondria, all our DNA is housed in the nucleus of our cells. Like mitochondrial DNA, it accumulates damage throughout our life, and this can theoretically lead to innumerable health problems. However, I believe that in practice only one of those problems—cancer—arises within what we currently consider a normal lifetime. Thus, if we could really thoroughly defeat cancer, nuclear mutations would be harmless. The most audacious component of SENS is just that—a way to defeat cancer altogether.

In a few places in earlier chapters, particularly Chapter 10, I've whetted your appetite with regard to telomeres and telomerase. I know you have that appetite, because when someone asks me what I do and I say I work on combating aging, the commonest response (apart from the just slightly predictable "Hurry up!") is "Ah, telomeres." And indeed, telomeres and telomerase play a very prominent role in SENS. But not the role that most of you are probably expecting.

As I've stressed throughout these chapters, the "engineering" approach to combating aging is fundamentally different from conventional thinking about aging and what to do about it, in that it focuses on the *actual damage* that the aging organism accrues, rather than the *metabolic processes* that cause that damage to accumulate.

This operational definition of aging makes the problem tractable. In the old-school, "gerontological" approach, the number of potential contributors to the aging process is legion, and getting control of *all* of them is a paralyzingly daunting task. It requires us to have a detailed understanding of an enormous number of complex pathways, interference with any of which is both difficult and bound to cause unwanted side effects as the normal function of those pathways is perturbed.

Anti-aging engineering sets us largely free of these problems. We leave metabolism to carry out its necessary but messy work, and find ways to undo or render harmless the relatively small number of fixed changes—molecular damage, in other words—that occur in the actual structure of the aging organism as a *result* of those processes. We are left with a field of just seven classes of damage to deal with—classes for which solutions are foreseeable, and whose repair is unlikely *in itself* to cause any negative side effects. All that we intend to remove is initially inert but eventually pathogenic (pathology-causing) damage—aspects of the aged body that the young organism does just fine without.

There is, however, one apparently gargantuan hole in this logic, and that is the question of damage to the DNA code in the cell's nucleus (as opposed to the DNA held in the mitochondria, which I discussed back in Chapters 5 and 6). Whereas mitochondrial DNA is responsible only for the production of the energy factories in which it is housed, nuclear DNA is the master blueprint from which our entire biological structure is built up and maintained over time. The proteins[1] that it encodes not only make up essential structural features of the body, from the lens of the eye to the pumping muscles of the heart and the miles and miles of arteries that carry blood to our cells, but also include tiny enzymatic machines that do everything from detoxifying poisons to building up fatty membranes and carrying chemical signals from one cell to the next. Damage the DNA, and you corrupt the code of our genetic program, or render perfectly good genetic instructions unreadable by the machinery that transcribes them into orders to be sent out to the body's protein-making "factories."

And your genes do suffer accumulating damage over time. The DNA in the nucleus is subject to a continuous assault on its structure. Each cell's nuclear DNA takes about a million damaging "hits" every day, caused by everything from ultraviolet radiation and environmental toxins to the free radical by-products of its metabolic processes. And even brand-new DNA isn't necessarily pristine: when the cell replicates itself, errors perpetrated by the machinery that copies the cell's genetic information often create production flaws of varying degrees of seriousness.

Much of this mischief is quickly fixed by the cell's elaborate quality control system for DNA, but some of it is irreparable by its nature. Some other damage is *potentially* reparable, but becomes indelible if the cell divides before repairs are made. Such permanent changes are *mutations,* and while mutations that occur elsewhere than in sperm and egg cells (and their progenitors) will not be passed on to the organism's progeny, they will be perpetuated in the cell in which they occur and in any of *its* "descendents."

On top of damage to nuclear DNA itself, there is damage to the so-called *epigenetic* structures of our chromosomes—the "scaffolding" that is anchored to our DNA. Epigenetic structures contribute important information by determining which genes are turned on in a cell and which are turned off, allowing the same overall DNA to be used to create cells as diverse as liver, heart, and kidney cells. Because of this, changes in the epigenetic scaffolding of a cell's DNA ultimately have the same range of *functional* effects on the cell as changes to the genes themselves: By turning on genes that should be turned off (or vice-versa), or increasing or decreasing their activity, these "epimutations" change the complement of proteins produced by the cell. Because they are operationally equivalent in terms of their impact on cell function, I'll allow myself a little terminological sloppiness to avoid belaboring the point with extra verbiage. From here on, I'll mostly be using "mutations" to refer to *both* these kinds of genetic damage—true mutations and epimutations.

Because they occur on an occasional, random basis and are permanent, mutations accumulate with age—and therefore, they qualify as "aging damage" by the definition embraced by the anti-aging engineer. The implication, then, is that we will have to either fix them or render them harmless if we're going to keep the body from progressively declining into pathology over time.

It *Is* Broke. We *Can't* Fix It

Having read this far, you may be expecting that I'll propose a fix for mutations similar to those I've suggested for AGE cross-links, unwanted cells, or lysosomes: just get rid of the junk. But a moment's reflection should show that this can't be done. Damaged genes may be dysfunctional, but we can't afford to just do without them: a cell with anything other than a *functional* gene is still damaged. And that presents a daunting challenge—so daunting, in fact, that for a time I despaired of its solution, and feared that mutations would act as ship-smashing cliffs to any ark that we might build to survive the deluge of metabolism and emerge into an ageless future.

So, you may think, *If we can't afford to* destroy *defective genes, can't we* repair *them instead?* Unfortunately, this is a technical near-impossibility in the foreseeable future, for the simple reason that there are so many different genes in the nuclear DNA. The human nucleus has two copies of nearly all our genes; the aggregate size of one copy (the "haploid genome") is about three billion "letters" of DNA. Exactly how much of this is really actual instructions for building and regulating the body is a matter of some debate, but certainly there are plenty of different places where damage to a "letter" could cause the misspelling of a "word," leading to a loss of information that could harm the function of the cell. That's a lot of potential damage that we would expect to need to fix.

The problem is how to do that for so many *distinct* genes. Any mechanism we might use to fix a damaged gene would have to somehow "know" how to distinguish an intact gene from a damaged one. True, the body already does this, by comparing the damaged DNA to its complementary strand in the double helix (or sometimes to the *homologous chromosome,* i.e. the other copy of the relevant stretch of DNA—remember I just told you that most genes are present in two copies in each cell), but it's not clear that this helps us much. It's hard to see how we could improve on our existing, inbuilt ability to repair DNA, which in humans is already amazingly effective.

We could, in principle, solve this problem if our DNA-repair system relied on a blueprint independent of the genetic information within the cell. But this would require a molecular-level tool that would carry *in itself* a master copy for each of the many tens of thousands of genes (each of which is typically a thousand or more DNA letters in length). It's conceivable that advanced nanotechnology might somehow be able to provide this level of detail to machines that could then carry out the repairs,[2] but only in the very distant future; we're nowhere near that level of proficiency today.

In fact, this summary actually *understates* the difficulty of the job we would face if our solution to mutations were to fix them one at a time. You're probably visualizing that these double-copy gene mutations are nice, clean breaks cutting across the two strands of the DNA helix—and sometimes they are. But sometimes, the DNA is either initially hit, or is mistakenly "fixed" by the body, in a way that leaves two *distinct* errors on either strand, separated by a dozen or so DNA letters. This leaves gaps whose repair, even given a proper blueprint, would entail first fixing the damage to both sides independently and then realigning the two strands and zipping them back together.

This one really had me stumped back in July 2000, at the conference where the engineering approach to developing anti-aging biomedicine first crystallised in my mind. If we can't just throw out damaged DNA, and there's no clear way to fix all of the possible forms of permanent DNA damage that accumulate with age, aren't we stuck with a mechanism of aging that we can't do anything about—and that will therefore kill us even if we repair or obviate every other molecular and cellular change that contributes to aging? At first I essentially ignored the problem, relying that other approaches to cancer would suffice, but I became increasingly doubtful that they would.

Who's Afraid of the Big Bad Mutations?

Throughout these chapters, I've always explained how a given kind of molecular or cellular change ("damage") probably contributes to aging pathology. But in most cases, it's not totally clear to what extent a given kind of alteration actually adds to the loss of function, increased disease risk, and exponential rise in death rates that characterise biological aging. In all the cases we've discussed so far, however, the answer to this question doesn't really matter. Aging damage is, by definition, not a part of a healthy, youthful body, so removing or obviating that damage certainly won't do us any harm—and, from what we can tell, will almost certainly do us significant good.

In this case, however, I couldn't see a clear way either to repair or to render harmless mutations in the nuclear DNA. So I had to take a step back and ask a more fundamental question: *Do we, as a practical matter, actually have to worry about nuclear DNA damage?*

Even to ask such a question sounds a bit crazy to some of my colleagues. Because the nuclear DNA is so clearly critical to cell structure and function, it seems beyond debate that mutations are a contributor to aging. This concept was first formally proposed in the late 1950s, before we really even understood what genes *were,* and it has become almost universally accepted by both scientists and the public at large.

But, however initially intuitive it may sound, this notion has never been justified with direct (or even good indirect, i.e., correlative) evidence. The only way to rule out something's involvement with aging is to speed it up and observe no effect on life span or age-related pathology.[3] This has been accomplished for nuclear mutations, but not yet well enough to be decisive. In mice, deleting a gene that normally fixes up free radical hits to genes before they have a chance to go on to become actual, fixed *mutations* greatly increases the steady-state level of such damage in the nuclear DNA. Yet these animals appear to suffer no pathology as a result, and have normal life spans.[4] However, we can't read too much into this result because the mutation rate is only elevated rather modestly. Similarly, a study was recently performed using four strains of mutant mice, each with a knockout of a *different* DNA-repair gene. In one such strain, mutations accumulated more than in normal mice—and yet, the effects on lifespan were unclear.[5]

A very good, and equally direct, test for showing that something *is* a key contributor to aging is to slow it down or arrest it and observe a direct anti-aging effect: an increase in the "natural limits" to the organism's life span and the preservation over time of youthful functionality. Of course, no one has ever done this with nuclear mutations, or indeed with anything else. Unfortunately for scientific clarity, all the successful anti-aging interventions that we currently know of in mammals change *many* things about the organism, from antioxidant enzymes, to maintenance of proteins, to the activity of the cell's garbage-disposal system. This prevents us from isolating any *one* of those changes as the dominant cause of the anti-aging effect, and thus isolating any particular type of damage as being the dominant contributor to aging. Calorically restricted animals, for instance, lose a significant amount of bone mass, but no one thinks that this loss is responsible for CR's anti-aging effect.

A close approximation of such a test, however, was published a couple of years ago, in the form of mice that had been given genes allowing them to produce extra amounts of an antioxidant enzyme (catalase), specifically targeted to different parts of their bodies.[6,7] I mentioned these mice in earlier chapters, but the study is worth a recap. Catalase detoxifies an abundant oxidizing molecule (hydrogen peroxide), so it can potentially protect things it's close to from at least one form of damage. Putting catalase into these animals' mitochondria, which significantly reduced the development of mitochondrial DNA deletion mutations, reduced their vulnerability to several age-related diseases, and extended their *maximum* life span by about 20 percent—the first unambiguous case of an antioxidant genetic intervention with an effect on this key measure of aging in mammals. Yet giving these organisms catalase targeted to the nucleus, which reduced nuclear mutations, provided *no* benefit in terms of lifespan.

Another type of evidence that is often raised in support of the idea of nuclear genetic damage as a contributor to aging is the existence of so-called "accelerated aging" models whose symptoms arise from accelerated rates of nuclear mutation accumulation. These are animals either with various inborn mutations, or subjected to outside assaults (like bombardment with toxic chemicals or X-ray radiation), that increase the accumulation of nuclear DNA damage as they age, either by increasing the rate at which it is formed or by knocking out the machinery that repairs it. This includes such human genetic diseases as Hutchinson-Gilford syndrome ("progeria") and Werner's syndrome.

Victims of these diseases, whether they walk on two feet or four, do often look in many ways like the elderly, suffering pathology that can be eerily parallel to that seen as animals get older, from bone diseases and failing hearts to scruffy fur and cataracts. But the fact that the *symptoms* of an abnormal pathology look a lot like the symptoms of "normal" aging doesn't prove that the *mechanisms* of one underlie the mechanisms of the other, any more than a wet lawn proves that the same mechanisms govern rainfall patterns and sprinkler systems. Almost anything that messes up normal equilibrium in the body but takes a while to kill you will look like "premature aging;" the question is what if any relationship a given change bears to aging in the rest of us. Evolutionary biologist Michael Rose, a geneticist at the University of California who has himself bred remarkably slow-aging, long-lived fruit flies, puts the point succinctly: "A lot of people can kill things off sooner, by screwing around with various mechanisms, but to me that's like killing mice with hammers: it doesn't show that hammers are related to aging."

Because we don't have this kind of direct evidence, we normally rely on more indirect, correlative types of evidence of a phenomenon's involvement with aging. One is to compare the rate of accumulation of some kind of damage in animals that age at different rates (that is, operationally: whose "oldest old" finally succumb to "natural" death, after first progressively losing youthful functionality, at different chronological ages). While all aging animals do accumulate nuclear mutations, the rate at which longer-lived animals suffer free radical damage to the nuclear DNA doesn't correlate well with their maximum life spans (unlike the corresponding rate in *mitochondrial* DNA, which does).

Frustrating, isn't it? But the good news is that, in my view anyway, the available evidence allows us to conclude with some confidence that nonspecific nuclear DNA damage is *not* meaningful contributors to aging. I briefly summarized the argument for this conclusion back in Chapter 4; here's the full story.

What gives force to the idea that nuclear mutations are a cause of aging is the fact that the mutations we inherit from our parents are major risk factors for a wide range of diseases, including age-related diseases from cancer to heart attacks to Alzheimer's. The same is sometimes true of mutations that occur very early in development, when there are so few cells in the little ball that will eventually transform itself into a human body that damage to the nuclear DNA of just one of them can infect almost our entire being with the same flaw.

But the situation is quite different after we're born, which is when *new* mutations occur and accumulate with the passage of time—and thus, where we have to think about an effect of nuclear mutations as a form of *aging damage* (as opposed to as a source of inborn *vulnerability* to aging diseases). This is because mutations only spread from one cell to another when the second cell's DNA is actually derived from the DNA of the first, as happens when (and only when) a cell divides and passes a copy of its DNA on to its progeny. Thus, whereas *inherited* mutations infect *all* of the cells in our mature bodies (because all of our mature cells derive from a single, mutated fertilized egg), *age-related* mutations happen in our mature bodies *one cell at a time,* as a result of random events like the radiation that comes into a plane at high altitude, or toxins produced by invisible mold spores in your food. Any such mutations can only affect the one cell in which they occur, and its descendents.

This fact greatly limits the potential of age-related mutations to become prevalent enough in our tissues to impair function. A lot of our cells—including the ones in tissues where the impact of aging is most clear, like the brain and heart—don't divide *at all* once we mature, so any mutations in such cells go no further. And even in cells that *do* divide—skin cells, say, or the cells lining your gut—the rate of cell division after we mature is balanced by the short life of the cells' descendents, so that an individual cell's mutations are present in only a few cells in the body at any instant and thus get little chance to "take over" the tissue in which the mutation occurs.

Also, even before we had hard numbers on the subject, we knew that complete "knockout" mutations would be relatively rare occurrences. For one thing, most mutations in DNA create errors in the encoded proteins that, if they affect proteins at all, only make them *less* effective, rather than *completely* dysfunctional. But even in cases where a gene is damaged so badly that either it can't be used as the basis for protein manufacture at all, or else the product that it produces will be useless—or even toxic—*still* generally no harm will ensue, because the damage will be to a gene that *isn't even being used* in that cell! Remember that the DNA of each cell includes not just the genes needed for that cell to do its specialised job, but the entire instruction manual for producing every cell type in your body. Individual cells acquire their specialized function—heart cell, liver cell, kidney cell, skin—by turning off most of their DNA, leaving active only those genes that they require to perform their particular function. In a typical cell, only about one tenth of a person's full gene complement is active. So about 90 percent of the genes in a cell could be irreparably damaged without affecting the cell's function one bit.

It gets better yet. Even though evolution is far too thrifty to allow cells to go around wasting their resources in producing proteins that aren't important to cell function, the cell can nonetheless suffer the loss of many proteins and still limp along without harming the body, even contributing in some degree to its internal economy. Remember, many people are born with such mutations affecting *every single one* of their cells, including all the cells where that protein is normally needed to do their jobs, and these people still live for decades. The same mutation occurring in only a small *fraction* of one's cells might not even be noticed in a normal lifetime, being compensated for by the fact that the rest of one's cells are fully functional.

Mutations: Few, and Not Far Between

And sure enough, the actual number of mutations that accumulate with aging does indeed appear to be rather low—too low to have any serious effect on the aging of the tissue in which they appear. We know this thanks in large part to the work of a team headed by physiology professor Jan Vijg, now at the Buck Institute for Age Research. Vijg's group came up with a very ingenious way to obtain a representative sample of both the *amount* of mutation that accumulates in different tissues with age, and the general *kind* of damage that occurs.

What Vijg's group found was surprising to them and to a lot of other researchers. They discovered, first, that there were not nearly enough relatively minor *point mutations* ("one-letter" mutations in DNA "words" or "sentences") to realistically have a meaningful impact on overall tissue function or on the aging of the organism as a whole.

More important, the number of mutations in a typical cell doesn't even increase between early and late adulthood in our most critical tissue (the brain), and the total burden of mutations only goes up by a factor of two to three in *any* tissue, even in old age.[8,9] This may sound like a lot, until you remember how many genes there are in such a cell and how few mutations are present in young people. Doubling or tripling a small number of initial mutations still leaves the cell with only a small number of them; and when you consider how few of those are occurring in genes actually used by the cell in which they occur, how few of that subset actually *disable* (as opposed to merely inhibit) a critical function of the cell in question, and how little the loss of a particular activity in a particular cell can really affect the overall function of a tissue, an organ, or an entire animal, it becomes clear that an increase like this is actually more of an inconvenience than a crippling blow to the aging organism.

But Vijg immediately appreciated that this was not a knockout blow to the involvement of nuclear mutations in aging. This is because some of the mutations identified by his group may be much more severe than the disabling of a single gene. Some of them, for instance, could take the form of *deletions*—the total removal of large stretches of DNA, annihilating many genes at once even though the event is, strictly speaking, only a single mutational event. Deletions can occur when two widely separated breaks occur at the same time in the same chromosome, and Vijg's data showed that a significant proportion of the mutations that accumulate with age do involve such breaks, which at first glance suggests there are a significant number of deletions amongst the modest increase in the total number of mutations that happens with aging. This would have a much greater impact than would be suggested by a simple tallying-up of mutation frequencies.

Luckily, however, it turns out that most of the cases in which a cell suffers two distinct breakages of DNA don't actually lead to deletions.[10] Fully half of them are actually cases of fractures happening on *two separate chromosomes,* which are not physically linked to one another and thus do not lead to a deletion—much as if you cut two separate pieces of string across their respective middles and switch the pieces' partners, rather than making two snips in the same length of material and discarding the portion between the two cuts. And even in the remaining cases, where both breaks occur on the same chromosome, many of the incidents are equally harmless, leading to mutations called *inversions* that shuffle chromosomes around while typically leaving them intact and functional.

For this and other reasons, not many of the events that at first look like deletions actually seem likely to have much more effect on actual genetic integrity than simple point mutations.

When we look at epimutations, the evidence available so far leads me to the same conclusion. The most well-studied kind of epimutation is changes in *methylation*—a chemical alteration of genes that prevents them from being expressed. Middle-aged mice[11] and humans[12] have indeed been found to have more alterations in their methylation patterns than immature ones do, but it's not clear that that trend continues further as the organism actually ages (as would be required of a true, stable form of molecular aging damage). Instead, as they go from adulthood into true old age, the change in methylation epimutation may stop accelerating and may even slow down. This could mean, for instance, that the observed methylation changes occur in the early life of the organism, but that their frequency is subsequently held to tolerable levels by repair mechanisms or even by the elimination of unacceptably aberrant cells, rather than accumulating with age the way that AGE cross-links or mitochondrial mutations do. Just as important, from our perspective as would-be interveners, there's no evidence that the observed methylation changes cause any actual functional problem over the life span.

Add it all up, and there seems to be remarkably little increase in mutations happening in aging cells—and of what there is, again, very few will actually affect the cell's functionality.

The Times, Not the Genes, They Are A-Changin'

Vijg's studies of that period, and those of a few of his forerunners, assessed the impact of mutations on aging by measuring the actual frequency at which genes are structurally altered during aging. But there's another kind of evidence that's often invoked to support the role of mutations in aging: gene expression studies. Starting in the late 1990s, scientists were able to use a new technology informally known as "gene chips" to measure the *activity,* rather than the structure, of nearly all the genes in the cells of tissues. This allowed for comparison studies of aging and younger animals—including humans.[13]

The results clearly showed that there are pretty substantial changes in gene expression in the cells of aging animals. This result has often been mistakenly thought to prove the importance of mutations in aging, because the gene expression shifts were assumed to be the result of mutations in the genes themselves. This idea was given some seeming support by evidence that the changes in gene expression happened alongside markers of "pre-mutagenic" free radical damage to genes—damage that can still be repaired as it stands, but that can go on to become full-blown mutations if not repaired properly.

But there's a much easier explanation for these shifts, and that is the very fact that gene expression *does* change—not just with aging, but all the time. Your cells are in a state of continuous dynamic adaptation to their environment, changing gene expression in response to new conditions at every moment. Every time your body needs to respond to its environment, the expression of genes involved in those responses is altered.

So as cells and tissues acquire aging damage that impairs their normal, youthful function, the body adapts to the changed circumstances as best it can. As oxidative stress climbs with the accumulation of cells that have been taken over by mutant mitochondria, cells ramp up the activity of genes that produce protective antioxidants and "heat shock" factors that help to repair some of the ensuing damage to proteins. When the increase in oxidative stress leads your arteries to become infiltrated with free-radical-damaged LDL, the surrounding cells produce more inflammatory factors to attract macrophages to clear the toxic stuff out. When your heart stiffens as it becomes riddled with cross-linked proteins, the body produces more "remodeling" enzymes that help to degrade the old tissue in hopes of clearing the way for fresh, undamaged replacement material.

Additionally, changes in the cellular environment imposed by aging can also interfere with normal gene expression. In some cases, such effects are easily observed: for example, oxidative stress directly inhibits the expression of some genes. But there are also more subtle effects afoot. For instance, you'll recall that free radicals, in addition to being produced as simple side effects of metabolic forces, are also produced *intentionally* as signaling molecules in various systems inside and outside cells. As free radical levels climb in the aging organism's cells, the excessive oxidative stress distorts these same signalling pathways, introducing "noise" into them. This can in turn result in inappropriate metabolic and gene-expression shifts as the cell responds wrongly to misheard, drowned-out, or counterfeit messages from (or superimposed on) the system.

And so on. The point is that the age-related shifts in gene expression seen in studies of tissues are not the *cause* of aging, but are an adaptive (and sometimes maladaptive) *response* to it. Changes in gene expression in cells subjected to free radical attack can occur not because that attack has damaged the genes whose expression levels change, but because an increase in oxidative stress requires the cell to change its metabolism to keep going in face of the challenge.

Now, you may be thinking that I've jumped the gun here. So far, I've explained how gene expression changes *can* be a compensatory response to aging rather than a cause of aging—but surely I have not shown that those changes *are,* overwhelmingly, responses and not causes. But it's easy to see that they must be responses, because they're *coordinated.* The only reason the studies I'm referring to were able to detect a change in expression of a given gene was because it was occurring in the *bulk* of the cells within the tissue being analyzed. Mutations and epimutations, by contrast, would affect one gene in one cell and a different gene in the next cell.

This coordinated change of gene expression in response to aging is shown most clearly in gene-chip studies that compare normally aging animals to those undergoing calorie restriction (CR),[14,15,16] which is, again, the only intervention now at our disposal that slows down aging in mammals, short of tinkering with their genes. These studies show that the greatest number of changes in gene expression that happen with aging occur in the blueprints for antioxidant, inflammatory, remodelling, and heat shock proteins—exactly the ones needed to respond to the damage inflicted by aging. They also show that CR not only slows down, but *reverses,* many of these changes when it's implemented late in life. This shows that the changes in gene expression can't be the result of mutations, because while CR can slow down the accumulation of mutations and other aging damage, it cannot undo damage that has already been done. Rather, it reduces the sources of damage that are leading to the changes in gene expression, cutting down free radical production in the mitochondria, reducing the accumulation of mitochondrially mutant cells, lowering blood glucose to prevent glycation cross-links, and so forth. Change the cells' pro-aging environment, and they scale back their gene-expression adaptations to that environment.

This sort of logic has led some, among whom Vijg's team are again prominent, to conduct much trickier studies assessing the variation of gene expression *from one cell to the next* in old animals as compared with young ones It was found that there is a dramatic increase with age in that variation.[17] On the face of it, this increased variability doesn't seem likely to be the result of the kinds of adaptation that are responsible for the dominant pattern of age-related gene expression shifts that we were talking about a moment ago, because the conditions that create them—and the adaptive responses needed to keep going under their influence—are more or less the same from one cell to the next in a given tissue.

But there are a lot of alternatives to explore before we jump to the conclusion that this increase in the amount of difference in gene expression between a given aging cell and its neighbors is the result of *mutations* (or epimutations). It may be due to some other age-related stressor, for instance. Vijg's team found that they could reproduce the increase in variability by exposing cells in culture to oxidative stress—and again, oxidative stress disrupts the ability of cells to relay signals within themselves. Such disruptions could lead to a failure to respond to signals coming in from the cell's environment—either local signaling molecules from its neighbors, or more "broadcast" signals like hormones and other factors. The effect of this would naturally vary from one cell to the next, because it results from noise in intra- and intercellular signaling, but if the oxidative stress were then removed, the variability might return to its original level.

Variability from one cell to the next can also result from *other* kinds of damage that have accumulated, randomly, to different degrees from one cell to another: telomere loss in one cell, high levels of AGEs in another, and the presence of mitochondrial mutations (without nuclear mutations) in still a third. Intuitively, the gene expression shifts required to increase the degradation of AGE-ridden proteins on the surface of heart muscle cells are quite distinct from those that occur in response to accumulated junk within them. When neighboring cells suffer different levels of different kinds of molecular damage, they mount different adaptive gene expression responses, both to counterbalance the effect of the damage on their ongoing metabolic processes and in many cases to attempt to repair it. Such effects have been observed at the cellular level in roundworms,[18] and preliminary studies suggest that similar factors could be at play in increases in gene expression variability in rats and humans, too.[19]

Also, such damage can send shock waves outside the cell in which it occurs, which will elicit changes in gene expression in nearby cells trying to cope with its impact. Consider cell senescence, which I discussed in Chapter 10. The activation of the senescence program will certainly change the expression of genes in the senescent cell as compared to its neighbors, but it will also change the expression profile of those neighbors, as they respond to the flood of growth factors, inflammatory signals, and remodeling proteins that it produces. Cells closer to the senescent one will be more affected than cells further away.

Importantly, these changes would distinguish the neighbors of a senescent cell *both* from the original culprit *and* from other cells not directly affected by its output. To pick the most obvious example: growth factors secreted by a senescent cell will trigger cell-replication programs in nearby cells. These programs, executed by gene expression changes, are ones that have been permanently shut off in the senescent cell itself, and that are quiescent in cells far enough removed from the senescent cell to evade its influence.

The same would be true of cells that had suffered any number of problems. And the variation can flow in the opposite direction, too. That is, in addition to *damaged* cells playing havoc with their neighbors by exporting their own internal woes, *healthy* cells can communicate with damaged ones to keep them functioning more normally. The best-researched case of this is cancer.[20,21] A single cell can harbor mutations that would by default lead to malignancy, but be held in check by its neighbors, which can do everything from inhibiting its proliferation (so-called *contact inhibition*) to inducing apoptosis. As with all other adaptive changes, the output required to impose this control requires shifts in gene expression that will distinguish cells working to keep their antisocial neighbor under control from the patterns of both the would-be cancer cell and noncancerous cells elsewhere in the tissue.

Hence, even the fact that there are more pronounced differences in the activity of genes from one cell to the next in the tissue of old animals (as compared with young) doesn't necessarily pin the blame on nuclear mutations. Moreover, much of the variability observed is a *good* thing—the distinct efforts, futile though they may ultimately be, of each cell to preserve its integrity under the unique burden of aging damage that each of them faces.

If, then, age-related gene expression changes don't flow out of a significant increase in nuclear mutations, but are instead the result of other kinds of damage (and cells' attempts, successful and otherwise, to carry on in the face of it), the proper engineering response is to leave nuclear mutations alone, and focus our attention on the *real* culprit in age-related gene dysregulation: the aging damage that forces our cells to flail about in increasingly desperate, disorderly, and panicked attempts to keep their heads above the waters of the aging process. Once our cells are no longer suffering under the onslaught of problems ranging from mitochondrial mutations to AGE cross-linking to visceral fat accumulation, their gene expression profiles can be expected to normalize, because both they and their environment will be normalized—returned to a youthful state of functionality.

Ironically, it may be precisely after we have cleaned up all of these sources of damage that nuclear DNA mutations may, eventually, begin to contribute to age-related death. The evidence shows only that cells accumulate relatively few nuclear DNA mutations, and those mutations do relatively little to impede cell function *within a normal lifetime*. But what happens when that life span is extended to centuries, with the same amount of metabolic damage to our DNA occurring, and with our DNA-replicating and DNA-repairing machinery no more perfect than it is today? It may well be that effects that are too subtle ever to reach a pathological stage within nine decades or so could build up into a real threat over the longer term.

However, we don't need to worry about that *now*. As I explained in respect of AGE cross-links in Chapter 9, one cornerstone of the engineering approach to developing anti-aging biotechnology is that we don't need to fix all possible forms of damage at once—we only need to do a good enough job of cleaning up those insults that meaningfully contribute to age-related frailty within today's life expectancies. Once this is accomplished, our bodies will remain youthful during the years in which they are now undergoing a slow descent into decrepitude. Some forms of molecular damage that aren't causing us problems now will then begin to reach the threshold beyond which we begin to meaningfully, functionally decay.

At that point, the next generation of SENS therapies will need to be brought in, to repair that newly pathological (and in some cases even newly identified) damage to the same standard. Fortunately, our extended lives will buy us the time to observe such damage accumulating in our bodies, and in those of shorter-lived animals whose lives have been extended in the lab, and whose youths renewed, using the same treatment—and as time goes by, we'll build better and better tools for identifying that new damage and for developing weapons to fight it. The key is to get past the barriers that hold us back *today,* and then be ready to break new ones as they approach. From all available evidence, nuclear mutations aren't yet such a limit—but they may well be one in the future.

The Exception That Creates the Rule

So far in this chapter, I've explained that one of the reasons not to worry too much about nuclear DNA mutations is that even the ones that are actually harmful to the cell are usually minor and, even when they are major, the damage that they inflict is mostly constrained to one or a few cells, so that any slack that their dysfunction creates can be picked up by other cells in the tissue. But there is, of course, one screamingly important exception to this rule: cancer. Cancer is famously a disease of nuclear DNA mutations (though, as we've seen, it usually takes more than *just* mutations to turn an aberrant cell into full-blown cancer). And the incidence of cancer clearly goes up with age.

Shortly, I'll discuss how I think we can really defeat cancer. But first, I'm going to explain why cancer is the reason for our apparently "unnecessarily" good defenses against the accumulation of nuclear mutations.

First, let's recall that one of the reasons cancer is so formidable an adversary is that any number of different mutations can contribute to the cancer process. A breakdown in any one of the many systems that protect against cancer—mutations that lead to the formation of a defective senescence or apoptosis "tumor suppressor" protein, or to the excessive production of a cell receptor for growth signals, or to the reactivation of a suppressed telomerase-coding gene—can be a key step along the broad and winding road that leads a healthy cell to become a renegade.

As with all other causes of age-related death, the evolved level of protection against potentially cancer-causing mutations is limited by the cost of creating and maintaining the machinery to achieve that protection. On the one hand, there's no sense in putting all of an organism's eggs into the basket of being so resistant to aging processes that it can stay young and healthy for two hundred years, if there are strong odds that it will freeze, starve, sicken, or become something's lunch in its third decade; those resources would be better spent on warmer fur, sharper claws, or simply a shorter gestational period. But on the other hand, the animal *does* need to remain internally intact for as long as it can reasonably be expected to survive those threats from the external environment, because every year of youth is another opportunity to spread one's genes around. If you're unsure about this, go back and reread Chapter 3.

Given those opposing priorities, natural selection will push hard to create machinery that protects against potentially cancerous mutations rigorously enough to keep cancer at bay for at least as long as the organism would likely get through winters, wars, and attacks from predators. But, precisely because of the nature of the cancer threat—that so many *alternative* mutations can contribute to the cancer—the body can't afford to cherry-pick the genes that it's going to protect. The only effective defense against cancer is to safeguard the integrity of *every single gene we possess.*

So this is how it comes to be that new, age-related nuclear mutations are so rare—why we possess such an elaborate system of oversight of DNA synthesis repair guarding our every gene, despite the fact that the great majority of mutations have negligible effect on the overall economy of the body. To protect the organism adequately against cancer, evolution gets the best bang for its buck from a system in which *every* gene gets the kind of gold-plated antimutation defense that you might think it would reserve for a privileged few, mission-critical, cancer-avoidance genes.[22,23]

A 2015 Challenge

This analysis shows that we don't need to address *all* mutations in order to develop a panel of interventions comprehensive enough to result in the first dramatic extensions of human life span. My realization of this was crucial to the nascent development of the SENS platform back in 2000, because it showed me that the sheer scope of the nuclear mutation problem was in one sense much smaller than I had at first feared. It had become clear to me that, for the most part, age-related accumulation of nuclear mutations are basically harmless over the course of a currently normal lifespan: the rate of their accumulation is quite insufficient to contribute significantly to age-related decline. But we emphatically do need to confront the one enormous exception to that rule: cancer. In principle, we can simply ignore nuclear mutations, *if and only if* we can find a truly effective way to protect ourselves against this one fatal disease.

The stakes here are high. Cancer is a deal-breaker for building an ageless organism. We can shatter the cellular manacles of AGE, free our brains and hearts of the webs of amyloid, clean out the dirty depths of our lysosomes, and all the rest—but if we fail to make a breakthrough against this one disease, we can still expect to be dead in our mid-eighties.

If you've been listening to the popular press accounts of progress in the War on Cancer, you may now be feeling much less alarmed than you should. The media, and also the scientists and bureaucrats on whom they report, love to trumpet every advance (indeed, every *hint* of an advance) in the treatment of cancer. There are so many reports of potential new cancer treatments that one might well think we were already far down the path to the day when we will have cancer mastered. This is especially so in light of the targeted cancer therapies that I reviewed in Chapter 10—therapies that generally are, or can be predicted to be, far safer and more effective than the knives, poisons, and radiation that have been the staples of cancer management for decades.

And you wouldn't be alone, nor even outside the scientific mainstream, in thinking this. In 2003, none other than Dr. Andrew von Eschenbach, the Director of the National Cancer Institute, famously put forward an ambitious but (so he claimed) realistic vision for his organization: to eliminate suffering and death from cancer by the year 2015. Dr. von Eschenbach wasn't just whimsically putting his dreams into words: he was putting forward his sober assessment of what the world scientific community, spearheaded by the NCI, could achieve in little over a decade. It became the formal agenda of the Institute: "Challenge Goal 2015." The timetable is now so embedded in the organization that it is routinely alluded to as simply "2015," with no further elaboration necessary—in the same way that we once talked about "Y2K." [24]

I think that this goal is utterly unrealistic—and that it only arose because of a failure to appreciate the flaws in the assumptions that are built into it. First, and quite explicitly, "it does not mean 'curing' cancer but, rather, it means that we will eliminate many cancers and control the others, so that people can live with—not die from—cancer."[25] If feasible, this is a perfectly legitimate medical goal: to have cancer under the same level of control that we today have over adult-onset diabetes or AIDS—in which the disease is still present but is so well managed that patients can lead nearly normal lives—would represent an enormous alleviation of human suffering and death from a terrible disease.

But cancer is fundamentally different from these diseases in a way that precludes its chronic "management." Diabetes and hypertension can be held at a safe, manageable level precisely because they are essentially stable diseases. By contrast, what makes cancer so fearsome a foe is that it's a constantly *evolving* disease, a hive of genetic inventiveness that continuously finds new and better ways to outwit our attempts to control it. Relegating cancer to the level of a chronic disease is an idea that could only ever be entertained by completely ignoring the basics of natural selection.

Cancer cells are characterized by an immense genetic instability, which results in large part from the fact that they nearly all get started from a mutation in one or more of the "guardians of the genome"—the genes that police mutations and direct either repair of DNA damage or the activation of senescence and apoptosis programs. Without this constant surveillance and maintenance, the random damage that cells suffer every day is allowed to develop into full-blown mutations, and the process feeds on itself as more regulatory genes are lost.

Many of these mutations are fatal to the cancer cell, but a few of them result in viable progeny that are just *different* from their parents and half-siblings. And that's where natural selection comes into play. Cancer cells, by definition, are reproducing themselves at an astonishing pace. They throw their bastard children out into the world and let survival of the fittest reign. The immune system or oncologists soon take their best shots at the tumor, exploiting the soft spots in the cancer cells' metabolism: their reliance on particular growth factors, for example, or their need for a reactivated telomerase gene, or their hunger for folic acid. But within a single tumor exists such an astonishingly varied population of cells, each with its own combination of normal and abnormal genes, that at least *some* of those cells nearly always have a way to survive any *particular* attack: a greater ability to detoxify a particular toxin, or an alternative way to keep their growth fueled even when a particular signal transduction pathway is shut down.

As a result, it ultimately just doesn't matter if a given therapy kills 99 percent of the cells in a tumor. Somewhere within its heart lurks the dark father of a "strain" of the cancer with a novel mutation that allows it to survive the drug that destroyed its cousins. This founding cell's furious growth continues even as we decimate its cousins, or resumes when the patient can no longer tolerate the stresses of the treatment. Its descendents remain standing after the assault, and are thereby *selected for survival by the very thing that killed their cousins.* When the ensuing tumor becomes large enough for us to detect, we attack what seems to be the same cancer in the same patient using the same treatment, but this time, the old tricks don't work. There really is plenty of truth in the saying that you can't outsmart evolution.

As I sat reflecting on all of this in the wake of my original SENS "Eureka!" moment at the turn of the millennium, a grim formulation of the problem crystallized in my mind. *It is not my goal,* I thought, *to buy time for cancer.*

Making Cancer Wilt Away

Soberly, I contemplated my own cancer challenge: to develop a cancer treatment that would keep us free of clinical cancer for just as long as the other SENS therapies would keep us free of other age-related maladies. The reasoning outlined above immediately ruled out all existing approaches, which leave us fighting one battle after another against an enemy with the implacable force of evolution on its side—a war in which we can win individual campaigns but must ultimately be defeated.

The answer came to me in March 2002 while nursing a beer in a café in Italy. As my mind had done late that night in California in 2000, it now leapt upon an insight that was in some ways completely obvious, and yet led inexorably to revolutionary conclusions. To defeat cancer, I saw, we would need a therapy that does not depend on *anything* that a cancer could escape through a mutation-driven change in gene expression. So any solution would have to have three key characteristics to be viable. First, it would necessarily involve denying cancerous cells access to some tool that is *absolutely indispensable* to *any* cancer's survival, so that they couldn't just make up for its loss by tweaking some other gene expression pathway through mutating its other genes. Second, we would have to take away that tool in such a way that no mutation could *restore* it, either. And third, this tool would have to be one that our normal, noncancer tissues could do without.

I quickly saw the tool that I wanted to lock up: telomerase. I mentioned this enzyme back in Chapter 10, when discussing senescent cells; now's the time to explore it in more detail. Our DNA comes equipped with a stretch of nonsense or "noise" DNA called the *telomere.* Telomeres are to our genes as the brief, silent stretch of "leader tape" at the beginning of a music cassette is to the songs on the tape: they give the "cassette player" (the DNA-replicating machinery) something to hold on to and advance over, so that it won't skip over the essential information at the beginning of the very first "song" (gene) on the tape.

One key difference between telomeres and cassette leaders is that leaders stay intact as long as the tape does, whereas telomeres become ever-so-slightly shorter every time the cell replicates itself or is hit by damaging agents like free radicals. If it weren't for telomerase, this gradual shortening would eventually lead to the complete loss of the telomeres in cells that replicate frequently during the lifespan, and thus the gradual erosion of the genes themselves. Telomerase periodically relengthens the telomere before it becomes critically short.

As with all of our other genes, the DNA that encodes the telomerase enzyme is present in all of our cells—but, because it's only needed after quite a few cell divisions have occurred, it's not needed in most cells for most or all of the time, so it's turned off. This widespread lack of the need for telomerase is used by evolution as a key component of our defense against cancer, because having a limit to the size and renewal of our telomeres prevents our cells from replicating themselves indefinitely—the crucial hallmark of cancer.

To become a full-blown cancer (as opposed to a cell with a single, *potentially* threatening mutation—a genetic risk factor for *becoming* a cancer) requires the accumulation of five to ten mutations, and statistically that requires multiple rounds of cell division and selection. The arithmetic is complex, but the consensus is that, to pose a health threat, cancers have to replicate at least two to three hundred times, even though a clinically relevant tumor contains "only" a million *million* (a "1" with twelve zeroes after it) cells, which could be achieved by "only" forty or so divisions if the originating cell had all the necessary mutations from the outset. And to be genuinely malignant (i.e., to be the founder of a colony of cancer cells that spreads its way throughout the body, as opposed to a localised tumor that could be simply removed with surgery and forgotten about), cancers must then be able to keep up the feverish pace of their replication even longer. The frenzied reproduction of cancer cells is also a key part of their ability to evade our assaults, because it is essential to their capacity to evolve new solutions to the challenges that we throw up against them.

It's no surprise, then, that mutations that unleash telomerase from the repressive strictures imposed on it in normal cells are found in over 90 percent of cancers. The remaining 10 percent also have a way to renew their telomeres—a little-understood mechanism called the "Alternative Lengthening of Telomeres" (ALT) pathway, which I'll discuss a little later on. Either way, without a way to renew their telomeres, the single-minded multiplication of potential cancer cells rapidly grinds to a halt as it reaches the end of its telomere "rope," and we wind up with a tiny (and generally short-lived) lump in our bodies instead of a life-threatening, malignant disease.

If, then, we could snatch this one tool out of the hands of cancers, we would cause any and all the aspiring cancers we developed to fizzle out before they became life-threatening—indeed, before many of them even became actual cancers, because they wouldn't get the opportunity to undergo the full spectrum of mutational events needed to give rise to the kind of renegade cell that can truly pose a threat to the body.

Of course, I was hardly the first to think the problem this far through. Several biotech companies—most prominently Geron, which first made a name for itself in telomere research—are working to develop anticancer drugs that would work by deactivating telomerase. But these pharmaceuticals suffer the same problem as all other approaches based on drugs that affect gene expression: they act as a force of natural selection against a disease with evolution at its disposal. A telomerase inhibitor would kill off those cancer cells in which it effectively turns off the enzyme (and in which ALT didn't take telomerase's place), but it would leave behind any cells that harbored mutations allowing them to keep on renewing their telomeres in the face of it. Different cancer cells might bear any number of variations that let them escape the drug's effects. Some would simply crank their telomerase activity up even further; some would enhance the activity of drug-metabolizing enzymes that degrade the inhibitor; still others would change their cell surface proteins in ways that would make it harder for the drug to penetrate into the cell. Whatever the mechanism, if even one cancer cell can evade the effects of such a drug, it can act as the seed for the tumor's renewed blossoming in a dark spring.

So, again, there was no sense in doing the job only halfway. If we're going to snatch telomerase out of the hands of cancer cells, I thought, we must *really* take it away. And there was only one way that I could think to do that reliably: by deleting the gene that encodes it.

Evolution can, of course, create whole new genes—but it takes a very, very long time to do it. Indeed, very little evolutionary change actually involves the creation of new genes, or even the removal of old ones, precisely because it's so hard to do: instead, evolution finds new ways to regulate old genes, or new functions for gene products other than the ones for which they originally evolved.[26] Thus, for example, the lens of the eye is made up of clear, flexible proteins called *crystallins,* which would seem to have no purpose other than to be used to focus light. Yet, there it is in the nervous system of the sea squirt, where it is part of an organ that keeps track of "down" by sensing gravity. The gene is of course present in every cell in its body, and a mutation in a proto-eye cell of one of our ancestors that carried this gene may have caused it to be expressed there, where previously it would have been turned off, making the protein available to let the light shine on in. And our genome contains no genes similar to the ones for telomerase, ready to mutate to replace it on cancer's demand.

So *deleting* the telomerase gene, unlike trying to *inhibit* it somehow, would be an almost sure-fire way to shut down cancer cells permanently. (Again: this logic ignores Alternative Lengthening of Telomeres but fear not—I will get to ALT shortly.) I had hoped back in 2000, and even speculated in the paper arising from the first SENS workshop,[27] that there might be some way that we could do this only in cancer cells. I became increasingly aware, however, that while it might well be possible to do this for *most* cancer cells, we'd never be able to do it for *all,* for the same evolutionary reasons that I keep hammering on: any mechanism that targeted cancer cells exclusively would have to have some mechanism for selecting a difference between them and normal cells—and of course, those differences would have a genetic basis, leaving the flap of the "tent" open for a mutant subpopulation of cancer cells to stick its evolutionary "nose" under.

Finally, fully eighteen months later in that Italian café, I stopped trying to run from this conclusion. The only way to be sure that we were denying telomerase to cancer cells would be to deny it to *all* cells. What we needed,

I realized, was to take the telomerase gene out of *every cell in the body,* along with the ALT mechanism whereby a small minority of cancer cells manage to lengthen their telomeres without relying on telomerase itself. I would soon term this therapeutic target the "Whole-body Interdiction of Lengthening of Telomeres" (WILT).

Removing telomerase from every cell in our body would preempt cancer before it got a chance to get started. But you can surely see why I took so long to explore this option—and why no one else had explored it before me. Deleting telomere elongation capacity throughout the body would also be life-threatening, because it would mean that our regular, proliferating cells (like those in the skin or the lining of the gut) would suddenly have iron limits on their ability to reproduce themselves and thus replenish tissue. From the moment that we denuded our cells of telomerase, a clock would be ticking. With each division the telomere would shorten by a notch from whatever it had been when we took telomerase out. We would be under the specter of a rather horrible death, as our stem cells went offline one by one under replicative senescence (see Chapter 10): with each failure of a stem cell responsible for supplying key functions, the tissue would fail to be renewed and would slowly degenerate.

So, the effect of telomerase deletion on frequently dividing cells would indeed be very serious indeed—fatal, in fact, in what I calculated to be around a decade from the point when telomerase was deleted.

But hang on, I immediately thought, *SENS already has a proposed solution to "normal," age-related cell loss: stem cells.* So we might just be able to deal with cell loss if we had a sufficiently sophisticated program of stem-cell replenishment—using cells engineered to lack the one linchpin function for cancer, namely telomere elongation.

Of course, these stem cells would eventually peter out, too, as *their* telomeres were worn down—but this is just the same situation that we face with *all* aging damage. The engineer knows that we don't have to root every last trace of cellular and molecular injury out of our systems in order to build a body that will not suffer age-related degeneration and death. The tissues of a twenty-year-old are already riddled with aging damage, and the level climbs every day, but you'd be hard-pressed to find much of a health difference between a basically clean-living person at twenty-five and the same person at thirty-five, because their level of damage at thirty-five is still beneath the threshold at which it causes functional deficits. As long as we keep it there, we will remain biologically young.

So if we introduced stem cells with nice, long telomeres in the first place, we could *let* them wind down and eventually be lost to apoptosis, senescence, or other sources of damage—and just *top our tissues up with more stem cells* before enough of those cells were lost to begin to impair tissue function. The need for regular treatments in this case would, ultimately, be no different from the need for regular rounds of AGE-breakers or of purges of anergic T-cell clones. Neglect your medicine and you will eventually suffer the consequences; keep up with your schedule, and stay young and healthy into a boundless future.

In this case, the engineer's logic is even stronger, because the same "damage" that might eventually kill us (in this case, the running down of our telomeres) is simultaneously the very thing that we need to ensure *does* happen, or else we will be killed by another means (the unchecked cell division at the heart of cancer). Putting an expiry date on all of our cells, but ensuring that they are regularly replenished with new ones, erases both problems at once.

In fact, the case for deleting telomerase was even stronger than this, because placing an absolute limit on the number of cell divisions that our stem cells (and the mature cells derived from them) could undergo would actually bring us an *additional* anticancer and anti-aging benefit. While we often are given the impression that most age-related nuclear DNA mutations are the result of damaging agents like free radicals, radiation, and mutagenic chemicals, the reality is that most nuclear mutations are the result of errors made in copying the DNA during cell division. And while it's almost never mentioned in the popular press, most cancers arise not in the mature cells of our bodies, but in our stem cells, where regular cell division and an active telomerase gene makes it relatively easy to take the brakes off cell growth.[28] (It was, in fact, precisely my increasing appreciation of this fact during 2001 that forced me along the line of thinking that led to the WILT concept.) By cutting down on the number of divisions that our stem cells can undergo before they die, we would simultaneously reduce the number of mutations that they would ever accumulate—and thus, the risk that they might suffer the combination of mutations that would turn them cancerous.

At that point, we'd have cancer licked. No cancer could reach a clinically significant stage. At worst, we would end up with a few little pebble-tumors, small balls of abnormal cells that have exhausted their ability to grow, no more life-threatening than a mole or a small cyst. And our normal tissues would be preserved intact, provided that we underwent regular rounds of replacement of stem cells. See **Figure 1**.

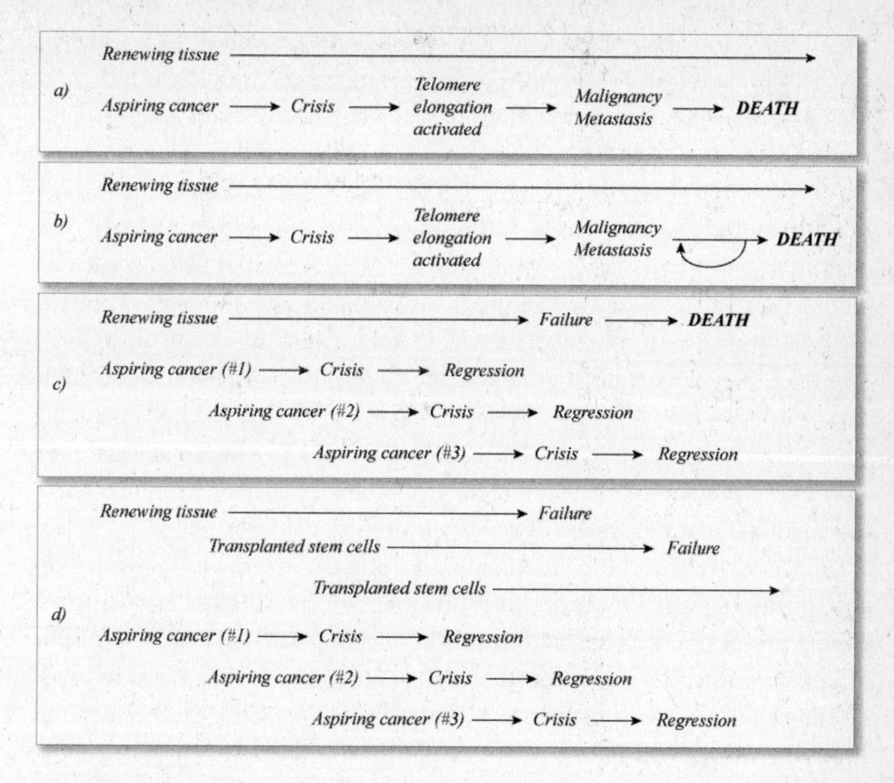

Figure 1. The effects of traditional (late-acting) cancer therapies (b), telomerase deletion (c), and telomerase deletion plus stem cell reseeding (d) on the prognosis for cancer (a).

So Crazy, It Just Might Work

I will not conceal that, when they first hear about it, virtually all my colleagues think the WILT proposal is utterly mad. Indeed, I myself, while not doubting my own sanity, initially worried that I must surely be missing some life-threatening side effect of the two-in-one intervention that was beginning to firm itself up in my mind. So I began consulting experts in all of the relevant fields—telomere biology, mutant humans, mice lacking functional telomerase genes, the ALT mechanism, stem cells, bone marrow transplantation, and of course cancer as a disease above and beyond its characteristic preservation of its telomeres—to confirm that I had my facts straight and hadn't neglected any actual showstoppers, and to ask them what they thought of the technical challenges facing the development of each of the biotechnologies that would be required to implement it.

Their reaction was interesting, and typical of my experience in networking with experts from different fields on interdisciplinary projects. Presented with the whole scheme, each of these experts thought the project as a whole was audacious at best, and something straight out of "soft" science fiction at worst. But to my moderate surprise, when I asked each of them to assess the feasibility and timescales for the development of the individual components that WILT would take from their own discipline, each and every one thought them achievable (albeit ambitious), and felt that nothing in the subfield of biomedicine in which they worked every day would pose an insurmountable challenge. It was in the areas that they *didn't* have intimate, working familiarity with the science or the biotechnology involved that they made *assumptions* of intractability.

Encouraged by these discussions, I held the third SENS workshop specifically on WILT, inviting many of the experts I'd already been consulting about the field. As with the 2000 meeting, my purpose was to put the participants in a room together and simultaneously put them to work fleshing out the way forward (or else the proof that none existed) to the complete, integrated intervention, thus showing them (I hoped) that the plan was sound in its elements and as a whole.

To my enormous satisfaction, it achieved both ends, and we published the results together in the *Annals of the New York Academy of Sciences.*[29] And of all the experts that were involved in the roundtable, only one scientist objected to being given the credit (and blame) for authorship of the paper—and her reasons were extremely revealing, as the Acknowledgements section of the paper, written with her approval, indicates. Dr. Nicola Royle, senior lecturer at the University of Leicester's Department of Genetics and an expert on telomeres (and especially the ALT mechanism), insisted that her name be withdrawn from its author list, not because she didn't think that WILT would work, but because she very much feared that it *would.* Her concern was that, as far as she could see, there was nothing stopping us from developing WILT as a final cure for cancer, and especially as part of a complete panel of SENS interventions that would finally free humanity from age-related degeneration, leading to indefinite youthful, healthy lifespans. But she was not ready to embrace that future: like so many others, her principled (but, I of course maintain, misplaced) fear about the potential drawbacks of unbounded human lifespans on the environment and on existing social structures was so great that she wanted no further part in promoting our progress toward any part of the SENS agenda.[30]

Let's look at some of the technical challenges and concerns that were discussed at this SENS roundtable—and, of course, in follow-ups with these scientists and other colleagues at other venues ever since.[31]

What Happens When We Take Telomerase Away?

This is an obvious one. We're talking about taking away a gene that is at least *present* in all of our cells, though it's permanently turned off in tissues where the cells never have to divide, like the muscles, the heart, and the brain. At least on the face of it, we wouldn't think that taking the gene (either or both of two genes in fact, as telomerase has two subunits) right out of the cell would cause these cells any problems. On the other hand, the enzyme is expressed and routinely used (under strict controls) by cells that need to undergo regular division—most notably stem cells. What would be the effects of taking it away from every last one of them?

Fortunately, we have some pretty reliable evidence on this point, thanks to two models: mice that have been genetically engineered to lack one or other of the two subcomponents of the telomerase enzyme, and a human inherited disease called *dyskeratosis congenita* (DKC). The picture here, overall, is a pretty optimistic one.

Unlike humans, mice *are* born with telomeres long enough to last them their whole lives without telomerase. (This makes mice a rather tricky species to use as models for human cancer, in fact.) Therefore, mice with their telomerase genes deleted have to pass their progressively shortening telomeres on through several successive generations before much of anything bad occurs. At that point, they develop the symptoms that you'd expect from a lack of stem cells, which appear first in the tissues that multiply the most. They become sterile as their sperm-forming cells run out of steam, and their guts and skin start to become depleted of cells and fragile.

They also, ironically, start developing high levels of cancer, which might at first seem to fly in the face of the entire program—but this is just one of the ways in which mice are a fraught model for human cancer. First, even though these mice telomeres are short enough to mess up stem cell proliferation, they're still long enough to let cancers grow to a size that's dangerous for a mouse, simply because mice's small size allows tumors that would be harmlessly small in a human to impede organ function and siphon off a fatally high percentage of their tiny bodies' resources. Second, mouse cells find it relatively easy to activate ALT, so the fact that these mice lack functional telomerase doesn't guarantee that they can't lengthen those telomeres anyway. Clever combinations of telomerase deletion with other mutations can largely sidestep these differences between mice and humans, though, and when this has been done,[32,33] cancer risk went down dramatically—in one case, to such low levels that *none* of the telomerase-lacking mice had died of the disease at a point when *all* of the animals from the same strain but with functional telomerase had been consumed.

DKC patients also give us reason for hope in the midst of their despair. They have a variety of mutations that prevent the effective functioning of their telomerase enzymes—either in telomerase itself, or in genes encoding proteins needed for its normal working. Patients do hold on to some limited telomerase activity, however. The likely reason we never see people *completely* lacking in the gene is that humans have to undergo a lot more cell division in the womb than mice do (and have shorter telomeres than mice at conception), so fetuses with a more severe telomerase mutation are probably aborted.

But telomerase activity in DKC patients is certainly very low. As a result, their telomeres are shorter than normal folks', and they develop predictable symptoms similar to those suffered by telomeraseless mice: mottled or web-patterned skin; patches of abnormal, thickened white cells in the mucous membranes, similar to those often seen in lifetime smokers; weak, thin nails with ridges and fissures; hair loss and lung problems; and bone marrow failure, causing problems with immunity, blood clotting, and delivery of oxygen and iron to their tissues.[34]

It used to be thought that the worst symptoms of DKC usually occur in people in their teens and twenties, but we now know that it isn't quite that simple—and the reason why that's so turns out to be very important. Dr. Inderjeet Dokal, who works extensively with DKC patients at the Department of Haematology at Imperial College in London, told me and the other participants back in 2002 at the WILT summit that he had noted that first-generation DKC patients of a particular type, the ones that have a mutation in one copy of a telomerase gene, don't develop symptoms until they're in their forties—but that, as they pass on their preshortened telomeres to their children and grandchildren, those that inherit the disease develop symptoms earlier and earlier. He and colleagues later confirmed this preliminary observation in rigorous studies of several families.[35,36]

This fact reinforces the principle that it is not the lack of telomerase per se, but the reaching of a threshold length by the telomeres themselves, that causes symptoms. This is an optimistic finding, because it implies exactly what we would expect (and hope) would be the case: that if we can periodically replenish the bone marrow and other stem cell pools with new stem cells whose telomeres are well above the critical length, we should be able both to cure DKC and also to avoid all the problems of the disease in people with *intentionally* extinguished telomerase. Indeed, the best treatment for DKC *today* is a bone marrow transplant, introducing new stem cells taken from people without DKC to replace the ones that are being depleted.

What About This Periodic Stem Cell Replenishment?

The Bone Marrow

Bone marrow transplants are, of course, already a common and nearly routine procedure, not only for DKC sufferers, but also for patients with a range of blood disorders, cancer patients who have lost their bone marrow to radiation therapy, and many others. Still, there are many complications in recipients today, and many technologies that we will absolutely have to master if we are to use bone marrow transplants for WILT.

One reason why bone marrow transplants often don't "take" is that they are not robustly incorporated into their niche in the bone while the original stem cells are still there. For the first round of WILT bone marrow replacement, we may have to perform chemotherapy to wipe out the native cells—but we'd want to do this anyway to minimize the cancer risk of having those old, telomerase-competent cells left behind. In subsequent rounds, the process will be easier, because we will intentionally wait to replace the cells from the first round of transplantation until the cells introduced in the previous round are beginning to die as their nonrenewing telomeres wear down.

How often will this stem cell replenishment have to be performed? It's looking good. People have done clever experiments to measure the average time between divisions of blood stem cells, and it's at least a couple of months in humans. Because it takes around fifty divisions before human cells not expressing telomerase start to feel the shortness of their telomeres, this rate of division should be slow enough to enable us to function just fine for about a decade between successive rounds of bone marrow replacement.

The Skin

Because of the pressure to supply skin grafts for burn patients, disfigured children, and cosmetic surgery, we've made remarkably fast progress in mastering the art of making new skin from stem cells. In mice, we can now peel the skin right down to the *dermis* (the layer of tissue beneath what we normally think of as "skin," which houses hair follicles, sweat and skin oil-secreting glands, and blood vessels), and fully reconstitute the old skin that we've removed. The cells of the dermis do not divide regularly, so we don't need to worry about their telomeres running out. Remarkably, the *epidermis* (the layer on top of the dermis, which is where we do need to replace stem cells) can be replaced using stem cells derived even from such different locations as the cornea of the eye, and the dermis orchestrates their rapid transformation into hair follicle stem cells that then expand outward, renewing the tissue.

Again based on how often skin stem cells divide under the status quo, a round of skin stem cell replacement should last us about ten years. Because the skin is so easy to access, it should be among the easiest and least invasive of WILT stem cell replacement routines.

The Lungs

The innermost layer of the lung, like the skin and the gut, is continuously sloughed off and is thus in continuous need of renewal. Unsurprisingly, lung complications are a major cause of death in DKC patients. Because the lung is in important ways similar to the skin, and is relatively easy to access, there is no reason to think that we won't make quick and relatively painless progress on this front once we put our mind to it. Indeed, some scientists are doing this already, mostly in hopes of treating cystic fibrosis patients. Better yet, the latest estimates are that lung stem cells divide considerably less often than even those in the skin.

Work to date has gone forward using similar approaches to those used in skin, though not yet from stem cells. Even so, progress is being made. In two different models of immunodeficient mice, scientists have purged the lung of its "skin" (epithelium), "scraping" it down to the underlying *basement membrane,* and then replaced the lost tissue using restructured cells taken from the innermost layer of the human lung. The next step will be to do it with stem cells.

The Gut

So far, there are still significant challenges facing the replacement of stem cells in the gastrointestinal tract in humans. Several years ago, Dr. F. Charles Campbell, now Professor of Surgery at the Queen's University Belfast's School of Medicine, made the first major advance. His team extracted stem cells from mouse intestinal tissue, wiped out the cells from small stretches of the colon, and then repopulated the tissue with stem cells, which differentiated into all the appropriate cell types and made fully functional new tissue. Progress has not been rapid since then, however. At the WILT summit Campbell explained that, in studies he has never reported in print, his team tried the same approach in pigs, but the result was a mass of dysfunctional scar tissue. Since then, however, considerable progress has been made by another group working in dogs,[37] and the same group has further advanced the technique in rats and mice.

Much work remains to be done here: not least, the opening-up of sections of the gut to remove the existing cells would be far too invasive for human use in WILT. Endoscopy, similar to what's currently in use to remove potentially cancerous colon polyps, may provide us with a more tolerable solution, and should be much more advanced in a few decades, when we'll actually need it.

A further question is how often we will need to replace stem cells in the gut. Previous estimates suggested that it would be much more frequently than the decade or so needed in these other tissues—more like a few times a *year*. Luckily, however, there is an easy way to see that those numbers must be wrong. If gut stem cells divided so much faster than those in other tissues, DKC sufferers and telomerase-lacking mice would suffer gut failure at a younger age than they suffer failure of the bone marrow or skin—but they generally don't. Rather, all these tissues fail at about the same age in any given patient; if anything, the blood is most often the first to go in humans, which is why bone marrow transplants are temporarily helpful. But still, the replacement of the stem cells of the gut seems likely to be a more invasive procedure than that in bone.

The Rest of Us

The rest of our bodies is, in the relevant respect, like the dermis: composed of cells that *don't* divide regularly. Some of these cell types (including the dermal cells, which are called *fibroblasts*) are *quiescent*: they can divide but they only do so when called upon to do so, such as to close a wound. Others are *postmitotic:* totally unable to divide, and instead renewed by incoming progenitor cells, if at all.

Postmitotic cell types are obviously no problem in cancer terms, but quiescent cells require a little more thought. The fact that they are not significantly affected in DKC or telomerase-knockout mice, and the relative rarity of cancers derived from them, gives us a bit more time to work on the job—and, once we have mastered it, probably a *lot* more time between each round of stem cell reseeding (as opposed to the more targeted replacement of "normal" and pathological cell loss to aging damage, which we will be tackling earlier on using ESCs, as outlined in Chapter 11).

But, precisely because they don't divide, these cells aren't going anywhere for a while either—and they do become cancerous occasionally. Thus, we do want to protect ourselves from cancer in them as much as possible. Removing senescent cells will reduce the already-low risk in these tissues considerably, but we will still want to cut it down much further.

The likely solution to this problem is targeted gene deletion. This is again a major challenge, because even in mice (where gene therapy is now routine) it's very difficult to target genes for insertion or deletion in cells that are already inside the organism (as opposed to putting genes into sperm or egg, or embryos, or into cells that have been taken out of the animal, modified, and replaced). We are, however, getting better at this. Hopes are high for using this technique for gene therapy generally, and it could potentially be used to remove telomere-renewing potential from tissues that aren't maintained by stem cells.

Over time, however, between replacing age-related cell loss using stem cells engineered to lack telomerase, and eventually replacing the original stem cells for these tissues with new ones engineered the same way, we will slowly "WILT-ify" these tissues a few cells at a time, progressively lowering the cancer risk they may pose.

Does WILT Underestimate Cancer's Evolutionary Ingenuity?

WILT is a highly complex, multifaceted proposal—but it rests, as I've explained, on one absolutely essential assumption. WILT will fail if cancers can figure out a way to grow indefinitely without the genes that we delete. Let's look at that eventuality in detail.

One formal possibility is that cells could develop a way to replicate their chromosomes without telomere shortening, and thereby avoid the need for any enzyme to reverse that shortening. It's been hypothesized that some stem cells effectively do just this. When stem cells divide, they typically produce one daughter that's still a stem cell and one "amplifying" cell that is set on the path to differentiation into its required function. (Indeed, the ability to do this is an essential feature of most people's *definition* of a stem cell.) It's possible that, by a clever system of controlling which DNA strands stay in the daughter stem cell and which ones go into the amplifying cell, stem cells could stop their DNA from shortening. But we don't need to worry about this possibility in respect of cancer, because it can only confer *linear* growth rates, not the exponential growth that is seen in cancers. With linear growth, a cancer would take thousands of years to grow large enough to kill us.

Another way to avoid telomere shortening is the way that bacteria, and indeed our own mitochondria, do it: not to have any telomeres! Bacterial and mitochondrial DNA are circles, so there's no end-replication problem. But it seems that there's no meaningful risk of this happening in humans. When telomeres get really short, the cell's DNA-repair machinery mistakes the raw ends of the exposed chromosomes for the result of a chromosome break. That causes the cell to make a misguided attempt at repair, joining the chromosomes together, end to end—which is rapidly fatal for a dividing cell, because the cell division machinery snaps the double chromosome apart again, not at the original join but at some random place, scrambling the genes into a dysfunctional mess.

The final possibility is altogether more real, unfortunately. I've mentioned it a few times in passing during this chapter: it is that cancers can occasionally relengthen their telomeres using enzymes other than telomerase. These enzymes have not yet been identified, so the system is given the uninformative name ALT, or Alternative Lengthening of Telomeres. The good news here is that ALT is quite rare—in some tissues it's almost never seen, and in those few tissues where it does appear it occurs in no more than about half of all tumors. This means that at least as much mutation is needed to activate it as to activate telomerase—and that tells us that ALT is almost certainly dependent on the turning-on of a gene that's usually very firmly off, rather than only the loss of activity of genes that are normally on. This is great news for WILT, because, if ALT is indeed dependent on the activation of some gene or other, rather than just on the inactivation of genes, we will be able to do the same for that gene (once we identify it) as WILT proposes to do with telomerase. There may be side effects, just as there are with deleting telomerase, but they should be addressable with stem cell or other regenerative therapies, just as telomerase deletion seems likely to be.

A SIDE EFFECT: *WILT*ING FERTILITY?

One potential side effect of the loss of telomerase from all of our cells might be eventual sterility for men. If having children is still a priority in a post-rejuvenation world, then men may choose to freeze their sperm in advance, as is presently done for sperm donors, for IVF. The sole sexual issue will be the actual making of babies, of course: nothing else will wilt from WILT.

Part Three

13

Getting from Here to There: The War on Aging

This book has presented, in as simple terms as possible, the biological details of what human aging is and how we can realistically set about defeating it. However, from what I've presented so far, you would be entirely justified in concluding that I've only made a case for the *eventual* development of therapies that could *modestly* delay aging. You might think that a realistic time frame for getting encouraging results in mice might indeed be a decade or so, but you're probably already thinking about the problems that there would be in negotiating FDA obstacles and such like in translating these therapies to humans. And you may be concluding that the sort of timeframe I've been predicting for the arrival of widely-available therapies—a few decades with 50 percent probability—is too short by a factor of at least three. You're probably also pretty skeptical about the *degree* of life extension that the techniques described in the last seven chapters could practically achieve, even in that extended timeframe: sure, they might be truly comprehensive if they worked absolutely perfectly, eliminating every scrap of their respective type of target damage, but we all know that no therapy is that perfect, and certainly not in its early versions. Thus, you're probably thinking that I have fallen drastically short of making my well-publicized case that a lot of people alive today may well live to be one thousand—and you'd be absolutely right. Thus, in this chapter and the next I'm going to put the scientific details to one side and directly address these two important and legitimate concerns. I'm going to start with the time frame for widespread availability of the panel of therapies described in this book.

I use the phrase "the War on Aging" to describe a specific phase in the process leading to the defeat of aging. I define it as the period beginning with the destruction of the pro-aging trance and ending with the widespread availability of therapies that can add a few decades to the life span of people who are already middle-aged. First I'll elaborate on this definition, then I'll explain why I call it the war on aging, and finally I'll explain why it has a good chance of lasting only fifteen to twenty years.

The pro-aging trance—the "rational irrationality" about aging that I described and critiqued in Chapter 2—will end only when its claim to rationality becomes unable to withstand even simple assaults, the sort that most people can understand. I believe that this will truly occur only when scientists obtain results in the laboratory—mainly with mice, I expect—that are so impressive that the majority of professional biogerontologists are finally prepared to say publicly that it's only a matter of time before we can postpone aging by at least a few decades in humans. Science is in a very real sense the new religion: what individual scientists say can be doubted, but the public scientific consensus is gospel. The result that I think is needed is something that I've called "robust mouse rejuvenation" or RMR.

RMR is a mouse life-extension result, and it's a rather precisely defined one. That's what's needed if we want scientists to set aside their paranoia about making predictions about what might happen in the future—we need to close all the major loopholes. By my definition, RMR will be achieved when:

- at least twenty mice of the species *Mus musculus,*
- from a strain whose natural average life span is at least three years,
- receive treatments starting only when they are at least two years old,
- and live to an average of five years of age, with all the extra time being healthy.

I thought about this definition pretty carefully before I publicized it, and it seems to be standing the test of time: no one has pointed out any way in which it could be achieved by "uninteresting" means, i.e., by means that would not convince knowledgeable scientists that a massive breakthrough had been made that was likely to be relevant to humans. The requirement to use at least twenty mice is so that we can be sure the age wasn't a fluke or a bookkeeping error. The requirement to use *Mus musculus* is because other mouse species already live longer than *Mus musculus* but are less well characterised by scientists. The requirement to use a strain of that species that naturally lives to three, which is unusually long for that species, is to avoid any possibility that the mice have some specific congenital problem, something that normally kills them rather young, and that the treatment merely alleviates that defect rather than comprehensively affecting aging. And of course the requirement that the treatment begin only when the mice are already two-thirds of the way through their natural life span is to ensure that it has potential relevance to people who are already alive, read the newspapers, pay taxes—and vote.

My reason for calling the period that begins with the achievement of this milestone the War on Aging arises from the initial, essentially immediate reaction that I expect society to have to it. In order to describe this reaction, I must first describe a side effect of the pro-aging trance that determines society's current reluctance to take aging seriously. I have a name for this, too: it's the triangular logjam. See **Figure 1**.

Experimental biology, like any other area of science, costs money—really quite a lot of money. Most biology isn't nearly as expensive as high-energy physics or astronomy, but it's expensive enough that professors have to spend

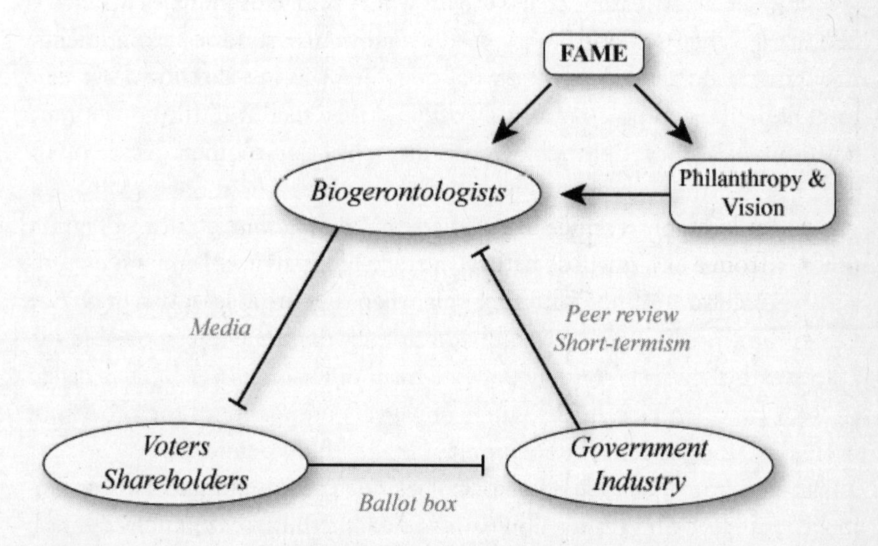

Figure 1. The triangular logjam impeding funding, and how philanthropy can unlock it.

a hideous amount of their time fundraising. The overwhelming majority of the funds that support experimental biology come from the public purse.

Biogerontology is typical in the above regard, but it's extremely unusual in one way: the public are absolutely fascinated by it, so biogerontologists get on the television all the time. I mean really, *all* the time. The difference in this regard between biogerontology and other biological fields—even really high-profile medical fields—cannot be overstated: even quite junior biogerontologists get more press attention than the world's leaders in other areas. And of course, when given that chance, biogerontologists are just as keen as any other scientist would be to talk about their own research—which, necessarily, is the research that they were able to obtain the funding to perform.

Consider, for a moment, what *else* a biogerontologist might choose to talk about to the media. In particular, consider the possibility of talking about research avenues that the public consider distinctly suspect: defeating aging, for example. What are the attractions of discussing such topics? Well, your name might get quite widely known to the general public, and you might get more media exposure. But hang on: What is the media exposure *for?* Scientists are intensely preoccupied, as I just mentioned, with the miserable business of maintaining a funding stream for their laboratories. How, exactly, would a high media profile achieve this—or, conversely, make it harder?

In order to explain the answer to that question, I must make sure you are aware of a key feature of the way in which public funding for science is allocated. When scientists want to do a particular series of experiments, they write a detailed description of what they want to do, how long they think it'll take and how much it'll cost, and they send it to the appropriate government agency. But the government agency doesn't then decide on its own whether the scientist can have the money. No: even though such agencies employ highly experienced ex-scientists as administrators of grant funds, those ex-scientists don't have anywhere near broad enough expertise to be able to tell the difference between a good idea and a poor one across the whole range of scientific disciplines that they're responsible for. So instead, they seek specialist advice from other scientists. This is called "peer review" and it's an absolutely universal component of the process of evaluating applications for government grants to do science.

Selection to evaluate your colleagues' ideas for experiments is an immense privilege and responsibility. It's not something that junior scientists get to do very often; generally the most senior scientists are the ones who do it most.

Do you see the problem yet?

Science is about the testing and refinement of hypotheses and theories. In principle, the most important quality of a scientist should be their ability to accept, with an open mind, evidence that challenges theories that they had believed for many years. But scientists are human, and moreover they know that the scientists that produced the new evidence are also human. In particular, they know that when a result is reported that contradicts established conventional thinking, the new evidence is often found later on to have been the result of experimental error. Thus, it is generally pretty hard to get senior scientists to change their mind about things, even if your evidence is really strong. The legendary physicist Max Planck famously remarked in the 1920s that "Science advances funeral by funeral" and this is only barely an exaggeration: It can take well over a decade for really fundamental changes of understanding of aspects of science to become generally accepted. A famous example in biology is the mechanism of action of mitochondria, cellular components that we've heard a lot about in this book.[1] And inevitably, this resistance to new ideas carries over heavily into how senior scientists evaluate grant applications.

So far, so unproblematic. After all, a modest degree of resistance to new ideas that you may not yet fully grasp is a good thing in some ways—we wouldn't want the scientific consensus to flit from one new idea to another too easily either, because (as I just mentioned) new ideas are often wrong. But the inertia that exists in scientific thinking is generally greater than this happy medium. And unfortunately, it's not just inertia of ideas, it's inertia of reputations. Senior scientists who have been appointed by the government to evaluate their colleagues' grant applications are, essentially by definition, members of the establishment. If they receive two applications of equal scientific merit, only one of which they have the resources to fund, and one is from a scientist who has a history of radical views about what science may shortly achieve, while the other is from a scientist who has never said anything outrageous in his life, you can bet that it's the latter who will get the money.

And that's not all. Reviewers of grant applications are, of course, given guidelines telling them what aspects of the applications' scientific merit or demerit they should consider particularly important. One aspect that is invariably high on this list, if not at the very top, is *feasibility:* perceived likelihood that the investigator will complete the proposed experimental program within the time and budget requested and obtain results that will merit publication in a reputable scientific journal. Sounds pretty uncontroversial, doesn't it?—but in fact this policy is a huge problem for science, because it is not (in practice) weighted by scientific significance. That's to say: A proposal for a study that will almost certainly tell us something whatever its outcome will fare much better in peer review than a study that may well tell us nothing, even if what the second study *might* tell us is far more important than what *any* outcome of the first study would tell us. This bias in favor of low-risk, low-gain research at the expense of high-risk, high-gain research pervades the whole of science and is extremely strong in biogerontology.

Well, I've spent quite a while in this chapter denouncing the stubborn closed-mindedness of senior scientists, but I hope you've taken on board that in the past couple of paragraphs I've explained that it's not entirely their fault: it's really the fault of their paymasters, the public funding agencies, a.k.a. the government. Grant reviewers are also grant recipients, by and large (though they don't review their own applications, of course). Thus, if the funding agency makes it known—either explicitly via written guidelines, or implicitly by their actions and off-the-record remarks—that they would prefer to fund middle-of-the road people to do dependable work than to fund troublemakers to do controversial work, the reviewers are hardly likely to dissent. It would be a very good thing if more such scientists did resist this sort of instruction, but realistically that's too much to ask.

But actually . . . it's not the government's fault either. The real culprit is you, the public.

This should not be a surprise to you. It's hardly a secret that governments in all democracies ultimately act to achieve one thing above all others: their own re-election. That will not be aided if the government spends, and is seen to spend, appreciable amounts of taxpayers' money on what the public regards as blue-sky pipe dreams: research that will probably lead nowhere. If the public were scientifically mature enough to appreciate that, in the long run, the rate of scientific progress is slowed by this overcautious approach, their elected representatives would be able to exercise similar vision and to instruct grant reviewers accordingly. But the public do not adequately understand how science works, so this doesn't happen. (A similar and even worse problem exists when it comes to medicine; I'll explore that topic later in this chapter.) This is basically because the judgment of how likely a hypothesis is to be true, or of how likely an experimental result is to lead to further knowledge, is actually one of the most sophisticated and unteachable aspects of doing science.

In biogerontology, however, there is potentially a way out—and this brings me back full circle (or triangle!) to the thing that distinguishes biogerontology from all other scientific fields in terms of its interaction with the public: the sheer extent of that interaction. Though there is no hope of turning the public into scientists sophisticated enough to understand the merits of highly ambitious experiments, there is ample chance simply to tell them that such-and-such an experimental approach is well worth pursuing. There's probably not enough chance of this for most scientific fields, but in the case of biogerontology it's abundantly possible. So, why don't biogerontologists do just that whenever they find themselves in front of a camera? I already told you: They don't want to gain a reputation for irresponsibility among their peers, because to do so would be to jeopardise their own chances of being funded even for unambitious work.

So there you have it—the triangular logjam. Biogerontologists are cautious in what they say to the public, in order to protect their funding, which is provided by the government, which are cautious in what and whom they fund, in order to protect their votes, which are provided by the public, who are fatalistic about what's even worth trying to achieve, because they see the biogerontologists saying only cautious things on the TV.

In order to defeat aging any time soon, I believe that an essential first step must be to break the triangular logjam. How can this be done?

Since I entered biogerontology, I've been chipping away primarily at one corner of the triangle: my fellow biogerontologists, especially the senior ones. Scientists are very politically aware, as described above, but they're also honest and sincere people. Moreover—and this is a key point—hardly any biogerontologists suffer from the pro-aging trance themselves. They know full well how horrific aging is, and with very few exceptions they want an end to it just as much as I do. Thirdly, there aren't very many of them, so personal contact is easy: I've known essentially everyone in the field personally for several years now. And finally, they're all smart enough to have earned doctorates. All in all, if I have a good strong case that we may be much closer to fixing aging than people have hitherto realized, shouldn't I be able to convince them—and even convince them to say so publicly?

Well, not quite—but nearly. As in any walk of life, what people say is important but what they don't say is also important. Those of my colleagues to whom I have presented the SENS panel in detail have mostly concluded that it is not fantasy, even though it's certainly very ambitious—but that hasn't translated into *explicit* public calls for SENS to be funded. What it has led to, however, is a variety of demonstrations of *tacit* support. It started with coauthorship by five senior colleagues of the first paper on

SENS, which arose from a workshop in 2000 (see Chapter 4); it's continued with refusal of several eminent colleagues to coauthor a denunciation of SENS published in the respected journal *EMBO Reports* and orchestrated by some of the less open-minded members of the community;[2] and most recently it's included the remarkable development that some people who did sign that denunciation have taken the initiative to divorce themselves from it by publishing constructive responses to the SENS agenda,[3,4] something that the *EMBO Reports* tirade specifically counselled against. While this may seem pretty tame as seen from the outside, I can assure you that it's an about-face as thorough as one ever sees in science.

It's obviously not enough, though. But it's all I'm likely to get from my senior colleagues in biogerontology for the time being, i.e., until I can piece together enough funding to give SENS research serious momentum despite its radical nature. Thus, in the (hopefully brief!) meantime I must address the other points of the triangle, too.

It's just conceivable that the government could be influenced directly. There are visionaries in government, and just occasionally they find a way to realize those visions. But in order to have any real chance on Capitol Hill or its counterparts in other countries, you really have to know the minds of the major players well—and that is something you don't pick up overnight. Thus, I've continued to leave that strategy to others—and I'm delighted to say that the ball seems to be slowly being picked up, most notably with the splendid "Longevity Dividend" initiative, a new effort spearheaded by the veteran lobbyist Dan Perry of the Alliance for Aging Research in collaboration with three gerontologists.[5] Whether they have much chance of success remains to be seen, but I emphatically wish them luck.

As I've become more prominent, however, I have been able to start addressing the third corner of the logjam: the public. You may recall that I started this book with a somewhat cantankerous complaint to the effect that if it weren't for the pro-aging trance I would be able to get on with the actual science and technology of defeating aging. Well, that's certainly true—but once given the chance, I have thrown as much energy into my advocacy and outreach efforts as I was already throwing into the science. Apart from anything else, I'm aware that the public are a source of funds in their own right, as well as a source of pressure on governments to alter their funding priorities.

The pro-aging trance dominates the nature of my interaction with the media, and via them with the public. The overwhelming majority of my time in interviews is occupied in discussions of the desirability of defeating aging, rather than the feasibility. But the good news, which I encounter mercifully often, is that it generally takes only a little bit of probing to reveal that the ultimate basis of my interlocutor's concern *is* their reluctance to accept the feasibility. It is this that convinces me so thoroughly that the achievement of robust mouse rejuvenation will consign the pro-aging trance to history in the twinkling of an eye.

The Intensity of the War on Aging, and Its Consequent Likely Duration

In order to give you a sense of what the world is likely to be like after RMR is achieved, I'm going to review some basic epidemiological and biomedical facts about three well-known viruses—HIV, CMV, and avian flu—and then examine an interesting scenario.

HIV has become one of the world's major killers. Belatedly, drugs that can suppress it and prevent it from progressing to full-blown AIDS are gradually becoming available in the developing world—still nowhere near in the quantities needed, but maybe soon even in those quantities. In the developed world, however, HIV is in a meaningful sense under control. It's possible to live with HIV for decades without any symptoms whatsoever, by the regular administration of expensive but (in the West) affordable drugs. What we still don't have, of course, are two key things:

- a drug to eliminate HIV from the body;
- a vaccine to stop it from infecting people.

CMV, cytomegalovirus, is not one of the world's major killers. Well, not obviously. In people with a normal immune system, it is completely suppressed and causes no symptoms at all. (My "not obviously" qualification arises from the fact that this suppression gradually wears down the immune system during aging, so that eventually people become more susceptible to more aggressive infections such as pneumonia; in this sense CMV is indirectly life-threatening. For more details on this and what we need to do about it, see Chapter 10.) But it is incredibly widespread: most Westerners are infected with it.

Avian flu is big news as I write these words (mid-2007), because for the past few years we have been somberly informed that it could soon mutate into a form that would cause a pandemic and potentially kill tens of millions or even hundreds of millions of people. All that needs to happen is for the avian flu virus to acquire genetic changes so that it can easily be passed from humans to other humans, the same as more familiar (and much less deadly) flu viruses can do. Such mutations are rare, but not astronomically rare; this could happen any time. Vaccines for avian flu are under development, but how well they will work depends on what mutations the virus picks up in the process of becoming human-to-human transmissible, and anyway vaccines often don't work so well on the elderly, who will be most at risk. Hence the possible death rates.

I've summarized these viruses as background for a scenario that I now want to explore in some detail, by way of an analogy with the situation after RMR has been achieved. I needed to lay out this background so that you appreciate that the scenario is reasonably realistic; I don't think I'll have any trouble convincing you that it's a valid analogy.

Let's suppose that HIV mutated to become transmissible by air, just like flu.

That's it. That's the situation I want to explore. Nothing else changes: the drugs to suppress HIV still work, they're still pretty expensive, and vaccines for HIV are still far away from development.

In this scenario, it's essentially certain that almost everyone in the world would have HIV within a couple of years. Not everyone gets flu in a pandemic, of course, but the difference is that once you get flu, if you don't die you mount an immune response that actually works, i.e., that eliminates the virus from your body. Thus, any given individual is infectious only for a rather short period (and after they've recovered, they can't be infected again, either). In the scenario we're looking at, once you get it you have it forever, and you're infectious forever. There will be no hiding place.

Pretty apocalyptic, isn't it? (Luckily, virologists think that in actual fact this scenario is vastly more unlikely than the corresponding mutation for avian flu.)

Well, hang on—is it so apocalyptic? We do have these drugs . . .

Let's look at a few round numbers. In the United States, roughly one person in every 250 has HIV, according to the Populations Reference Bureau—that's about one million people. The drug treatment to keep HIV under control costs about $30,000 per year, so that adds up to about $30 billion per year. Thus, if everyone in the United States had HIV, we'd be talking about $30 trillion per year. But the actual cost of production of these drugs is far, far lower: generic forms of them are being synthesized in

India and sold (still at a profit, mind) for only $300 per year, and even lower prices are in the offing. So actually we're looking at only $300 billion per year—$1 billion per day—to keep everyone in the United States healthy even if they all have HIV.

Now, a couple of points. First, you might think that "only $300 billion per year" is a pretty curious use of the word "only." Well, think again, because that's almost exactly what the United States is currently spending on the war in Iraq. (I'm not commenting here on the relative merits of these expenditures, you understand—I'm just pointing out that we have a precedent of an unexpected expense of the same size that is not bankrupting the nation.) Second, you might be against the infringing of patents, so you might object to my slashing the cost by a factor of a hundred. But is your belief in the patent system stronger than your belief in stopping your neighbors—or yourself, or your family—from coming down with AIDS and dying horribly? Ask yourself honestly: If this scenario actually happened, and one major party campaigned on a manifesto to raise taxes by $300 per year for the average person and to spend that money on generic drugs to prevent AIDS, and the other party campaigned on a commitment either to raise taxes by $30,000 per year for the average person or not to provide the drugs at all, who do you seriously think would get elected?

I hope I've convinced you what would happen in the above scenario: we would find the resources to treat everyone. We'd probably find the resources to treat everyone in the developing world, too, just as we're now stirring ourselves to treat everyone who needs such drugs in the developing world today.

Now, let's look at society's post-RMR view of aging in the same way. I suspect you can quickly see the similarities. Everyone has aging. The therapies we'll be looking to make available will be suppression therapies, which we will have to take for as long as we live (though much less frequently than those with HIV need to take their drugs). Within that limitation, however, the therapies will work: people's aging won't progress. But the therapy will be very expensive. (In the first instance that expense will be mainly for funding of research, training greatly increased numbers of medical personnel, building additional drug synthesis facilities and such like. The same figure I discussed above, $1 billion per day, is as good an estimate as any.)

So let's ask the opposite question: what are the *differences* between a post-RMR world and a universal-HIV world? I would say that there are really only two:

- the therapies won't yet exist at the time RMR is achieved;
- our acceptance that human aging can probably be defeated fairly soon will be new, while the universality of aging is age-old: this is the reverse of the situation with HIV in the above scenario.

I would argue that neither of these differences has any real chance of causing society to behave any differently in the aftermath of RMR than in the universal-HIV scenario. The nonexistence of the therapies is really no different than the nonexistence of enough antiretroviral drugs, which would certainly be the initial situation in the scenario I've described: we will work to develop those therapies as fast as possible, just as we would work to scale up production of antiretrovirals as fast as possible. The idea that the order of events could matter seems equally far-fetched; if everyone has a life-threatening health condition and we have a shot at making it no longer life-threatening, we'll clearly strive to do so.

Why "the War on Aging"?

I think you can probably see by now my reasons for calling this period "the War on Aging." In the early 1970s, there was an initiative called "the War on Cancer" that involved a sharp and sustained hike in the funding for cancer research fueled by the hope that cancer could be cured within as few as five years.[6] The war on cancer was not as abject a failure as some people tend to suggest—without that funding we would not have advanced nearly as fast as we have in our understanding of cancer, so there's little doubt that that initiative will have brought forward the true defeat of cancer quite substantially—but it was a complete misnomer, for one simple reason: the amount of money involved was really quite small, imperceptible to the U.S. taxpayer. As summarized above, the war on aging will be extremely expensive—not imperceptible at all. And yet, it is clear that the public will embrace the necessary tax rises: it will be quite obviously impossible to get elected except on a manifesto commitment to attack aging with all available resources. This is a mind-set that has previously been seen only at one type of stage in a wealthy nation's history: wartime.

Hippocrates and Gelsinger

In closing this chapter, I want to touch on one final aspect of the war on aging—one which completes my case that it may well last only 15-20 years.

In 1999, a teenager named Jesse Gelsinger died of anaphylactic shock in a trial of a gene therapy procedure at a hospital in Philadelphia.[7] This was the first such incident of its kind, and it sent the gene therapy world into its own form of anaphylactic shock. The bottom line was that essentially all gene therapy trials worldwide were suspended for about a year. We don't know how much delay that will eventually turn out to translate into in terms of the development of safe and effective gene therapy, but the chances are pretty good that it'll be a few weeks at least, and given the enormous breadth of applicability of gene therapy, that could mean thousands of lives lost—maybe even hundreds of thousands if it delays the defeat of aging. Bearing this in mind, was the suspension of trials for so long a proportionate response?

The U.S. Food and Drug Administration (FDA) would answer in the affirmative, as would their counterparts around the world. Regulation of experimental drugs and therapies, whether it be in terms of what results are needed or how they are obtained, is based on one abiding principle above all others: the minimization of risk that the therapy might make the patient worse. Specifically, this minimization of risk explicitly counts for more, much more, than maximizing benefit. In this way, the FDA is following a principle that has existed since medicine's earliest days: the famous edict of Hippocrates, *primum non nocere,* or "first do no harm." (Note that this phrase is actually not part of the Hippocratic Oath, the set of principles by which medical professionals swear to abide as part of their qualification process.)

I take the view, quite simply, that Hippocrates has had his day. The avoidance of harm was a rational strategy to adopt during the early days of medicine, when people very often recovered spontaneously from what their doctors thought were fatal conditions simply because the doctors had inadequate diagnostic tools. In such a situation, the *psychological* effect of possibly causing harm, whether it be the effect on the doctor or on the patient's loved ones, legitimately skews the objective cost-benefit analysis of a given treatment. In the modern world, however, such recoveries are relatively very rare. I therefore believe that the 10:1 (at least) ratio of lives lost through slow approval of safe drugs to lives lost through hasty approval of unsafe drugs[8] is no longer acceptable.

Furthermore, I believe that in the turbulence of the War on Aging, the general public will also come to the view that it is unacceptable. This will lead, in a matter of months from the achievement of RMR, to a root-and-branch revision of the laws and regulations governing clinical trials and approval of drugs and therapies. A fair guess is that drugs will be approved for universal use (via prescription) after a degree of testing that approximates today's Phase 2. People will die as a result; the 10:1 ratio mentioned above will probably reduce to 2:1. And people will be happy about this change, because they'll know that it's wartime, and the first priority—even justifying considerable loss of life in the short term—is to end the slaughter as soon as humanly possible. I am the first to acknowledge that, without such a change of priorities, my prediction that the war on aging may well last only fifteen years would be totally absurd. But with that change, only the pace of research will be limiting.

14

Bootstrapping Our Way to an Ageless Future

I have a confession to make. In Chapters 5 through 12, where I explained the details of SENS, I elided one rather important fact—a fact that the biologists among my audience will very probably have spotted. I'm going to address that omission in this chapter, building on a line of reasoning that I introduced in an ostensibly quite circumscribed context toward the end of Chapter 9.

It is this: The therapies that we develop in a decade or so in mice, and those that may come only a decade or two later for humans, will not be perfect. Other things being equal, there will be a residual accumulation of damage within our bodies, however frequently and thoroughly we apply these therapies, and we will eventually experience age-related decline and death just as now, only at a greater age. Probably not all that much greater either—probably only thirty to fifty years older than today.

But other things won't be equal. In this chapter, I'm going to explain why not—and why, as you may already know from other sources, I expect many people alive today to live to one thousand years of age and to avoid age-related health problems even at that age.

I'll start by describing why it's unrealistic to expect these therapies to be perfect.

Evolution Didn't Leave Notes

I emphasized in Chapter 3 that the body is a machine, and that that's both why it ages and why in principle it can be maintained. I made a comparison with vintage cars, which are kept fully functional even one hundred years after they were built, using the same maintenance technologies that kept them going fifty years ago when they were already far older than they were ever designed to be. More complex machines can also be kept going indefinitely, though the expense and expertise involved may mean that this never happens in practice because replacing the machine is a reasonable alternative. This sounds very much like a reason to suppose that the therapies we develop to stave off aging for a few decades will indeed be enough to stave it off indefinitely.

But actually that's overoptimistic. All we can reliably infer from a comparison with man-made machines is that a truly comprehensive panel of therapies, which truly repairs everything that goes wrong with us as a result of aging, is possible *in principle*—not that it is foreseeable. And in fact, if we look back at the therapies I've described in this book, we can see that actually one thing about them is very unlike maintenance of a man-made machine: these therapies strive to minimally alter metabolism itself, and target only the initially inert side effects of metabolism, whereas machine maintenance may involve adding extra things to the machinery itself (to the fuel or to the oil of a car, for example). We can get away with this sort of invasive maintenance of man-made machines because we (well, some of us!) know how they work right down to the last detail, so we can be adequately sure that our intervention won't have unforeseen side effects. With the body—even the body of a mouse—we are still profoundly ignorant of the details, so we have to sidestep our ignorance by interfering as little as possible.

What that means for efficacy of therapies is that, as we fix more and more aspects of aging, you can bet that new aspects will be unmasked. These new things—eighth and subsequent items to add to the "seven deadly things" listed in this book—will not be fatal at a currently normal age, because if they were, we'd know about them already. But they'll be fatal eventually, unless we work out how to fix them, too.

It's not just "eighth things" we have to worry about, either. Within each of the seven *existing* categories, there are some subcategories that will be easier to fix than others. For example, there are lots of chemically distinct cross-links responsible for stiffening our arteries; some of them may be broken with alagebrium and related molecules, but others will surely need more sophisticated agents that have not yet been developed. Another example: obviating mitochondrial DNA by putting modified copies of it into the cell's chromosomes requires gene therapy, and thus far we have no gene therapy delivery system ("vector") that can safely get into all cells, so for the foreseeable future we'll probably only be able to protect a subset of cells from mtDNA mutations. Much better vectors will be needed if we are to reach all cells.

In practice, therefore, therapies that rejuvenate sixty-year-olds by twenty years will not work so well the second time around. When the therapies are applied for the first time, the people receiving them will have sixty years of "easy" damage (the types that the therapies can remove) and also sixty years of "difficult" damage. But by the time beneficiaries of these therapies have returned to biologically sixty (which, let's presume, will happen when they're chronologically about eighty), the damage their bodies contain will consist of twenty years of "easy" damage and eighty years of "difficult" damage. Thus, the therapies will only rejuvenate them by a much smaller amount, say ten years. So they'll have to come back sooner for the third treatment, but that will benefit them even less . . . and very soon, just like Achilles catching up with the tortoise in Zeno's paradox, aging will get the better of them. See **Figure 1**.

Back in Chapters 3 and 4, I explained that, contrary to one's intuition, rejuvenation may actually be easier than retardation. Now it's time to introduce an even more counterintuitive fact: that, even though it will be much harder to double a middle-aged human's remaining life span than a middle-aged mouse's, multiplying that remaining life span by much larger factors—ten or thirty, say—will be much *easier* in humans than in mice.

The Two-Speed Pace of Technology

I'm now going to switch briefly from science to the history of science, or more precisely the history of technology.

It was well before recorded history that people began to take an interest in the possibility of flying: indeed, this may be a desire almost as ancient as the desire to live forever. Yet, with the notable but sadly unreproduced exception of Daedalus and Icarus, no success in this area was achieved until about a century ago. (If we count balloons then we must double that, but really only airships—balloons that can control their direction of travel

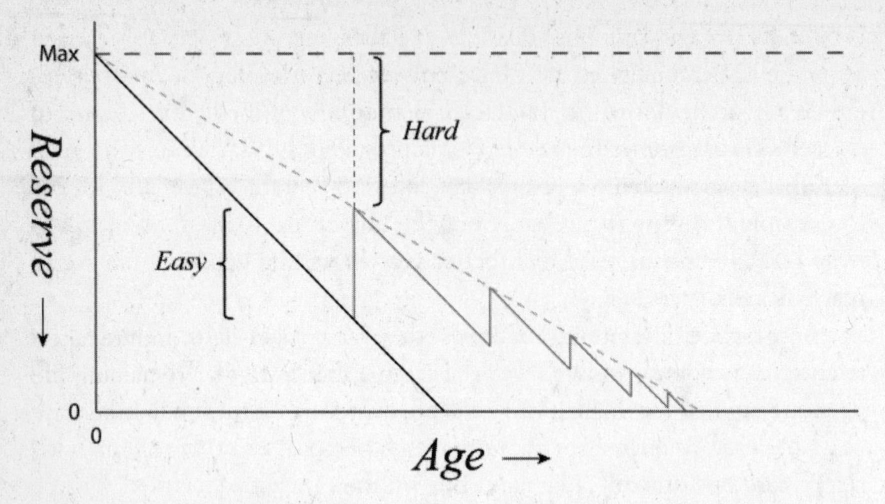

Figure 1. The diminishing returns delivered by repeated application of a rejuvenation regime.

reasonably well—should be counted, and they only emerged at around the same time as the airplane.) Throughout the previous few centuries, engineers from Leonardo on devised ways to achieve controlled powered flight, and we must presume that they believed their designs to be only a few decades (at most) from realization. But they were wrong.

Ever since the Wright brothers flew at Kitty Hawk, however, things have been curiously different. Having mastered the basics, aviation engineers seem to have progressed to ever greater heights (literally as well as metaphorically!) at an almost serenely smooth pace. To pick a representative selection of milestones: Lindbergh flew the Atlantic twenty-four years after the first powered flight occurred, the first commercial jetliner (the Comet) debuted twenty-two years after that, and the first supersonic airliner (Concorde) followed after a further twenty years.

This stark contrast between fundamental breakthroughs and incremental refinements of those breakthroughs is, I would contend, typical of the history of technological fields. Further, I would argue that it's not surprising: both psychologically and scientifically, the difficulty of bigger advances is harder to estimate.

I mention all this, of course, because of what it tells us about the likely future progress of life extension therapies. Just as people were wrong for centuries about how hard it was to fly but eventually cracked it, we've been wrong since time immemorial about how hard aging is to combat but we'll eventually crack it, too. But just as people have been pretty reliably correct about how to make better and better aircraft once they had the first one, we can expect to be pretty reliably correct about how to repair the damage of aging more and more comprehensively once we can do it a little.

That's not to say it'll be easy, though. It'll take time, just as it took time to get from the Wright Flyer to Concorde. And *that* is why, if you want to live to one thousand, you can count yourself lucky that you're a human and not a mouse. Let me take you through the scenario, step by step.

Suppose we develop Robust Mouse Rejuvenation in 2016, and we take a few dozen two-year-old mice and duly treble their one-year remaining life spans. That will mean that, rather than dying in 2017 as they otherwise would, they'll die in 2019. Well, maybe not—in particular, not if we can develop better therapies by 2018 that re-treble their remaining life span (which will by now be down to one year again). But remember, they'll be harder to repair the second time: their overall damage level may be the same as before they received the first therapies, but a higher proportion of that damage will be of types that those first therapies can't fix. So we'll only be able to achieve that re-trebling if the therapies we have available by 2018 are considerably more powerful than those that we had in 2016. And to be honest, the chance that we'll improve the relevant therapies that much in only two years is really pretty slim. In fact, the likely amount of progress in just two years is so small that it might as well be considered zero. Thus, our murine heroes will indeed die in 2019 (or 2020 at best), despite our best efforts.

But now, suppose we develop Robust Human Rejuvenation in 2031, and we take a few dozen sixty-year-old humans and duly double their thirty-year remaining life spans. By the time they come back in, say, 2051, biologically sixty again but chronologically eighty, they'll need better therapies, just as the mice did in 2018. But luckily for them, we'll have had not two but *twenty* years to improve the therapies. And twenty years is a very respectable period of time in technology—long enough, in fact, that we will with very *high* probability have succeeded in developing sufficient improvements to the 2031 therapies so that those eighty-year-olds can indeed be restored from biologically sixty to biologically forty, or even a little younger, despite their enrichment (relative to 2031) in harder-to-repair types of damage. So unlike the mice, these humans will have just as many years (twenty or more) of youth before they need third-generation treatments as they did before the second.

And so on . . . see **Figure 2**.

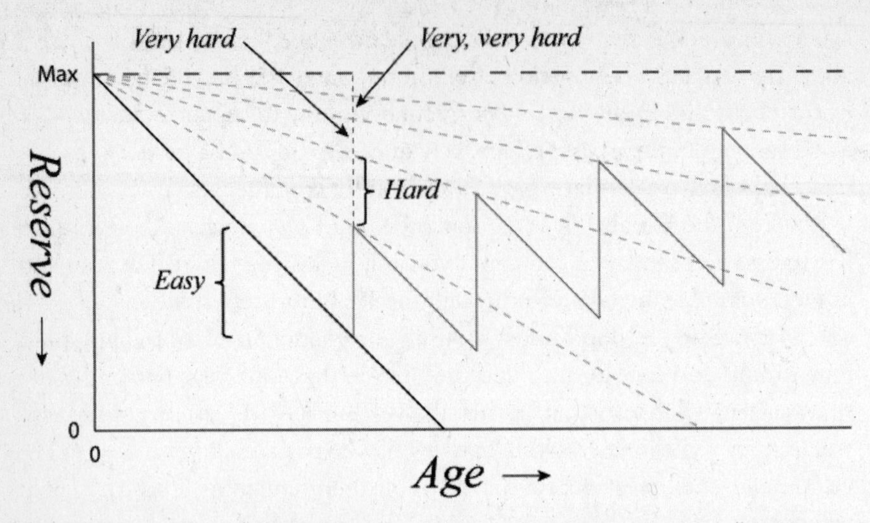

Figure 2. How the diminishing returns depicted in Figure 1 are avoided by repeated application of a rejuvenation regime that is sufficiently more effective each time than the previous time.

Longevity Escape Velocity

The key conclusion of the logic I've set out above is that there is a threshold rate of biomedical progress that will allow us to stave off aging indefinitely, and that that rate is implausible for mice but entirely plausible for humans. If we can make rejuvenation therapies work well enough to give us time to make them work better, that will give us enough additional time to make them work better still, which will . . . you get the idea. This will allow us to escape age-related decline indefinitely, however old we become in purely chronological terms. I think the term "longevity escape velocity" (LEV) sums that up pretty well.[1]

One feature of LEV that's worth pointing out is that we can accumulate lead time. What I mean is that if we have a period in which we improve the therapies faster than we need to, that will allow us to have a subsequent period in which we don't improve them so fast. It's only the *average* rate of improvement, starting from the arrival of the first therapies that give us just twenty or thirty extra years, that needs to stay above the LEV threshold.

In case you're having trouble assimilating all this, let me describe it in terms of the physical state of the body. Throughout this book, I've been discussing aging as the accumulation of molecular and cellular "damage" of various types, and I've highlighted the fact that a modest quantity of damage is no problem—metabolism just works around it, in the same way that a household only needs to put out the garbage once a week, not every hour. In those terms, the attainment and maintenance of longevity escape velocity simply means that our best therapies must improve fast enough to outpace the progressive shift in the composition of our aging damage to more repair-resistant forms, as the forms that are easier to repair are progressively eliminated by our therapies. If we can do this, the total amount of damage in each category can be kept permanently below the level that initiates functional decline.

Another, perhaps simpler, way of looking at this is to consider the analogy with literal escape velocity, i.e. the overcoming of gravity. Suppose you're at the top of a cliff and you jump off. Your remaining life expectancy is short—and it gets shorter as you descend to the rocks below. This is exactly the same as with aging: The older you get, the less remaining time you can expect to live. The situation with the periodic arrival of ever better rejuvenation therapies is then a bit like jumping off a cliff with a jet pack on your back. Initially the jet pack is turned off, but as you fall, you turn it on and it gives you a boost, slowing your fall. As you fall farther, you turn up the power on the jet pack, and eventually you start to pull out of the dive and even start shooting upward. And the farther up you go, the easier it is to go even further.

The Political and Social Significance of Discussing LEV

I've had a fairly difficult time convincing my colleagues in biogerontology of the feasibility of the various SENS components, but in general I've been successful once I've been given enough time to go through the details. When it comes to LEV, on the other hand, the reception to my proposals can best be described as blank incomprehension. This is not too surprising, in hindsight, because the LEV concept is even further distant from the sort of scientific thinking that my colleagues normally do than my other ideas are: it's not only an area of science that's distant from mainstream gerontology, it's not even science at all in the strict sense, but rather the history of technology. But I regard that as no excuse. The fact is, the history of technology is evidence, just like any other evidence, and scientists have no right to ignore it.

Another big reason for my colleagues' resistance to the LEV concept is, of course, that if I'm seen to be right that achievement of LEV is foreseeable, they can no longer go around saying that they're working on postponing aging by a decade or two but no more. As I outlined in Chapter 13, there is an intense fear within the senior gerontology community of being seen as having anything to do with radical life extension, with all the uncertainties that it will surely herald. They want no part of such talk.

You might think that my reaction to this would be to focus on the short term: to avoid antagonising my colleagues with the LEV concept and its implications of four-digit life spans, in favor of increased emphasis on the fine details of getting the SENS strands to work in a first-generation form. But this is not an option for me, for one very simple and incontrovertible reason: I'm in this business to save lives. In order to maximize the number of lives saved—healthy years added to people's lives, if you'd prefer a more precise measure—I need to address the whole picture. And that means ensuring that you, dear reader—the general public—appreciate the importance of this work enough to motivate its funding.

Now, your first thought may be: Hang on, if indefinite life extension is so unpalatable, wouldn't funding be attracted more easily by keeping quiet about it? Well, no—and for a pretty good reason.

The world's richest man, Bill Gates, set up a foundation a few years ago whose primary mission is to address health issues in the developing world.[2] This is a massively valuable humanitarian effort, which I wholeheartedly support, even though it doesn't directly help SENS at all. I'm not the only person who supports it, either: In 2006 the world's second richest man, Warren Buffett, committed a large proportion of his fortune to be donated in annual increments to the Gates Foundation.[3]

The eagerness of extremely wealthy individuals to contribute to world health is, in more general terms, an enormous boost for SENS. This is mainly because a rising tide raises all boats: once it has become acceptable (even meritorious) among that community to be seen as a large-scale health philanthropist, those with "only" a billion or two to their name will be keener to join the trend than if it is seen as a crazy way to spend your hard-earned money.

But there's a catch. That logic only works if the moral status of SENS is seen to compare with that of the efforts that are now being funded so well. And that's where LEV makes all the difference.

SENS therapies will be expensive to develop and expensive to administer, at least at first. Let's consider how the prospect of spending all that money might be received if the ultimate benefit would be only to add a couple of decades to the lives of people who are already living longer than most in the developing world, after which those people would suffer the same duration of functional decline that they do now.

It's not exactly the world's most morally imperative action, is it?

Indeed, I would go so far as to say that, if I were in control of a few billion dollars, I would be quite hesitant to spend it on such a marginal improvement in the overall quality and quantity of life of those who are already doing better in that respect than most, when the alternative exists of making a similar or greater improvement to the quality and quantity of life of the world's less fortunate inhabitants.

The LEV concept doesn't make much difference in the short term to who would benefit from these therapies, of course: it will necessarily be those who currently die of aging, so in the first instance it will predominantly be those in wealthy nations. But there is a very widespread appreciation in the industrialised world—an appreciation that, I feel, extends to the wealthy sectors of society—that progress in the long term relies on aiming high, and in particular that the moral imperative to help those at the rear of the field to catch up is balanced by the moral imperative to maximize the average rate of progress across the whole population, which initially means helping those who are already ahead. The fact that SENS is likely to lead to LEV means that developing SENS gives a huge boost to the quality and quantity of life of whomever receives it: so huge, in fact, that there is no problem justifying it in comparison to the alternative uses to which a similar sum of money might be put. The fact that life span is extended indefinitely rather than by only a couple of decades is only part of the difference that LEV makes, of course: arguably an even more important difference in terms of the benefit that SENS gives is that the *whole* of that life will be youthful, right up until a beneficiary mistimes the speed of an oncoming truck. The average quality of life, therefore, will rise much more than if all that was in prospect were a shift from, say, 7:1 to 9:1 in the ratio of healthy life to frail life.

Quantifying Longevity Escape Velocity More Precisely

This chapter has, I hope, closed down the remaining escape routes that might still have remained for those still seeking ways to defend a rejection of the SENS agenda. I have shown that SENS can be functionally equivalent to a way to eliminate aging completely, even though in actual therapeutic terms it will only be able to postpone aging by a finite amount at any given moment in time. I've also shown that this makes it morally just as desirable—imperative, even—as the many efforts into which a large amount of private philanthropic funding is already being injected.

I'm not complacent though: I know that people are quite ingenious when it comes to finding ways to avoid combating aging. Thus, in order to keep a few steps ahead, I have recently embarked on a collaboration with a stupendous programmer and futurist named Chris Phoenix, in which we are determining the precise degree of healthy life extension that one can expect from a given rate of progress in improving the SENS therapies. This is leading to a series of publications highlighting a variety of scenarios, but the short answer is that no wool has been pulled over your eyes above: the rate of progress we need to achieve starts out at roughly a doubling of the efficacy of the SENS therapies every forty years and actually declines thereafter. By "doubling of efficacy" I mean a halving of the amount of damage that still cannot be repaired.

So there you have it. We will almost certainly take centuries to reach the level of control over aging that we have over the aging of vintage cars—totally comprehensive, indefinite maintenance of full function—but because longevity escape velocity is not very fast, we will probably achieve something functionally equivalent within only a few decades from now, at the point where we have therapies giving middle-aged people thirty extra years of youthful life.

I think we can call that the fountain of youth, don't you?

15

War Bonds for the Campaign Against Aging

It is, as you now know, my bold but solidly grounded contention that there is a strong chance that *you*—the reader of this book—will live to experience the rejuvenation of your body into a flesh that is years or even decades younger, biologically, than your chronological age, leading ultimately to an endless summer of literally perpetual youth. But that is a prediction about what *could* be—not about what *must* be. Make no mistake: Once the War on Aging begins, it must end in victory, and the future of indefinite health will be ours. But whether that process begins in time to save our parents, or only ourselves, or only our children, or even *their* children depends entirely on when the first bomb of that war—the achievement of robust mouse rejuvenation (RMR)—is finally dropped. So with that understanding, the key question for each of us is, *What am I going to do about it?*

Granted that the War on Aging *could* begin in as little as a decade, it makes sense to take care of yourself and your family, so that you don't miss out on the arrival of robust human rejuvenation by just a few years. There is plenty of well-established science on this front: eat more fruits and vegetables, get your essential fatty acids, exercise, and maintain a healthy weight. But a much more effective way to increase your personal odds of seeing your body rejuvenated—and one that has the decided advantage that it also increases the odds of survival for those close to you, and for untold thousands that you have never even met—is to hasten the day when the battle is well and truly joined. Fortunately, there *are* some very powerful things that you can do, *today,* to help ensure the saving of lives, again of *tens of thousands of lives a day,* possibly including your own or your most dearly beloved. The most immediately obvious actions would be to lobby for more funding for rejuvenation research, and for the crucial lifting of restrictions on federal funding to embryonic stem cell research in the United States, by writing letters to your political representatives, demanding change.

But an even more powerful thing you can do is to donate to the Methuselah Foundation.

Let's step back a moment to remind ourselves of where we are today, and see how we can affect our future.

Remember the logjam that I outlined in Chapter 13? The reason why the investments necessary to bring forward robust mouse rejuvenation—the first and critical benchmark for the instigation of a total life-and-death struggle against biological aging—are so hard to obtain lies in a mutually reinforcing ring of politically directed funding restrictions, scientific overcaution in public statements and grant requests, and public opinion. The fastest way out of this vicious circle is to create an independent source of funding, pouring the needed billion or so dollars *directly* into work designed to achieve RMR. Unfortunately, it would be very difficult to raise the funds required to do this in a short time frame, precisely because of the widespread pessimism engendered in the public by scientists afraid for their careers in a world where the logjam exists.

There are two plausible ways to change that, and the Methuselah Foundation is the worldwide spearhead for both of them.

The first is to support SENS research directly. You can do that by donating to the Methuselah Foundation, because we are a standard science funding organization. Just like the National Institutes of Health or the National Science Foundation, our scientific team, headed by me as Chief Science Officer, evaluate scientific publications and research findings and provide funds to professors around the world. The difference between us and other agencies, of course, is that we are focused on a particular goal and we are not afraid to fund projects that may take a long time to succeed.

Or . . . that may not succeed at all. And that's why, by way of hedging our bets, we have the second strategy.

I believe that SENS is far and away the most promising way to achieve RMR soon and corresponding human therapies thereafter—but, like any scientist, I could be wrong. Really hard technological goals, whether in medicine or elsewhere, vary a lot in terms of how confident the experts in the relevant field are that their preferred approach will work. At one extreme, in some cases there is almost no doubt about how to proceed—all that holds the project back is availability of resources. The Apollo project was a fine example of this: once national pride freed up the necessary cash, the project went from start to finish faster than Boston's Big Dig. But at the other extreme—powered flight before 1900, for example, or essentially all medicine before 1800—people have ideas about what might work that are still extremely speculative, and attempt after attempt fails or never even gets tested. Testing everything that might work just costs too much for any organization, even a highly motivated government, to afford when there are so many comparably plausible possibilities.

Luckily, there is one time-tested strategy that has been successfully used, again and again, to solve engineering challenges of this sort without having to raise even a small fraction of the full sum required to complete the project. That strategy is the *research prize.*

Research prizes with a specific benchmark have a long and illustrious track record of producing the development of effective prototype technologies with only a small investment of funds. Charles Lindbergh's famous transatlantic flight in 1927; John Harrison's invention of a way to fix the longitude location of a vessel at sea (crucial to successful navigation away from a coastline); and the dramatic photo-finish race for the first privately funded suborbital flight with human cargo that was driven by the Ansari X-Prize, are all examples of the power of such prizes to spur daring technological innovation.

What is so powerful about research prizes is that they only reward success. Not a single dollar of the prize goes out until someone achieves the goal laid out by its creators. As a result, the existence of a *single* prize motivates *many* teams of independent scientists and engineers to go after it—each of them using a different approach, and each of them investing *their own dollars* independently. As a result, the money that is mistakenly invested in approaches that are ultimately shown to be unsuccessful does not deplete the prize jackpot by a single penny. In the end, about ten to twenty dollars is ultimately laid out by the contestants for every dollar raised to put in the jackpot of a research prize—even though *each* contestant usually gambles less private money than is sitting in the pot—and the goal of the prize is achieved at a cost to the prize's principals that is far less than would be required for a monolithic "Apollo Project" approach.

You can see where this is going.

The Methuselah Mouse Prize (or Mprize, with a tip of the hat to the recent success of the X Prize) is a project that was dreamed up independently around the turn of the century by a few biogerontologists (starting with Gregory Stock's concept of the Prometheus Prize) and by long-time humanitarian visionary David Gobel. When David and I discovered each other, our complementary talents allowed us very rapidly to bring the prize to fruition. The purpose of the project is to break the logjam that we've been discussing using a research prize structured similarly to the X Prize model. Enjoying the support of X Prize Chair and CEO Peter Diamandis as a chief advisor, the Mprize's chief project is the Rejuvenation Prize for the greatest extension of life span in mice that are already elderly—in other words, for progress toward fully-fledged robust mouse rejuvenation.

The Mprize has the potential to remove the stumbling blocks that currently restrict scientists in government and industry from taking on the aging process as a curable disease. For the scientists in academia, it creates an incentive to write the right grant proposals, in hopes of obtaining more funding directly and greater prestige for their institutions by winning the prize—prestige that itself tends to attract more funding from outside sources.

But not only that: it's a popular win, too. The prize concept, by its nature, captures public imagination and provides a dramatic way to inform the public and the media that scientists are working on extending healthy life spans in mammals. This increases the credibility of any similar reputable efforts, and wins acceptance for the idea that it can be done in humans. In turn, changes in public opinion ease political constraints on awarding public funding for such projects, and may even generate active pressure for such awards to be made *before* RMR has occurred.

And the Mprize also reorients the incentives for industry. Right now, there is no specific motivation for private researchers to perform life span studies in mice: at most, they are a stepping stone toward long, expensive, human trials. When a significant financial reward—and the promise of substantial publicity—is on the table, however, a business case is created for spending a few years rather than a few months in testing a compound in mice. Should a startup company succeed in rejuvenating mice, you can bet that Big Pharma will be beating down its door for the rights to translate their intervention to humans.

Thus, the Mprize can bypass the vicious circle that has put the chill on serious biomedical gerontology in academic research. More than this: it can *reverse its course,* putting its converging, self-reinforcing mechanisms to work in a new, *virtuous* circle. Scientific results will drive public optimism, in turn driving political acceptability, leading to more public and private funding. Those investments will eventually—maybe very rapidly—lead to robust mouse rejuvenation even if the SENS approach were to falter; and then, the War on Aging will begin in earnest.

Eat well, exercise, and support the Methuselah Foundation, and I shall look forward to shaking your hand in a future where engineered negligible senescence is a reality: where we can enjoy dramatically extended lives in a new summer of vigor and health, the dark specter of the age plague driven away by the sunshine of perpetual youth.

Afterword

Michael and I did a lot of agonizing over the manuscript of *Ending Aging* in the winter of 2006–2007. We were at once cutting out of the manuscript precious details to make sure that readers wouldn't get bogged down in excessive detail, and restraining our natural impulse to add in *more* detail, as labs all over the world continued to report results that were advancing the science underlying SENS. We also knew, of course, that there would be months of editorial and publishing work between our submission of the manuscript and its ultimate publication, so that the book would be out of date when it hit the shelves, no matter how up-to-date we tried to make it.

The ensuing months would bring home to us the fact that, as with computing power, biomedical gerontology is now advancing on an exponential curve. The release of this new paperback edition gives me the opportunity to share with you the huge strides that have been made in just one year. I've structured this afterword according to Part Two of the book, with a final section that extends Chapter 14.

Meltdown of the Cellular Power Plants

Part 2 began with a chapter that differed from the rest, in that rather than describing a solution to an aspect of aging it focused on evidence for whether the phenomenon in question—accumulating mitochondrial mutations—is important in aging at all. This question is still wide open. However, some important findings have been published recently that bring an answer closer. I'll summarize just two, both of which involve the "mitochondrial mutator mouse" and related models that were developed independently in a number of laboratories in recent years.

A particularly interesting finding, from the laboratory of Nils-Göran Larsson at the Karolinska Institute in Stockholm, was that mice in which some brain cells were mitochondrially mutant and others were not exhibited a pattern of mitochondrial dysfunction that was not restricted to the mutant cells. To be precise: mutant cells were indeed unhappy, as a result of "suffocating" (being unable to use oxygen to extract energy from nutrients), but cells that were *adjacent* to mutant cells were *also* unhappy.[1] This is, to my knowledge, the first evidence for "collateral damage" inflicted by a mitochondrially mutant cell on a mitochondrially healthy one in a living organism. The mechanism is probably not related to my "reductive hotspot" proposal, but interest in all possible intercellular mechanisms for the pathogenicity of mitochondrial mutations is sure to rise as a result of this finding.

Secondly, Larry Loeb's group at the University of Washington in Seattle have examined what *types* of mitochondrial mutations accumulate in the mutator mice and which of them are harmful. By comparing not only the mutator mice with normal mice, but also examining "heterozygous" mice that carried one mutator mutant and one normal gene, the group was able to demonstrate that elevated levels of deletions, and not of point mutations, were the cause of these mice's shortened lifespan, and in particular that the pathogenicity appeared to correlate with the clonal expansion of these deletions and the concomitant loss of normal mitochondrial DNA from the affected cells.[2] There are still reasons to be cautious about the fidelity with which these mice recapitulate the problems in normal aging, because the cell types that are most affected seem to be different. Nonetheless, this study constitutes important support for the idea that the clonal expansion of deletion mutants, making their host cells incapable of using oxygen, is the originating mechanism for why mitochondrial mutations are bad for us.

Getting off the Grid

As you'll recall, mitochondrial "power plants" have their own DNA, which is uniquely vulnerable to damage because of its proximity to the cell's major source of mutagenic free radicals. The solution I favor is called *allotopic expression*: moving backup copies of this DNA into the "bomb shelter" of the cellular nucleus. But this approach has been frustrated by a property of the proteins encoded by the mitochondrial DNA: their *hydrophobicity,* which makes them curl up into a form that the mitochondrion cannot import.

Setting Up a Branch Plant

Historically, as described in Chapter 6, the most promising approach to overcoming the hydrophobicity problem has appeared to be to look for changes to the relevant proteins' sequence—changes that somewhat lower the protein's hydrophobicity without impairing its function. We're still very keen on that, but we're now pursuing it in parallel with a complementary approach. This second approach does not involve changing the sequence of the proteins to be imported. It has been developed by Dr. Marisol Corral-Debrinski, of the Quinze-Vingts National Center of Ophthalmology in Paris, who hopes to use allotopic expression to cure inherited mitochondrial diseases that commonly cause blindness.

It had been known for decades that some of the mitochondrial proteins that are already naturally nuclear-encoded in our cells are actually *produced* very close to the mitochondrial surface itself, thanks to having their mRNA, "messenger RNA" (the instructions, copied from the original template in the nuclear DNA, that provide the direct working orders for the cell's protein production machinery) effectively "addressed" to protein "factories" located close to mitochondria—just as a multinational corporation might have branch plants in different countries tool up to create product lines customized to the local market. Various researchers had speculated that proteins synthesized on the mitochondrial surface would be importable even if they were viciously hydrophobic, simply because the curling-up process is not instantaneous: If a protein is being produced right next to the mitochondrion, it can be imported as it's built ("cotranslationally," to use the technical term), and before it has the chance to curl up at all.

Corral-Debrinski, working initially in yeast but more recently in mammalian cells, was the first person to turn this speculation into reality. She discovered the "address tag" that caused the cell to transport some mRNAs to the mitochondrial surface was in the very back end of those mRNAs' sequence—a sequence called the "three-prime untranslated region." The operative word here is "untranslated"—this region is an ostensibly useless chunk of sequence, because it comes after the point where the protein factory (the ribosome) that reads the mRNA has fallen off on encountering the code that says, "Here's the end of the protein." This means that, in principle, one should be able to take the DNA that encodes a hydrophobic protein, put a sequence on the front that tells the cell that the protein is destined for mitochondria (see Chapter 6), and put one on the back that tells the cell that the mRNA is destined for mitochondria, and the protein should be imported efficiently despite its hydrophobicity.

Corral-Debrinski and her team did just this in mammalian cells, using the gene for ATP6, which is the one on which most published studies of allotopic expression have focused. They were pleased to see high levels of protein successfully imported into the cell's mitochondria—and so were the editors of *RNA,* the journal that published the result.[3] But that still left open the question of whether the protein was *functioning* as part of the ATP-synthesizing machinery once it got there.

Her group's next experiments proved that it was. When they loaded the same specially directed ATP6 gene construct into skin cells harboring the same ATP6 gene mutation that causes NARP in human subjects, the result was "a long-lasting and complete rescue of mitochondrial dysfunction." When provided with various fuel sources under conditions that force them to rely entirely on oxidative phosphorylation for ATP (blocking off the less efficient glycolytic pathway—see Chapter 5), NARP skin cells' peak energy output was reduced by 61%–74 percent compared to normal cells. But when the same cells were given the doubly targeted but otherwise normal ATP6 gene construct as a nuclear backup copy, energy output rose to fully *98 percent* of a normal cell. And to prove that it wasn't a fluke—and that the approach could be used to get around other mitochondrial mutations—she successfully repeated the same trick with a different mutation (LHON) in a different mitochondrial gene (ND4), which encodes a subunit of a different complex of the oxidative phosphorylation chain (Complex I)—and even used a different native double-targeting system to do it (this one copied wholesale from COX10, whose mitochondrial protein helps to assemble Complex IV, instead of from SOD2).[4]

As soon as I heard of these results, I invited Corral-Debrinski to present her results at SENS3, the third in the series of interdisciplinary conferences I've hosted in Cambridge (beginning with IABG 10) to bring together researchers from all of the disparate fields of science and technology that will be needed to implement the full SENS platform. She agreed, and by the time the conference rolled around she had even more exciting news to report: She had shown that the system works not just in cells, but in *whole, living rats.*

Corral-Debrinski's group did not *fix* an existing mitochondrial mutation in the animals—no one has yet developed a colony of LHON or NARP lab animals. Instead, they used the double-targeted allotopic expression system to insert a copy of ND4 bearing the most common LHON mitochondrial mutation into the eyes of rats. The gene was taken up and expressed at high levels in about 40 percent of the *retinal ganglion cells* (RGCs—a kind of nerve cells that relay information from the retinal light receptors to processing centers in the brain, defects in which induced by LHON cause the associated blindness). As a result, RGC viability plummeted, and many of the cells died outright. When raised in culture, neurons taken from such animals failed to properly extend the branches (*dendrites* and *axons*) that they use like telephone lines to communicate with other neural cells.[5]

It's important to note this result only shows *that* the allotopically expressed protein is toxic to the RGC—not *why* it is. Allotopically expressed proteins might themselves cause problems just by being in the cell, even if they didn't actually integrate into the relevant Complex of the oxidative phosphorylation system. But that concern was partially addressed (and the general progress of the allotopic expression technology was further advanced) by another study coming out of the University of Florida's medical campus in Gainesville, published just as we were sending *Ending Aging* to press.[6] These scientists reported having used a completely different sequence to target the ND4 protein to the mitochondria, and weren't even relying on localizing the protein's mRNA to the mitochondrial surface. In this study, some rats' eyes were targeted with LHON-mutant versions of ND4, just as Corral-Debrinski's group had done—but they *also* used a second group of rats, in whom they allotopically expressed the *healthy* protein using the same system. The encouraging result: allotopic expression of the normal protein was harmless—but that expressing the defective protein caused toxicity to the RGCs similar to that observed by Corral-Debrinski's group in Paris.

While both Michael and MitoSENS researcher Mark Hamalainen have pointed to some weaknesses to this study, it at least demonstrates that an allotopically expressed ND4 is not itself harmful to the cell. And the Paris group now plans to use their allotopic expression system to combine both arms of the Gainesville experiment into one, elegant proof-of-concept. Her group will first *cause* LHON in rats by allotopically expressing the mutant ND4 into their RGCs—and then *cure* it, by inserting the healthy one afterward.

Combined, the two labs' results told me how Mark's Methuselah Foundation grant could best be used. Mark has now shifted base to Paris, where I've arranged for him to work with Corral-Debrinski's team directly, learning their methods, using their cells, and refining their protocols to see whether her system will overcome the problems the Holt lab has had with ATP6. The Foundation has also expanded the resources of the Paris lab by funding technician Sébastien Augustin to provide essential research support to Mark and future members of the group. If Mark's efforts are successful, he'll go on to collaborate on allotopic expression of yet more mitochondrial genes.

Teach a Man to Fish . . .

Dr. Samit Adhya, of the Division of Molecular and Human Genetics at the Indian Institute of Chemical Biology, is pursuing a completely different approach to mitochondrial resuscitation. His work focuses on providing mutant mitochondria not with finished proteins (as in allotopic expression), but with RNA. Not the *mRNA* we've been talking about, however, but *transfer RNA* (tRNA), which are tools that deliver the protein building-blocks (amino acids) to our protein factories (the ribosomes) for incorporation into a protein. A given tRNA can only handle one of the twenty types of amino acid that occur in proteins.

Proteins encoded in the mitochondrial DNA are built by mitochondrial ribosomes with the help of mitochondrial tRNAs. Mammals encode all their mitochondrial tRNAs in mitochondrial DNA, but some single-celled organisms instead encode some of them in nuclear DNA, and then *import* them after production in the cytosol, just like proteins. One species, a skin parasite called *Leishmania tropica,* does this for *all* its mitochondrial tRNAs, using a protein complex (appropriately called the *RNA import complex,* RIC) that it has evolved for this purpose. Adhya reasoned that if he could introduce RIC into mammalian cells, mammalian mitochondria might take up the RIC-bound tRNA in the same way that these simpler organisms do, bypassing the need for a mitochondrial copy of the DNA to make tRNA *within* the mitochondria. This would have potential therapeutic use, since mutations in the genes for tRNA are responsible for several inherited mitochondrial diseases, such as *Myoclonal Epilepsy with Ragged Red Fibers* (MERRF), whose victims suffer loss of mitochondrial function and death of muscle fibers.

And indeed, it worked. Adhya first showed that mammalian cells would indeed take up RIC-bound tRNAs. Then, in 2006, he took cells whose mitochondria harbored the same tRNA mutation that causes MERRF, and cured them of their mitochondrial defects by delivering the relevant tRNA using the RIC system.[7]

But how to show that the system would work in an intact animal? A normal mitochondrion manufactures its own tRNA, so the functionality of such cells would not be improved by adding more. And Adhya faced the problem that plagues research into mitochondriopathies: a lack of a viable animal model of the disease. So, like Corral-Debrinski, Adhya decided to work *backwards* as a preliminary test. You may recall studies discussed in various places in the book using a molecular tool called "antisense RNA": a mirror-image copy of a corresponding functional RNA, which binds to it and thereby inactivates it. So instead of delivering more of the tRNA itself, Dr. Adhya used the RIC to force the mitochondria to take up antisense RNA, thereby preventing protein production and fouling up mitochondrial function. Sure enough, mitochondria took up the RICs and the relevant antisense RNA, and the cells also had the expected defects in mitochondrial function and energy production. Now Adhya was in a position to test an intact animal. His team injected RIC-bound antisense RNA for ND1 (a subunit of Complex I) into the rear left paws of normal, healthy rats, to show that the cells of intact animals would respond in the same way that cells in a vat had done. Sure enough, the animals' feet started suffering muscular degeneration eerily similar to human mitochondrial disease. After initial swelling and inflammation, the animals exhibited a severe slackening of the muscles in the paw, and were clearly hobbling, dragging the affected paws along behind them. The effects were tracked down to the muscle cells taken from biopsies from the affected feet, which documented the extensive necrotic death of the local muscle cell tissue.[8]

There's an obvious potential to use this RIC biotechnology to deliver functional tRNAs to cure victims of mitochondriopathies linked to tRNA mutations, such as MERRF. But also, there's a way in which, conceptually at least, this borrowed technology might one day be used to overcome the problem of the large deletion mutations that progressively colonize more and more of our long-lived cells with age, driving the rise in oxidative stress.

This could be done if it could be used to deliver not only tRNAs but also actual *mRNA* to replace the mRNAs for which knocked-out mitochondrial DNA can't serve as a template—something that no known species does, and that might be frustrated by the greater length of mRNAs than tRNAs. And sure enough, Adhya informs me that he is now successfully using the RIC to deliver normal mitochondrial mRNA into mitochondria, in work that will probably be in print by the time you read this.

The next step will be to introduce functional RNA into animals with dysfunctional mitochondrial genes. If this restores normal mitochondrial function and blocks the symptoms and pathology associated with the disease, there's no reason to think that it couldn't do the same thing for the human equivalents; if so, the technology should allow us to sidestep not only the mutations in the mitochondrial DNA of those rare and unfortunate souls who suffer with congenital mitochondrial diseases, but also those responsible for the universal mitochondrial failures of aging.

And More Progress . . .

These two advances were certainly the most important to have been reported since the hardcover edition of this book was written—but there's more. One group, headed by Dr. Tonio Enriquez of Spain's University of Zaragoza, is working on a solution to the hydrophobicity problem for allotopic expression using a method that I first proposed in 2000:[9] the protein "bracing bars" called *inteins,* that (as described in Chapter 6) would keep allotopically expressed mitochondrial proteins flexible until after mitochondrial import. And in another novel solution, Dr. Volkmar Weissig of Northeastern University reported the import of *whole new mitochondria* into mitochondria-deficient cells[10]—an effect which, as it turns out, has a lot of independent support. By coincidence, another group reported shortly before Weissig that "[T]he active transfer [of mitochondria] from adult stem cells and somatic cells can rescue aerobic respiration in mammalian cells with nonfunctional mitochondria."[11]

Ultimately, I'm not attached to allotopic expression as *the* solution to this problem: As engineers of aging, our focus is *endpoints,* not procedural niceties, and the more trails that are being blazed to our destination (functional mitochondria, impervious to their self-created mutagenesis), the faster we will start saving lives, and alleviating the sickliness of aging bodies made sluggish by the wastes generated by failing power plants.

Upgrading the Biological Incinerators

A lot of work has been underway this year on identifying new enzymes to clear out the recalcitrant wastes that build up in our cells with aging, and the Methuselah Foundation has been at the heart of it. We currently support two labs working, using different approaches, to identify genes, enzymes, and degradation pathways used by a variety of microorganisms to degrade two key lysosomal wastes that contribute to aging and age-related disease.

One group, at the Center for Environmental Biotechnology at Arizona State University, is headed by Ph.D. candidate John Schloendorn under the guidance of Dr. Bruce Rittmann, and has enjoyed extensive intramural assistance from young scientists devoted to curing the age plague. To date, the lab has benefited substantially from the volunteer efforts of Kent Kemmish at the University of Arizona in Tucson, Justin Rebo at St. George's University, Grenada, and (earlier in the project) the multitalented Mark Hamalainen. The team will soon expand to include Tim Webb of Bishop's University in Quebec and (if everything works out) Lauri Tontson from the University of Tartu, Estonia, both of whom have previously done some related work at the Center and who have now gone to significant trouble to make arrangements to contribute directly to this project. The other lab, at Rice University's Environmental Engineering lab, is being spearheaded by grad student Jacques Mathieu, under the supervision of department head Dr. Pedro Alvarez.

Drano Meets Bioremediation Meets Cardiology

One key area of focus to date has been *7-ketocholesterol* (7-KC), which I mentioned in Chapter 7 as the constituent of oxidized cholesterol that appears to be the most important contributor to lysosomal dysfunction, foam cell formation, and ultimately, heart disease. Both groups start by identifying which of a variety of microorganisms can digest 7-KC by putting them in a medium in which this waste product is the only energy source available for food. These studies have identified a variety of bugs with hearty appetites for the waste, including species of the genus *Nocardia,* a group of rod-shaped bacteria that (unsurprisingly) are found widely dispersed in soils rich in organic matter.

The ASU team then tracked which degradation products were produced as the bugs eat their way through the material, looking for key metabolites that would give insight into the biochemical pathways by which they sequentially break 7-KC down—and, thereby, the properties of the enzymes responsible. So far, they've identified two likely key intermediates, and some of the chemistry and energy sources that the enzymes involved probably require. They've also shown that these species can degrade modified forms of 7-KC that probably exist in lysosomes and contribute to its toxicity, but that our macrophage lysosomes are especially unlikely to be able to handle, suggesting that the species in question really are doing a job that our own cells can't. The Arizona team has also identified a specific enzyme that can initiate the breakdown of 7-KC; unfortunately, it also breaks down *unmodified* cholesterol, so it's not clear if we can use it as it occurs in nature. If not, though, understanding the action of this enzyme should help us to identify (or design) a variation of it that has a more pinpointed activity. John gave an update at SENS3,[12] and the group are now working to characterize the reactions and the enzymes responsible further.

At Rice, Jacques Mathieu has focused his attention on a bacterial species that happens to be very extensively used in bioremediation. The species (called RHA1) looked particularly promising, because its use in bioremediation has led to its genes being well characterized. Importantly, "gene expression microarrays" have been constructed for these organisms, which scientists can use to tell which genes are turned on at a given time.[13] Genes that are expressed more heavily when the organism starts feeding on 7-KC are likely to be involved in producing the enzymes required to keep the feast going. Therefore, performing expression microarrays should allow Jacques to identify the key genes—and thereby, ultimately, the actual enzymes—that allow these humble bacteria to do the job that our own cells can't.

In a collaboration with Dr. William Mohn at the University of British Columbia, Jacques has identified two clusters of such genes that are turned on when the bacteria are feeding on 7-KC—but *not,* importantly, when they are given unmodified cholesterol. Moreover, analysis of these genes suggests that they're involved in the degradation of the right family of chemicals. Like the ASU team, Jacques has also done some work in studying the pathways of these reactions. They've now been able to make copies of the DNA responsible for making the enzyme and inserted them into the common bacterium *E. coli,* which will allow them to produce enough of the 7-KC-digesting enzymes to allow for extensive further study.

Cast the Aggregate from Thine Eye

One of the other major compounds the program at Arizona State is targeting is A2E, the mangled vitamin A derivative that is the responsible for an inherited form of blindness called *Stargardt's macular degeneration,* and is generally accepted to be behind "wet" macular degeneration in the rest of us. John's team had spent much of 2006 and 2007 trying to use the kind of microbial dietary restriction exercises that had worked so well for 7-KC to identify something that would eat A2E, but with no success.

Frustrated by this, our researcher Kent Kemmish began to wonder about the *kind* of enzyme that they were ultimately looking for, and the relationship that it would bear with enzymes already known to break down some of A2E's close chemical cousins. Was it possible that a *known* enzyme, already characterized, might be redirected (perhaps with a little molecular engineering) into clearing out the vision-stealing junk? He started to trawl through the relevant literature and, surprisingly, he quickly identified two enzymes or enzyme classes (let's call them X and Y),[14] that seemed to fit the bill. Enzyme X was so common that it seemed a surprising candidate for a cure for a prevalent cause of blindness, but there was no scientific reason to rule it out—and it was easy enough to find a commercial supplier. Meanwhile, Kent got in touch with a lead researcher who had published some studies on enzyme Y. After helpfully answering Kent's questions, the researcher ultimately disclosed that in addition to the enzymes characterized in the published reports that had led Kent to call, they also had a much larger collection of similar enzymes in their lab.

When the commercial enzyme X arrived, Kent tested its ability to degrade A2E. It's reasonably easy to get a quick answer to this question because as a relative of vitamin A and the carotenoids that give most orangey vegetables their color, A2E is dark red; as it is degraded, it slowly loses pigmentation, typically winding up a pale yellow. Sure enough, he found that exposure to the enzyme for an afternoon, especially at higher concentrations of aggregate (where the pigment was darkest to begin with) did begin to lighten the color of the plate. Using purer A2E and waiting for longer (six days) led to even better results.

The fruits of Kent's collaboration to study enzyme Y were even more bountiful. Working with the graduate student who had done much of the work on the already published members of the lab's enzymatic library, he performed the same simple visual study to see which of them might be able to degrade the high-grade A2E that he'd brought along. To his delight, one of them not only had an effect, but a very powerful one: squirting a titer of enzyme into the vial totally bleached the red pigment away after an overnight incubation.

Flushed with this success, the group put in a call to Dr. Janet Sparrow of Columbia University's Department of Ophthalmology, who is one of the most distinguished scientists working on macular degeneration today, and has a particular focus on retinal pigment cells and the role of A2E in macular degeneration. Along with Rittmann and Alvarez, Sparrow had participated in the National Institute of Aging workshop that I'd put together to explore the LysoSENS concept, and had helped to prepare the report detailing and endorsing the proposal.[15] She repeated and confirmed John's and Kent's results, and expressed interest in going further.

We are delighted to have an investigator with Sparrow's expertise and resources working on the project, so she's now using grant money from the Methuselah Foundation to identify the substances produced when the enzymes degrade A2E, with an eye to figuring out the essential chemistry. Next, she intends to deliver them into retinal pigment epithelial cells preloaded with A2E, to see if the same thing happens in living cells as under glass—and if so, how well and how safely. If the test in cells is successful, we'll ask her to test any viable-looking enzymes out to see if they are safe in—and can restore vision to—mice with a congenital form of macular degeneration similar to Stargardt's disease. If all goes well, we could begin to move them (or a biochemically "tweaked" version of them) through other animal models in a few years, with the ultimate intention of using them in clinical trials to cure human patients.

Taking a Splice out of Life?

Before concluding my update on "intracellular junk" I want to touch briefly on a discovery that might, at first sight, seem to fit rather poorly into the seven-point SENS classification. You'll be relieved (as I was!) to discover that this can now be confidently described as a false alarm.

There's a good chance you've seen images of children with *Hutchinson-Gilford progeria syndrome* (HGPS, or sometimes just "progeria"): the small children whose wrinkled skin, hair loss, and pinched nose make them look "old before their years." Beyond these visible hallmarks, the mutant gene that causes these defects also gives these children thin, weak bones, and a kind of heart disease, so that they indeed die young (typically in their teens), making it seem all the more as if their bodies had been turned onto biochemical fast-forward. And indeed, there are plenty of people who will happily refer to, or think of, HGPS as a kind of "accelerated aging"—or will, at least, assert boldly that the disease tells us something important about "normal" aging.

In fact, it's not at all clear that it does. The mechanisms underlying some of the symptoms of HGPS are clearly not the same as seen in the same organs during normal aging, and many of the other consistent aging pathologies are absent entirely: sufferers' eyesight and hearing are fine, they are not at any particular risk of cancer, and (small mercy) their minds remain sharp throughout their short lives. The mechanism underlying their tragic symptoms was identified back in 2003: an unusual sort of mutation called a "splice variant" that messes up the production of the mRNA used to make a protein called *lamin A,* which is part of the scaffolding that supports the membrane of the cell nucleus. The fact that the rest of the population doesn't harbor this mutation constituted further evidence that HGPS might have little to tell us about "normal" biological aging.

But many investigators couldn't let go of the idea, and in 2006 the notion that there might be a connection was given new legs by a study[16] showing that cells in laboratory culture that were originally taken from healthy people not harboring the HGPS mutation nonetheless did occasionally produce small levels of the defective mRNA—not because of a mutation, but by random *errors* in stitching the mRNA together. More ominously, the cellular defects characteristic of the disease were also present—and when the researchers compared cell cultures derived from very young (three- to eleven-year-olds) and very old (aged eighty-one to ninety-six) donors, they found that while the levels of mangled mRNA *produced* by cells cultured from the two age groups were similar, the level of the *damage* attributable to it was much higher in the old-derived than the young-derived cultures.

On the face of it, this looks like the operational definition of true aging damage: The cell is continuously producing this HGPS-like damage due to random events in metabolism that it is unable to repair, and which therefore builds and builds over time, eventually reaching pathological levels. However, there are compelling reasons not to worry about this finding.

First, to prove that the HGPS-like problems seen in the older cells were indeed the result of the defective mRNA and not another, independent age change, the researchers inserted a construct into the old-derived cells that prevents the early-stage mRNA from making the occasional splicing mistake that causes the production of defective lamin A protein. This treatment not only prevented further worsening of the associated abnormalities in the cell nucleus—it actually *reversed* those problems.[16] This would not have occurred if the defective protein were simply being produced at a continuous rate, incorporated into the nuclear membrane, and never removed: in that case, blocking the splicing mistake would simply prevent the addition of *more* such protein into the membrane, and would not have undone the preexisting damage. Moreover, the same preventive exercise, when done in young-derived cells, had no measurable benefits when the cells were still fresh: It was only as the cells were artificially aged in cell culture—when additional damage would have built up from other sources—that adding in the blocking molecule allowed them to recover. And finally, the researchers eliminated the possibility that the cells' recovery might be due to nonphysiologically rapid dilution of damage, due to the much more rapid cell division that typically occurs in the culture dish than in the intact organism: They introduced conditions that prevented any cell division and the recovery was still seen.

This tells us pretty unequivocally that the youthful cells actually *can* undo this damage quite nicely, and that the key difference between young-derived and old-derived cells is that the old cells have accumulated *other* aging damage that interferes with clearance mechanisms that work just fine in young cells. Thus, if otherwise restored to their youthful state by the proposed SENS therapies, these old cells would again be able to remove the defective lamin A.

Cutting Free of the Cellular Spider Webs

The time since Michael and I finalized the manuscript of *Ending Aging* has seen an explosion of research into vaccines targeting the amyloid beta protein as a therapy for the main cause of the neurodegenerative process of Alzheimer's disease. There are too many, in fact—too many reports, of too many different kinds of vaccines, using too many different vectors, with too many different but exciting outcomes, at too many different stages of development—to allow me to do justice to the scope of the field, and any attempted review would anyway be badly outdated in the months between turning these words over to our publisher and their appearance on the page in front of you.

Leaving aside novel or extremely promising concepts given initial proof-of-principle in robust animal studies, then, let me focus your attention on the most advanced-stage research: human clinical trials. These include the ongoing follow-up of the original trial of Elan Pharmaceuticals's AN1792 vaccine, which continues to fuel optimism and new research into the vaccine approach; in addition, however, there are at least six *new* amyloid vaccines being tested in human clinical trials at this writing: ACC-001

(Elan/Wyeth), LY2062430 (Eli Lilly), RN1219 (Pfizer), CAD106 (Cytos), and above all AAB-001 (otherwise known by the equally, er, memorable code name of "Bapineuzumab"), also from Elan/Wyeth, which I'll discuss in more detail below.

The Trial's Not Over Yet . . .

You'll recall that Elan was the first out of the blocks on this front, advancing their AN1792 vaccine into a relatively large and rigorous, but still limited, "Phase II" clinical trial—one that is intermediate between a small, simple study designed entirely to give a range of possible doses to test and to give some preliminary information about safety, and the advanced "Phase III" trial, designed to provide the robust proof of clinical benefit required for FDA approval. And you'll recall that that trial was stopped early because a small number of patients developed brain inflammation, but that the mining of what data was available suggested that, in patients whose immune systems mounted a robust antibody response, brains were rapidly cleared of plaque and minds were preserved, while patients taking a placebo or failing to respond to the vaccine continued down the harrowing road into oblivion that is Alzheimer's.

Researchers have continued to mine that data, and also to collect *new* data, following up on the patients in the trial over the years since the plug was pulled. Since our book came out, two of the remaining patients who mounted a strong antibody response have died (of problems unrelated to Alzheimer's disease or any of the observed vaccine side effects) and been autopsied.[17, 18] Like the three previously studied patients, their brains were found to be remarkably clean of amyloid plaque, even though they died three[17] and four[18] years (respectively) after taking their last dose of vaccine. The female patient's case is especially remarkable: Her score on the Mini-Mental State Examination (a test of cognitive function) remained stable—a stability that is all the more remarkable since, while her brain had been cleared of plaque by the vaccine, it was still full of neurofibrillary tangles (see Chapter 7), and she had also developed *dementia with Lewy bodies* (another brain disease associated with deposits of indigestible junk inside the neuron, which will require its own, lysosome-based clearance strategy to overcome).

Moreover, the benefits gained by people from successful vaccination at the end of the trial appear to have continued amongst the twenty-five treated patients still alive 4.5 years after immunization. Of that number, seventeen *still* have measurable levels of active antibodies in their system—

and, even all these years on, the group as a whole continues to show "significantly slower decline on the Disability Assessment for Dementia than placebo patients."[19]

Excitingly, moreover, neither of the recent autopsied patients shows any signs of the brain inflammation that had shut down the original trial, suggesting that such a reaction is neither a prerequisite nor an inevitable result of achieving plaque removal or stabilization of brain function (although it's quite possible that any *initial* inflammation might have cooled down in the intervening years since their last injections).

In fact, in a head-slapping twist, there is now evidence that the brain inflammation observed in the aborted trial may not have been even an inherent risk of the vaccine itself! As I write this (March 2008), a report has just appeared[20] comparing the antibodies present in the blood of patients from the Phase II trial with those in the blood of patients from the earlier, smaller, Phase I study. The researchers found that while, as we noted in Chapter 7, the T-cells of patients in the last AN1792 trial mounted a "Th1" (inflammatory, cell-attacking) response, patients in the earlier trial mounted a "Th2" response, which would be expected to not lead to inflammation and to focus more on the antibodies that drive the actual *benefit* of the therapy!

It is possibly crucial that there was a seemingly trivial change in the material used in the two trials. In the Phase II trial, in which the brain inflammation reared its head for the first and final time, the vaccine had an added emulsifier called polysorbate 80, which had never been used in previous studies and which the investigators argue might well be responsible for the inflammatory response.[20] That would offer a frustratingly simple explanation for the fact that no such reaction had been observed in either animal studies or the initial rounds of clinical testing. In other words, a common *ice cream firming agent* might have killed several people, scuttled a perfectly good vaccine, and sent researchers scrambling needlessly for ways to develop an alternative that would lack a deadly side effect that could have readily been avoided with a simple formulation change.

Bumped Up to Executive

Whatever the explanation for the closure of the AN1792 trial, scientists have subsequently been investigating numerous alternative vaccines, with a strong focus on approaches that would have similar or better plaque-busting abilities, but lower risk of a destructive inflammatory side effect.

One of the most obvious approaches under investigation is the use of *passive* vaccines: active antibodies, generated *outside* the body, which are then infused directly into patients, rather than injecting them with a foreign protein and then relying on the body to generate its *own* antibodies in response—a response which, as I've just outlined, was unreliable in many people, and may (or may not) have been toxic in a few.

Among many teams pursuing such an approach, Elan itself (in collaboration with Wyeth) had a first-mover advantage, and used the in-house expertise and relationships it had built during the research on AN1792 to develop a passive vaccine called *bapineuzumab.* This trial moved its way through preliminary Phase I (rough-and-ready safety) testing into a Phase II (more careful safety, and preliminary efficacy) clinical trial. The Phase I trial tested three doses of vaccine versus a placebo (injected salt solution). Encouragingly, despite the fact that Phase I trials aren't really designed to test for clinical benefits, patients receiving the lowest dose appeared to enjoy benefit on scores of cognitive function—a benefit which was statistically significant for those getting the medium dose. Moreover, all patients showed elevations in their plasma levels of beta-amyloid, which is consistent with mouse studies, in which antibodies administered as passive vaccines appear to pull amyloid out of the brain. The Phase II trial began in December 2006, and will conclude in mid-2008.

But in early 2007, there was a prescheduled "interim peek" at the data. We don't know exactly what the investigators saw: The data are still being kept under blinded conditions, and the companies have been dutifully keeping their lips sealed. But it must have been good. Soon thereafter, Elan pooled *all* of the patients from their ongoing Phase II trials into one megastudy, in which *all* patients—even those previously receiving a placebo—would be given one or another dose of the real thing. And by May 2007, Elan was announcing that they had sought—and been given approval for—the commencement of a Phase III clinical trial,[21] spread out over more than 350 sites across the world, and including a remarkable 4,000 total patients with mild to moderate Alzheimer's disease. They will be looking at brain imaging scans, and analysis of levels of biomarkers such as beta-amyloid and tau in the spinal fluid, but the basis for demonstrating clinical benefit and seeking regulatory approval will be scores on tests of cognitive and functional outcomes.

The first patient received his first injection in December 2007; the trial is expected to wrap up at the end of 2010. Obviously, I'm very optimistic at this time, and will follow the interim reports as they become available. Indeed, the progress with bapineuzumab has stimulated me to explore aggressively the possibility of recommending that the Methuselah Foundation sponsor immunotherapy work against other amyloids, once a suitably qualified research team can be identified and a suitable research protocol developed.

Breaking the Shackles of AGE

Unfortunately, to the best of my knowledge, there has been little if any progress made on AGE breakers during 2007. Research on alagebrium, the first taste of a cross-link breaker, has been languishing, in part due to the reorganization of the company that owns it (formerly Alteon, now Synvista)—though they did announce a Phase II study late in 2007, involving patients with chronic heart failure. I've had little success in encouraging researchers to work seriously on the problem, and especially on the issue of breaking the key AGE cross-link, glucosepane: The Foundation is gearing up to try some highly exploratory approaches to this, but we're not expecting rapid progress, and most external scientists interested in glucosepane are not yet interested in working on destroying it.

This stasis has led Michael and me to think hard about alternative approaches to the cross-linking problem, and I'm pleased to say that we're increasingly of the view that it may be *easier* than it seemed. Not because we've had a blinding insight into the chemistry of glycation, but because we've stood back and looked at the problem like engineers. In this case, tissue engineers.

As I described in Chapter 9, the problems caused by extracellular cross-links are intrinsically *mechanical*. As such, any systemic effects they may have are downstream of this mechanical damage. That means we have a much better chance of thoroughly solving cross-linking-derived pathologies by tissue-specific therapies than we might expect in the case of most of the seven deadlies. Two tissues in which cross-linking has major effects during aging are the lens and the kidney—but we are already quite proficient at replacing lenses, and work is proceeding apace to develop artificial kidneys.[22]

But there's one other tissue that clearly we can't ignore: the major arteries. As they stiffen and their buffering capacity is progressively lost, more and more of the surge of the squeezing heart is transmitted further and further down the arterial network, and increasing amounts of this pulsatile assault reaches target organs unmitigated. This appears to be substantially responsible for, inter alia, the age-related rise of kidney damage and stroke.

Well, I'm pleased to report that the solution I just mentioned for the lens and kidney—wholesale replacement—has already been given proof of principle in the artery too. In a recent animal study,[23] transplant surgeon John Mayer, tissue engineer Frederick J. Schoen, and others demonstrated that the bioengineered artery patches they had developed were able to graft into some of the lower-stress arteries in the body, leading to at least suggestions of partial regeneration. Clearly these are early days, and much more investigation will be required to determine the full scope of the problem and the options for its repair. However, it's a highly promising start.

This means that I now consider the accumulation of extracellular cross-links to be the SENS strand most likely to be addressed by tissue engineering, rather than by stem cells, gene therapy, or traditional drugs. A major additional advantage of tissue engineering, of course, is that it delivers a much more complete rejuvenation of the relevant tissue than any other. In this case, rather than grinding through each major species of cross-link only to encounter another, we'd be rid of them all in one fell swoop. Not only that, we would remove mechanical problems *not* related to cross-linking, such as the simple fraying and breakage of elastic connective tissue: While not proven to be life-threatening, this is a form of damage in the major arteries that may eventually contribute to arterial stiffening and hypertension.[24, 25]

Putting the Zombies to Rest

Following the Smoke Trail

When I discussed the various kinds of toxic cell that accumulate with aging back in Chapter 10, I outlined a general strategy of identifying proteins that the target cells manufacture in exceptionally high quantities, and then either vaccinating against such proteins and letting the immune system clear the cells out, or using toxic drugs that could be activated in (and only in) such cells using a variety of new and emerging selective drug-delivery systems. In some cases, you'll recall, we had a rather good idea of which markers we will need to target, but in other cases we didn't. The situation most up in the air was visceral fat.

Our first problem in understanding visceral fat is its composition. While fat may look, to the naked eye, like a single, amorphous blob, in fact it contains not only the energy storing *adipocytes,* but also (among others) *fibroblasts, endothelial* cells (which also line our blood vessels), and cells from the immune system. Not even knowing which cells we need to target, we were clearly handicapped in identifying the distinctive cell-surface markers that would allow us to remove them from the body without also damaging innocent bystanders.

But a convergence of research has been closing in on a particular kind of cell within visceral fat as the culprit behind its progressive metabolic dysfunction: the *adipose tissue macrophage,* or ATM. Like the macrophages that engulf bacteria or that attempt (sometimes to ultimately disastrous ends, as we have seen) to clear out toxic cholesterol products from our arteries, ATMs are specialized immune cells that are recruited to the visceral fat depot and become toxic when engorged.[26, 27, 28]

The year 2007 saw the publication of a potentially landmark animal study apparently identifying the marker we seek. The researchers fed one group of mice a normal diet, in amounts that let them get a little more than svelte but not frankly obese, while another group was given free access to a high-calorie, high-fat diet that led to actual obesity. A third group of mice was fed a normal diet, but which they grossly *overate* because of a genetic defect that makes them perpetually hungry however much they eat. The key finding was in the abundance of a cell surface protein named *CD11c.* Barely more than 1 percent of the ATMs taken from lean mice displayed CD11c; this was to be expected, since CD11c is characteristic of an entirely different class of cells from the immune system called *dendritic cells,* and only rarely appears on the surface macrophages. But close to *one third* of the ATMs from mice fed the high-calorie diet displayed this same surface marker—as did an astounding *74 percent* of those from the genetically obese mice.

In later test-tube studies, immune cells engineered to mimic this profile were found to pump out the same inflammatory factors that are so central to the metabolic disruptions of visceral fat, and to pump out even *more* when awash in free fatty acids (FFAs)—as they are in bloated visceral fat. Under such conditions, the model toxic ATMs inhibit the action of insulin in muscle cells—and, worse, it appears that they may engage in a vicious feedback loop, releasing factors that encourage adipocytes to dump even more of their stored fat out as FFAs. By contrast, FFAs have very little metabolic effect on normal adipose-tissue macrophages.[29]

These cells will now need to be further characterized, and their equivalents in human visceral tissue sought out. Such work is not trivial, but it requires no new tools to carry out—just the time and money to do the studies. With those results in hand, the happy combination of a cell-surface marker signature to home in on, and the localized nature of the visceral fat depot, should make targeting these cells one of the easiest tasks we face—and just in time, as the obesity crisis threatens to overwhelm our society with diabetes, heart disease, and organ failure.

Calling Old Soldiers in from the Cold

Staying within the "toxic cell" category, another source of great concern is the aging immune system, and the role of the accumulation of anergic T-cells in that process. Various ways to eliminate these cells have been proposed, but there is currently no concerted effort to do it. My preference, as you will recall, is a vaccine-based approach, which would remove such cells and allow the expansion of a wider range of cells and of cells with full disease-fighting capacity. I am pleased to report that a very prominent researcher in the immunology of aging has expressed great interest in a project to develop such a procedure in mice and (if it shows promise) potentially to test it out in nonhuman primates. Most exciting, the proposed study will involve not only the attempted removal of the anergic cell population, but also the simultaneous attempt to reconstitute the thymus, which (as mentioned in Chapter 11) suffers particularly badly from cell loss with age. As soon as we finalize priorities and protocols, this investigator will receive funds from the Methuselah Foundation to begin the project. As always, any research program that emerges from these discussions will be announced on the Foundation web site at http://www.mfoundation.org/ and elsewhere.

New Cells for Old

In contrast to the lack of progress on tackling AGE cross-links in over a year, a disappointing laggard in an era of accelerating scientific progress, many of us have had the wind positively forced out of our lungs by the amazing surge in stem cell research. There were two particular areas in which progress was astonishingly fast and promising, and which I won't spend too much time on as they received wide coverage in the media and their therapeutic importance is widely recognized: "therapeutic cloning" in nonhuman primates, and the use of genetic manipulation to reprogram mature cells into cells which appear to have all the power of embryonic stem cells (ESCs).

Cell Rejuvenation Climbs Evolution's Tree

"Therapeutic cloning" (more properly termed *somatic cell nuclear transfer,* SCNT) remains, theoretically, amongst the best possible sources for cells for therapy, generating cells that would have all the useful properties of embryonic stem cells, but would have the added benefit of being perfect genetic matches for the recipient. Aside from the confused political hullabaloo which has slowed down progress in the field, there have also been some legitimate technical difficulties, as the outrageous case of Korean fraud Hwang Woo-Suk illustrates. Even though we've been doing quite well in mastering SCNT in rodents and in farm animals, work in primates has been slow, leading some to fear that there was something about the fundamental biology of humans and our close relatives that would not allow such reprogramming to occur, or would require decades of work to untangle.

Late in 2007, however, Dr. Shoukhrat Mitalipov's group at the Oregon National Primate Research Center reported the successful derivation of functional ESC-like cells via SCNT from Rhesus macaque monkeys.[30] The procedure was still very inefficient compared to the state of the art in mice, requiring 305 monkey eggs to make just a tenth as many blastocysts, and then ultimately leading to only two lines of cells—and one of those is genetically abnormal. Still, the success of each of the several steps along this path was itself a breakthrough.

Despite their outward expressions of caution, few of my colleagues in the field doubt that Mitalipov's methods can be refined and improved just as we have already done with their four-legged predecessors, and then translated into the human case. Indeed, the establishment of a reliable regulatory framework for work on human SCNT here in the United Kingdom has allowed Mary Herbert's stem cell group at Newcastle University to set up a collaborative project with the Oregon scientists aimed at doing just that, using 500 high-quality surplus human eggs donated by women undergoing fertility treatments. And there's reason to think that it may ultimately prove *easier* or *more efficient* to perform SCNT using human eggs than one might anticipate based on the work in monkeys. This is because the art of using hormones to coax nonhuman primate ovaries to release more eggs is relatively crude, and the treatments may have compromised the quality of the eggs that Mitalipov ultimately used, whereas human fertility clinics have long since worked out less stressful and more reliable methods of doing the same thing for in vitro fertilization in humans.

Wrapping Hands Around the Levers of Cell Rejuvenation

Of course, what ultimately differentiates an ESC—with all of its infinite possibility for self-replication and transformation into specialized cells and tissues for medicine—from a mature skin, muscle, or kidney cell is not the passage of time, but the differences in gene expression. Remember, all of the cells that ultimately make up the adult body are derived from ESCs, and the secrets of what makes (for example) lung and liver cells different not only from one another, but also from the embryonic cells that were their umpteen-great cellular grandparents, are locked up in the different ways that these various cell types regulate the expression of their commonly held genetic library. As a result, most researchers have long felt confident that, at some point, we'd be able to use the techniques of molecular biology to take ordinary, mature cells and turn back their clocks by turning some genes on and some genes off.

Still, I and many others were greatly surprised when scientists at the Department of Stem Cell Biology of Kyoto University, led by Dr. Shinya Yamanaka, reported the successful genetic cell reprogramming of adult mouse skin cells into an ESC-like state called *induced pluripotent stem* (iPS) cells as early as 2006.[31]Indeed, many were quite skeptical: it seemed too simple a protocol, involving just four widely studied genes, and the researchers had used just one potentially unreliable marker as evidence of the cells' successful reprogramming. I mentioned the paper briefly in Chapter 11, when it was still quite a recent report, expecting verification—or refutation, as the case might be—to take some time.

But I was wrong. Just one year later, Yamanaka would repeat the experiment, using two additional markers to verify his initial report[32]—and labs at MIT (Dr. Rudolf Jaenisch) and Mass General (Dr. Konrad Hochedlinger) would simultaneously publish independent results confirming it.[33,34] Needless to say, these were tremendously exciting developments, opening up the promise of a way to produce cells with the full therapeutic potential of ESCs without the confused, pseudo-moral hand-wringing over the need to use a human blastocyst, and also without the logistical problems of securing enough viable eggs to derive cells by SCNT.

Everyone expected researchers to pile on the result and start working on demonstrating the cells' full powers, with the ultimate hope of controversy-free research leading to real-world cures; but, even taking that into account, subsequent progress has been quite unbelievably fast. Just *five months* after those confirmations, Yamanaka had generated a line of *human* iPS cells,[35] and Dr. Jamie Thompson's prestigious stem cell lab at the University of Wisconsin, Madison was saying it had achieved the same. Even more excitingly, Thompson's group had done it by switching on two (out of four) genes *different* from the ones that all previous labs had used.[36] This not only indicated a diversity of possible routes to the ultimate development of iPS cells that could be used for human therapy, but also relieved another concern: c-Myc, one of the genes used by Yamanaka, is a gene whose overactivation contributes to the progression of cancer, and indeed both Yamanaka and Jaenisch had reported that when they had made living mice out of mixed conventional and iPS-derived cells, there had been widespread tumor formation. But Thompson's new method doesn't rely on c-Myc at all, having replaced it (and KLF4) with two different genes.

And that wasn't all. Just weeks later: Jaenisch pushed the envelope *again*, giving proof-of-concept of the therapeutic use of iPS cells! Starting with mice with a mutation equivalent to the one responsible for sickle-cell anemia in humans, his group used the iPS technique to transform their skin cells into an embryonic stem–like state. Then he performed in vitro gene therapy on them to correct the mutation, differentiated the corrected cells into blood stem cells (you can do that with ESCs, remember), and used these cells to perform a bone marrow transplant. From those transplanted cells—a perfect immunological match for the mouse in question, because each mouse acted as its own donor—the blood was quickly repopulated with new, healthy blood, ultimately curing the disease.[37]

Well, developments couldn't proceed any faster, could they? Secure in this belief, Michael and I delivered this afterword to our publisher in early April 2008. Well . . . you guessed it: Within a week I had to tell him that he needed to wait for an update.

First of all, news broke of a still-unpublished advance: a group of Harvard Medical School scientists, led by Dr. Willy Lensch, revealed that they had used the iPS technique to make new ESC-like cell lines using skin cells taken from patients with several specific congenital diseases, apparently including type 1 diabetes, Down's syndrome, Huntington's disease, muscular dystrophy, and possibly Parkinson's disease. While we would obviously not use these cells—which, by definition, carry the very flaws that underlie these terrible diseases—as the basis for new cells and tissues to replace those lost to age-related or other disease, stable lines of such cells that can reproduce themselves and maintain their essential characteristics when differentiated into the cells most affected by the disease have significant biomedical importance, both as models in which to study the origins and development of their donors' disease, and as possible tools with which to do greatly improved early screening for drugs.

Later that week, the journal *Nature Medicine* published an even more important result: scientists from the Memorial Sloan-Kettering Cancer Center had successfully used SCNT-derived cells, differentiated into the *dopaminergic* neurons that fail in Parkinson's disease, to treat an animal model of the disease.[38] Four years earlier, they had shown that such SCNT-derived cells could improve the abnormal motions seen in the model animals when they were taken from *different* animals than the ones receiving the transplant—cells that were, in other words, given all the powers of ESCs, but were not genetically matched to the recipient—but that, as might be expected, they didn't work optimally, and caused the predictable autoimmune reaction, leading to very poor survival of the transplanted cells and inflammation in their brains. By deriving the SCNT cells from *the same mouse* that was suffering from the Parkinson-like disease, they overcame the problem of injection, and got an impressive restoration of function. When you consider the promising but limited results of similar strategies using fetal stem cells in recent human trials, this in itself confirmed further the promised superiority of SCNT as a source of therapeutic cells—a significant advance.

Unfortunately for these, they had the bad timing to have their results come out at the same time that Rudolf Jaenisch once again rocketed the iPS field forward. Jaenisch's group first generated semi-specialized neural precursor cells from iPS-derived cells from mice, and then in turn nudged some of *these* to become dopaminergic neurons. The neural precursor cells were transplanted into the brains of fetal mice, and he demonstrated that they not only survived, but differentiated into various kinds of specialized neurons, following the body's normal developmental cues, and integrated into functioning contributors to the mice's brain circuitry. The dopaminergic neurons were transferred into a model of Parkinson's disease similar to the one used in previous experiments. Just like their SCNT cousins, these cells successfully helped normalize the diseased mice's motions—and, again, without needing to trouble an embryo to create the customized, immune-matched cells. Moreover, in previous experiments with both conventional ESCs and their own early experiments with iPS cells, the neurons in question had occasionally transformed themselves into tumors; Jaenisch considerably reduced the incidence of this problem by doing a better job of sorting *which* cells they would use for transplant, choosing only those cells that had fully differentiated into their dopaminergic neural form—another technique we'll need to master for human therapeutic use.[39]

We also learned that a private research institute in Japan claims to have used Yamanaka's mouse iPS protocol to derive *human* iPS cells, not only independently, but *months* before Yamanaka and Thompson had done so—and using a different virus to insert the reprogramming genes, somewhat further documenting the flexibility of the basic recipe.[40] Moreover, they documented that the genes used to induce ESC-like pluripotency in their iPS cells were shut down in the cells that resulted, easing worries that the genes might lead to cancer. The company (an arm of the pharmaceutical firm Bayer AG) claims to have filed a patent at the time on this basis; we will have to await the details, and hope that this does not create the kind of legal wrangling that early work in ESCs and genetic diagnostics created in the last decade.

While there is still a lot of work to be done—first to validate these cells' full pluripotency and therapeutic powers, and then to get the protocol working in humans—I am simply astounded by the rate of progress to date. With many other labs now turning to work with these cells (including, famously, Ian Wilmut of the Roslin Institute and Dolly the sheep), one can dream of a rate of progress that seemed impossible just months ago. The calls of religious conservatives to abandon work on true ESCs and SCNT *at once* in favor of the iPS technique are still premature, but the evidence to date all points to them being the real deal.

There is a lesson here, too, for those who are too quick to jump to ethical conclusions: iPS cells are not obviously ethically superior from a bioconservative point of view. The main reason why they aren't is that the most problematic biotechnology of all—*reproductive* cloning, the creation of a new live organism, possibly a human being, by means not involving fertilization—seems to be much easier to achieve using iPS cells than by other means. It turns out to be quite easy to double the number of chromosomes in the single-cell embryo, making a "tetraploid" cell that has four copies of each chromosome. Embryos like that can develop for a short while, but only the placenta develops properly: the cells that would normally become the embryo itself die off. It has now been established that iPS cells can be injected into the five-day-old tetraploid mouse embryo and successfully hosted by the tetraploid placenta, resulting in live mice that are entirely derived from the iPS cells.[41]

Harnessed Healing

In many cases, we will want not just to replenish lost cells with fresh new ones, but also to engraft functioning tissue and even whole organs. There has been remarkable progress in this regard recently, including work on how raising cells on scaffolds, and how exposing them to mechanical forces similar to the ones that are present in the organs in which the cells normally develop, can allow them to develop into mature forms and structures.

One such study caught the public imagination like wildfire in 2007: the regrowth of a functional heart out of a stripped-down frame of proteins. Dr. Doris Taylor of the University of Minnesota developed a protocol to progressively strip away the muscle cells from the underlying protein scaffold in a heart taken from a rat, and then used that scaffold as a frame onto which to seed another, newborn animal's heart-precursor cells. Over the course of a week, these living cells attached themselves to the frame, integrating with it and with one another—and then began to contract. Ultimately, the cells were able to start this shell of a heart beating again.[42] If the proto-organs developed in this preliminary study can be further developed into fully functional hearts, it could form the basis for customized organ creation. For instance, accident victims could donate their hearts or other organs, which could then be stripped of the donor's heart cells and filled back in with the recipient's stem cells, creating a heart that was both youthful and functional, and also an almost perfect immunological match for the recipient. It's even possible that nonhuman hearts could be used as the starting material. This is, therefore, a major breakthrough in tissue engineering, where real success has thus far been limited to tissues with little need for a robust blood supply, such as the skin and the bladder.

A different approach is to attempt to harness the body's own regenerative powers, facilitating the work of stem cells in regrowing tissues and organs at the site of injury. Dr. Alan Spievack and Dr. Stephen Badylak, of the McGowan Institute for Regenerative Medicine at the University of Pittsburgh, work with extracellular matrix scaffolds for a variety of tissue engineering applications, repairing defects in the chest, esophagus, and urethra of dogs. These are the same sort of scaffolding that was used by Taylor as the frame on which she would regrow a heart. Along with many researchers, Spievack and Badylak have found that such scaffolding tissues are not just passive frameworks awaiting an enveloping layer of living cells, but contain factors that actively help to guide the process of building developing tissues, and of rebuilding them after injury. Years ago, Spievack developed a series of proprietary extracts derived from the extracellular matrix of pig intestine and bladders and, after experimental work, was able to quickly move it into veterinary applications. The extracts attract progenitor and immune cells, and also stimulate the growth of blood cells to nourish the regrowing tissue. But what has really caught the public imagination have been reports of regrown human fingers in two men, one of them being Spievack's brother Lee. These results are not yet published formally in scientific journals, but press reports indicate that, after accidentally severing the tips of fingers, the two men received the newer powder extract and grew back *intact fingertips*—not just a scar, but skin, blood vessels, and in Lee's case even the nail—in just four to six weeks.

Stimulation of latent regenerative capacity has also been shown recently in other tissues, by exposing them to an environment resembling younger tissue. As earlier, I can't hope to give a comprehensive account here; instead I provide a summary of the most conspicuous advances.

Ovary rejuvenation has been for some years the focus of Dr. Jonathan Tilly, director of the Vincent Center for Reproductive Biology at Massachusetts General Hospital. It's long been "known" that mammals lay down all the egg cells they'll ever have very early on in life—before birth, in fact. In 2004, however, Tilly carefully looked at the ovaries of both immature and adult mice, and found that both groups' ovaries had a layer of tissues with egg precursor cells that were actively dividing, quite similar to the ones from which eggs are known to develop in embryos, but that the older animals' supply of such cells was on the decline. And he found that the existence of such an egg-replenishing cell population was actually *necessary* to explain the size of the pool of eggs in the ovarian follicle at a given age, granted the rate at which mature eggs were constantly being lost not only from the menstrual cycle, but from much larger numbers of eggs being lost to random molecular damage. Moreover, giving immature animals a drug that is selectively toxic to *dividing* germline cells wiped out the pool of eggs by the time they reached adulthood—a result that should not have occurred, if the animals were already born with all the eggs they would ever have.[43]

Michael and I didn't tell you about that in the original 2007 edition of this book, largely because there turned out to be a number of plausible alternative explanations for Tilly's results. Now, however, he has closed off the major loopholes and also discovered something that's arguably even more exciting than his original finding. It now seems that new oocytes are indeed generated from egg precursor cells, but that this is somehow stimulated by bone marrow stem cells infiltrating the ovary. Bone marrow from young mice does this particularly well. The body *does* still contain within itself the ability to make new eggs throughout the life span—but as Tilly's first study had shown, aging slowly robs of this ability. But infusing new bone marrow into the aging organism provides some factor that rejuvenates this capacity, awakening dormant cells and restoring the potential that they had become progressively less able to manifest as the body that they inhabited grew inhospitable with age.[44]

In parallel work, Dr. Irina Conboy's group at UC Berkeley has shown a similar effect in muscle. They'd shown previously that aging muscle slowly loses the ability to mobilize muscle precursor cells in order to repair damage, and that this ability can be reawakened by exposure to factors present in the youthful systemic environment.[45] Now it transpires that these factors can be provided to the local muscle stem cell niche by infusion of embryonic stem cells.[46]

And finally, Dr. Howard Chang's group at Stanford have shown that *nuclear factor kappa-B* (NF-κB), a protein that regulates the response to oxidative stress, ultraviolet radiation, many immune challenges, and inflammation, and that has been intensively studied because of its role in chronic inflammation, autoimmune diseases, and cancer, also plays a wide-ranging role in determining the age-related changes to gene expression in mouse skin. By turning off the expression of the gene in the skin of mice, they were able to reverse a large number of age-associated gene changes, and also to make the skin *look* and *behave* more like young skin—within just two weeks.[47]

Nuclear Mutations and the Total Defeat of Cancer

The kind of approach pursued by Badylak, Conboy, Tilly, Chang, and others is not the prototypical "engineering" solution, as it relies on manipulating the metabolic pathways underlying the regenerative system inside the body—an exercise always fraught with risk of perturbing the rest of the system, as we have learned countless times in the past. Viewed that way, this clutch of results seems too good to be true—too simple. And it probably is . . . for now.

It's now widely accepted that our cells and tissues don't lose the regenerative capacity of early life merely because of accumulating damage to the regeneration machinery: If they did, simple tricks like turning off one gene could never succeed in reversing those changes. No—the reason why that reversal is turning out to be so easy is because the original decline is a coordinated, programmed change—and programmed changes have evolved for a reason. In this case, the reason is very probably to defend against cancer. Cancer needs mutations, and young people have not had time to accumulate many mutations in their cells, so they can get away with playing fast and loose with regeneration, but older people don't have that luxury. It's generally very hard to prove for certain that restoring regenerative capacity will encourage cancer, because the cancers may only become big enough to be detectable decades later—but that's bad enough if we're expecting to live a great deal longer. That's another reason to work as hard as we can to develop a truly comprehensive approach to defeating cancer, which I described in Chapter 12. So, how is that quest going?

"General Cellular Malaise"

First, let's look at the question I explored early in Chapter 12: Is cancer really the only thing we need to worry about in regard to nuclear DNA damage? Intuitively, a gradual increase in mutations across a substantial number of cells with age might lead to enough cells in a given tissue being just a little bit sick to impair the functioning of the tissue as a whole. In Chapter 12, I gave my reasons for doubting this intuition. But to some extent I advanced those arguments out of an excess of caution: If there *were* no age-related increase in mutations (aside from those involved in cancer, cell death-resistance and cell loss, for all of which SENS has other solutions), it just wouldn't *matter* much what else those mutations might do, and the most important tissue of all—the cerebral cortex—was indeed shown to have no accumulation at all of a wide variety of mutational classes during the whole of a mouse's adult life.

But I'd still like to be even surer. In particular, there are certain types of mutation that have still not been checked out for whether they accumulate in the brain; we also have no solid information concerning epimutations in any tissue other than the heart. Therefore, I am hoping to arrange for a top expert in this field to have a serious look at this question. By happy coincidence, this researcher is about to take over chairmanship of a department at a major research institution in the United States, with all that that entails regarding freedom of research resources. Once we hammer out a plan for tests of "general cellular malaise" that are likely to yield sufficiently informative results, I will make Methuselah Foundation funding available to seek more robust answers to the question.

ALT: More Like Telomerase Than it Seems?

ALT, you will recall, is a system of telomere elongation that about 10 percent of our cancers use instead of the obvious one, telomerase. There has been considerable concern that ALT will prove to be the undoing of WILT, my proposed anticancer therapy, because in order to close ALT down we need to identify a gene that we can *preemptively* shut off in *all* our cells, not only in the cancer, without unmanageable side effects. The side effects of doing this to telomerase are very considerable, and arguably only just manageable; ALT might be even worse.

I've long been convinced that this concern is misplaced, however. I feel certain that, of the genes necessary for ALT (and there are probably quite a few, some of which have already been found), there must be at least one that is robustly turned off in almost all our cells. If that were not true, ALT would be much easier to activate than telomerase (for which—as you will recall—the "TERT" gene, encoding the protein component of the telomerase enzyme, is the gene with this property), and 90 percent of our cancers would use ALT, rather than only 10 percent. But this is only a theoretical argument; until we find such a gene, therefore, there will always remain doubt.

Well, I can't tell you that the gene has been found. But I can tell you that we have identified a very promising suspect, and the Methuselah Foundation has now arranged for two researchers, at different labs, to test it out—one in mouse ALT cells and one in human ALT cells. This should be a relatively straightforward study, so we may even have the results by the time you read this.

What's even better is that the gene in question may not, as has always been presumed, be one that exists for a nontelomeric purpose and is mutagenically coopted by ALT cancers: Rather, there are strong hints that it may have a physiological role in elongating telomeres in a few tissues where cell division is so rapid that telomerase is just not fast enough to keep up. That would be really terrific news, because it would argue against the fear that there may be a wide variety of different possible mechanisms of telomerase-independent telomere elongation that each require about the same number of mutations to activate them. A mechanism that already exists for the express purpose of elongating telomeres, and thus that cancers merely need to turn on, is likely to require considerably fewer mutations to activate than one that has to be reconfigured for a function it normally lacks.

Remote-controlled Precision Engineering of Our Genes

In parallel to the work required to characterize the cells themselves, we will need to develop ways to deliver the needed genetic tweaks. As you'll recall, this is not likely to be all that difficult in respect to the continuously renewing tissues like the blood or the skin, because (so long as the work I referred to a paragraph ago pans out) we can perform the manipulations outside the body and prescreen the treated cells so that we only put back cells with exactly the desired genetic change. Not so, however, for cells like glia or muscle satellite cells, which divide very rarely in normal circumstances and are not maintained by a stem cell population. We'll need to use bona fide somatic gene therapy there, and moreover it'll need to be gene targeting, the extra-difficult form of gene therapy that modifies specific existing DNA rather than just inserting new DNA into an unspecified place in the genome. In this regard, I've been very excited by recent progress in a gene therapy technique using enzymes called *zinc finger nucleases*, which allow for targeting of new DNA into precise locations in the genome. I featured talks by the leading group in this area at my 2005 and 2007 conferences, but recent work has further improved this method, reducing the potential risk of making inappropriate additional cuts into the recipient's DNA.[48]

Those Side Effects: Manageable or Not?

As you'll recall from Chapter 12, the preemptive elimination of telomere-elongation capacity from all our cells will have potentially devastating side effects on our continuously renewing tissues, causing us to die of anaemia, intestinal failure, and lung atrophy after an estimated ten to twenty years. Therefore, it can only be contemplated if we can simultaneously avert these side effects by periodically replenishing the stem cell populations in the blood, skin, gut, and lung, thus making those tissues immortal even though their stem cells no longer are. Importantly, the new stem cells will lack the genes for telomere elongation. Will they still perform as fully functional stem cells?

One might think, well, of course they will—they are identical to normal stem cells except for the absence of a function that they don't need (and won't need for another decade or so). But in recent years there has been an unnerving series of reports that telomerase may have additional cellular roles, over and above its telomere-elongation function. These roles must be pretty minor, given how healthy telomerase knockout mice are until their telomeres shorten over multiple inbred generations—but it's still a concern.

Therefore, I'm delighted to say that another professor whom I brought to SENS3 has been looking at this sort of question. Dr. Lenhard Rudolph, now at the University of Ulm, works on the hematopoietic system of telomerase knockout mice, and he has identified important features of the interaction between blood stem cells and the cells nearby in the bone marrow (the cell "niche," as it's known) that seem likely to allow us to test the competence of telomerase knockout stem cells in realistic circumstances very soon.[49]

Obviating WILT: Let's Not Give Up Yet

Let's make no bones about it, though: WILT is scary. Even though the evidence available to us today suggests that we will indeed be able to develop stem cell therapies sufficient to avert the consequences of having every cell type in our body be mortal, the lingering doubt legitimately remains that there will turn out to be some tissues for which stem cell replenishment is unfeasible. Because of this, no one would be happier than I if a therapy for cancer were developed that was as effective as WILT promises to be, but without the complications. It is, therefore, appropriate to highlight here a couple of recent advances in combating cancer that have attracted both public and scientific attention for their apparently remarkable promise. While I do not currently expect them to be so effective as to make WILT unnecessary, I devoutly hope to be proven wrong.

The first is an extraordinarily simple intervention, based on a feature of cancer that has been known for over seventy years: the Warburg effect. Warburg discovered that cancer cells, which must tolerate an unusually low concentration of oxygen because cancers generally have a poor blood supply, actually *avoid* using oxygen very much even when it's made available to them in cell culture. For some time it has been suggested that this apparent adaptation might be an avenue for therapy: that if cancer cells could be forced to use more oxygen than they seem to want to, they would suffer in some way. But this has not been thoroughly explored, partly because there is no obvious way in which the cells *would* suffer, and partly because forcing them to use more oxygen is easier said than done. But now, a very simple way to do just that has been discovered. Dr. Evangelos Michelakis and his group at the University of Alberta in Edmonton found that a cheap and simple compound called dichloroacetate does just this, by biasing the cell to import pyruvate (the end product of the non-mitochondrial component of carbohydrate metabolism) into mitochondria rather than export it from the cell.[50]

The other advance that I want to highlight was probably the most talked-about presentation at SENS3: Dr. Zheng Cui, of Wake Forest University. In 1999, Cui serendipitously discovered a heritable trait in mice that confers exceptional cancer-resistance. For several years, Cui had made frustratingly slow progress in characterizing the basis and mechanism of this trait, but in 2006 he determined which specific type of immune cell was responsible—an unexpected one, as it turned out, called *granulocytes.* At SENS3, he reported the first results of a trial exploring the effects on human cancer cells of granulocytes from cancer-resistant humans.[51] These results were highly encouraging, and also provided preliminary evidence for the involvement of other factors including vitamin D, and he is now proceeding to a phase II trial involving the infusion of granulocytes from highly cancer-resistant people into cancer patients. It has long been known that the human population exhibits a wide divergence of susceptibility to cancer, and this is potentially the biggest breakthrough yet in determining the basis for that variation and in exploiting it in the clinic.

Bootstrapping Our Way to an Ageless Future

In Chapter 14, I alluded to a project that I was working on with computer programmer Chris Phoenix to create a computer simulation of both the model of aging as damage accumulation that underlies SENS, and the effects of the implementation of increasingly effective SENS biotechnologies on mortality risks, as a way to put the intuitive concept of "longevity escape velocity" (LEV) onto a firm mathematical footing. This work was duly completed and published.[52] I am delighted to say that it does all that I wanted it to do, and more.

The way that biogerontologists, evolutionary biologists, and demographers think about aging has been profoundly influenced by a British actuary named Benjamin Gompertz, and his eponymous "function." Gompertz showed that there was a surprisingly regular acceleration of death rates as we age: at any given adult age, a person's risk of dying in the next year is *double* what it had been just seven years earlier. This, of course, creates an exponential relationship.

This exponential decay function—which also holds robustly across nearly all organisms, and especially mammals—intuitively matches the general model that is accepted by nearly all aging theorists, and that forms the basis for an "engineering" approach to aging: that aging is the result of several distinct, but *mutually synergistic, progressive,* and *deleterious* changes that happen in the body, feeding back upon one another to generate an incessant acceleration of biological decay. As each mechanism progresses, it makes some aspect of the body as a homeostatic whole less robust: less adaptable to change, more vulnerable to every fresh insult from the internal and external environment, whether it takes the form of a blood clot shooting toward the coronary artery, a flu bug that you picked up at your last handshake, or a slip on the ice on a wintry morning. And each form of damage feeds back on the other kinds, so that the body doesn't just become progressively more frail each year at a steady pace, but rather loses a *greater absolute amount* of biological reserve in any given year than it did the year before, making us frail faster. Eventually, the system is unable to recover from some assault, and it kills us—even though the same blow might have only badly shaken us a couple of years earlier, and have been brushed off with a laugh a few decades previously.

Chris and I were able to build not only this general outline, but the specific understanding of an aging process driven by multiple categories of accumulating damage, each of which can potentially be fatal on its own (in other words, the "Seven Deadlies"). Within these categories, we modeled the fact there are numerous individual contributors (such as the various different *kinds* of AGE cross-links within the cross-link category). Moreover, we incorporated the fact that each mechanism has both an inherent, ongoing rate of increase, and a feedback function working upon it, so that the rate of frailty that it adds to the body increases more rapidly over time. There are also factors built in to accommodate variations between individuals in the rate of aging. Then, finally, a given individual modeled by the simulation is subjected to a challenge each year, whose severity is randomly generated. If the level of that challenge exceeds the resilience of an affected system, the system fails, and the person dies.

The first piece of good news was that our model worked on real-life data. Our data for a model population fitted real-world actuarial numbers from the contemporary United States—in fact, fitted them even better than the Gompertz formula did.

Our main purpose, though, was to model what happens if we apply "engineering" anti-aging therapies to people within the model. To simulate this, we modeled the arrival of each new SENS intervention as the elimination of just *half* of a single category of damage in its recipient, thereby acknowledging that real-life therapies will not be totally comprehensive. We then explored different scenarios: ages at which a person is when she or he receives the first therapy; rates at which new interventions are developed; effectiveness of new rounds of intervention at further reducing overall category damage; and so on. And thankfully, just as I'd predicted, we saw that a complete SENS platform does indeed offer a highly realistic path to escape velocity.

Let's start with what I view as a highly pessimistic scenario, in which we halve the burden of damage within one or another of seven categories of damage every six years. For instance, we first cut down the burden of vascular stiffness created by AGE cross-links by 50 percent, and then six years later remove half of the total amount of accumulated extracellular junk (amyloids) by the same amount, and six years after that we successfully incinerate half of the burden of long-lived aggregates in our lysosomes—and so on, until, forty-two (six times seven) years after our first therapy arrives, a new therapy is rolled out that cuts in half the *remaining* damage in one or another category. Then the process continues, with additional improvements in one or another category again rolling out every six years. As I say, I consider it extremely unlikely that we will take as long as forty-two years to achieve so modest an advance in the removal of damage in a given category, once the first round of therapies has arrived—but nonetheless, let's see where that leads.

Well, the bad news is that we cannot save anyone who is already eighty biological years old when the first SENS therapies reach the clinic: such a person will live a little longer, on average, than he or she would have without such a therapy, but even the luckiest such people will only obtain an additional twenty or so years of life, simply because they were too close to the precipice when the treatment began. (I emphasize "*biological* years old," however: put more graphically, that means within about nine years of death given only today's medicine. Unusually robust eighty-year-olds, who are chronologically eighty but biologically maybe only sixty-five, are not the people I'm talking about here.) But the terrible logic of the self-feeding synergy of aging damage works both ways: benefiting from the first age-reversing therapy gives even a seventy-year-old a one in ten chance of surviving long enough that the chance of dying from the forces of aging (and not from a skateboarding mishap) will finally become negligible. And for sixty-year-olds, the odds are even better: almost even odds of indefinite

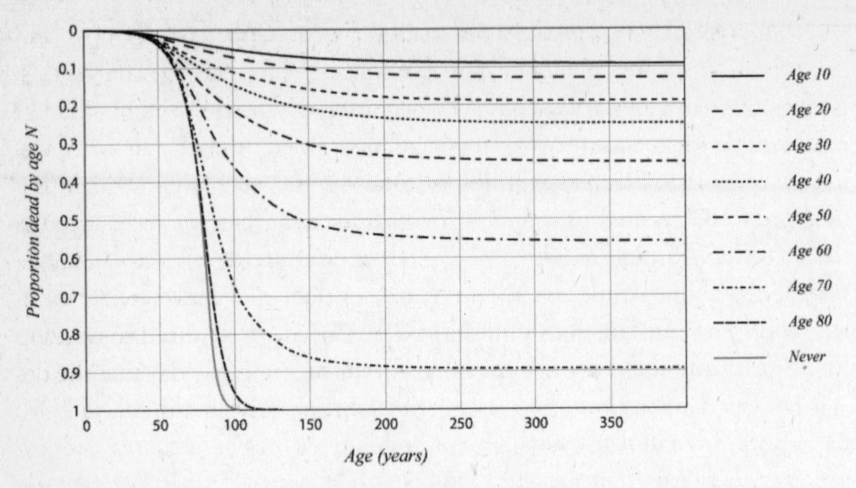

Figure 1. Survival curves for cohorts aged ten, twenty, . . . eighty when the first therapy arrives, under the schedule of therapies described in the text. The black line is the survival curve in the absence of any such therapies at any age.

youth—and in the meantime, a strong rejuvenation, a much longer healthy life even if you don't go on to LEV. And so on—see Figure 1.

And the faster we roll out the therapies, the model affirms, the more of their own parents and even some grandparents will be with them, too. At SENS3, Chris showed that a more realistic rate of progress—halving each category of damage every twenty-one years, for example—would dramatically improve the survival chances of those who were already elderly when the therapies began to arrive. Better yet, I've also shown that this rapid rate of progress is really only needed for a couple of cycles: thereafter, the necessary rate of progress to keep people biologically youthful actually *slows*—good news for those of you who are skeptical about the applicability to medicine of the concept of "accelerating change" championed by my friend Ray Kurzweil.

Instead of facing a future in which the lying promise of "golden years" is a shared doublespeak for a dark brooding winter sunset, these people will live in the glow of the summer sun of youth for as long as they want to have it, and enjoy life itself for as long as a contingent world allows. People will still sometimes die in mountaineering accidents and the like, but never again will anyone ever need to suffer an increasing burden of age-related suffering—or watch in pity and horror as their loved ones do.

Afterword . . . and Forward!

There is certainly much more of import that has been uncovered in the last year, detailing the contributions of various metabolic processes to the damage of aging, the mechanisms linking that damage to pathology, and the foreseeable biotechnologies through which I am now more confident than ever we can bring an end to aging. But the point must be made that the foreseeability of real anti-aging biotechnology does not imply its inevitability; more important, it does not dictate the timescales required to develop that technology and move it into clinical practice. For every day that the development of a comprehensive panel of such interventions is delayed, another 100,000 men and women will die of the molecular rot that has accumulated over their too-short lifespans in their essential biomolecules and cells—and many times this many more will suffer terribly, as they either slide down the fearful path into age-related degeneration, disease, and dementia, or watch their loved ones be taken by those same forces.

Please do what you can to help. Talk to your friends and family about the real humanitarian crisis of aging, and pass on your copy of this book—or buy them (and perhaps your local library) a copy for their own use.

If you're a journalist, shout it from your columns.

If you're a young student with an aptitude for science, or a scientist bold enough to change the course of a career, consider taking the achievements of Pasteur and Salk well beyond what they would have dared dreamed.

Contact your political representatives, and demand that legal restrictions on the use of embryonic stem cells and somatic-cell nuclear transfer be done away with. Insist also that funding be made available for research into a *cure* for the age plague, and not just the academic study of the aging process, or myopic efforts to make the best of its assaults on our health and independence.

And finally, consider joining the thousands of individuals who have directly contributed to the advancement of the science that will give us the rejuvenation biotechnologies of tomorrow, by directly supporting the Methuselah Foundation (http://www.mfoundation.org).

The defeat of aging can only be done by a heroic effort. And with every life at stake, every life *has* a stake—and a part to play—in curing the plague that has crippled and killed so many, and looms over all of our heads.

Notes

2. Wake Up—Aging Kills!

1. There are many available sources of the number of deaths from various causes in various parts of the world. A reasonably concise one from which the approximate numbers I give here can be derived is: Lopez AD, Mathers CD, Ezzati M, Jamison DT, Murray CJ. Global and regional burden of disease and risk factors, 2001: systematic analysis of population health data. *Lancet* 2006;367(9524): 1747–1757.
2. The name "Tithonus error" derives from this reaction's similarity to a Greek myth, in which a goddess (Eos) fell in love with a warrior (Tithonus) and asked her father (Zeus) to make Tithonus immortal. In the myth, Zeus agreed to this, but it all ended in tears, because Eos had forgotten to ask Zeus to make Tithonus permanently youthful. Tithonus thus became frailer and frailer, until eventually Eos had to turn him into a grasshopper. The survival of this myth through so many centuries, despite the blatant arbitrariness of its central assumption, indicates very powerfully how comforting people find it to pretend that defeating aging would not be such a good thing after all.
3. All the material relating to the SENS Challenge—*Technology Review*'s original profile and editorials about me and SENS, the terms of the Challenge, the entries, my rebuttals, the judges' verdict and the furious dissent from that verdict written by the authors of the main entry—are available on the *TR* Web site at http://www.technologyreview.com/sens/index.aspx. If you have access to hard copy, you may like to know that the original profile and editorials appeared in the February 2005 edition and the result appeared in the July 2006 edition.

4. Most of my Web site is written for the nonspecialist, so as to be useful to as many people as possible. My academic publications are also available there, in preprint form. The site covers the science described in this book (though not nearly so thoroughly as this book does, so if you're thinking of putting it back on the shelf, don't!) and also the social context.

3. Demystifying Aging

1. This was pointed out in characteristically splendid language by Leonard Hayflick in his 1994 book *How and Why We Age* (Ballantine, New York, 377pp). Hayflick's book initiated something of a procession of senior biogerontologists writing about their field for a popular audience. (What you won't find in any of those books, of course, is a plan for defeating aging.)
2. In 1972, Nixon vetoed the first bill to create the NIA, as a result of lobbying from other NIH institutes and government officials. It passed at the second attempt because, with the Watergate waves breaking around him, Nixon was less available to be lobbied. The strength of feeling against biogerontology in that era gives me a decided sense of déjà-vu.
3. As this book is being written, four of my colleagues are spearheading a new push to get this message across to U.S. politicians, under the catchy moniker "The Longevity Dividend." I sincerely applaud their persistence in this effort, and I enthusiastically joined one hundred or so of our colleagues in publicly endorsing it—but, if the truth be told, it will be a near-miracle if this new initiative is any more successful than those in whose frustrating wake it follows.
4. Rose MR. Can human aging be postponed? *Sci Am* 1999;281(6):106–111.
5. McCay CM, Crowell MF, Maynard LA. The effect of retarded growth upon the length of life span and upon the ultimate body size. *J Nutr* 1935;10:63–79.
6. Friedman DB, Johnson TE. A mutation in the age-1 gene in *Caenorhabditis elegans* lengthens life and reduces hermaphrodite fertility. *Genetics* 1988;118(1):75–86.
7. Kenyon C, Chang J, Gensch E, Rudner A, Tabtiang R. A *C. elegans* mutant that lives twice as long as wild type. *Nature* 1993;366(6454):461–464.
8. The number of such reports is now in double figures.
9. Migliaccio E, Giorgio M, Mele S, Pelicci G, Reboldi P, Pandolfi PP, Lanfrancone L, Pelicci PG. The p66^{shc} adaptor protein controls oxidative stress response and life span in mammals. *Nature* 1999;402(6759):309–313.
10. Mitsui A, Hamuro J, Nakamura H, Kondo N, Hirabayashi Y, Ishizaki-Koizumi S, Hirakawa T, Inoue T, Yodoi J. Overexpression of human thioredoxin in transgenic mice controls oxidative stress and life span. *Antioxid Redox Signal* 2002;4(4): 693–696.
11. Schriner SE, Linford NJ, Martin GM, Treuting P, Ogburn CE, Emond M, Coskun PE, Ladiges W, Wolf N, Van Remmen H, Wallace DC, Rabinovitch PS. Extension of murine life span by overexpression of catalase targeted to mitochondria. *Science* 2005;308(5730):1909–1911.
12. de Grey ADNJ. The unfortunate influence of the weather on the rate of aging: why human caloric restriction or its emulation may only extend life expectancy by 2–3 years. *Gerontology* 2005; 51(2):73–82.

4. Engineering Rejuvenation

1. de Grey ADNJ. The mitochondrial free radical theory of aging. 1999; Austin, TX: Landes Bioscience. (ISBN 1-57059-564–X).
2. Barja G. Rate of generation of oxidative stress-related damage and animal longevity. *Free Radic Biol Med* 2002;33(9):1167–1172.
3. Rhee SG. Redox signaling: hydrogen peroxide as intracellular messenger. *Exp Mol Med* 1999;31(2):53–59.
4. Actually, my early expositions of SENS also listed age-related changes in the immune and endocrine (hormonal) systems as separate items to be repaired. They are no longer included, because they both arise from changes that come under some of the other SENS headings. Immune decline is ultimately the result of having too few of certain types of cells that perform key roles in our immune defenses, together with the overabundance of immune cells that are no longer working well. Endocrine changes are similar: we have too many of some types of cell, especially fat cells in the abdomen, and too few of others (such as in the ovary in women); also, some important endocrine cells become encumbered by intracellular garbage that slows them down.
5. I used the phrase "engineered negligible senescence" in my 1999 book (see note 1 above), which was mostly written in 1998. I first prefaced it with the word "strategies" in the article arising from the workshop I ran just three months after my California revelation: de Grey ADNJ, Ames BN, Andersen JK, Bartke A, Campisi J, Heward CB, McCarter RJM, Stock G. Time to talk SENS: critiquing the immutability of human aging. *Ann N Y Acad Sci* 2002;959:452–462. The title of that article was also the first use of the acronym "SENS."

5. Meltdown of the Cellular Power Plants

1. This terminology actually makes me bristle, because so many substances that *behave like* "free radicals" and are often *called* free radicals don't, strictly, meet this definition—and conversely, some molecules that strictly are free radicals are not harmful at all. The same goes for "reactive oxygen species," another popular term used interchangeably with free radicals by many people. For instance, *reduced iron ions* (Fe^{2+}) make significant contributions to free radical chemistry, but are not molecules, are not oxygen-based, and do not contain an unpaired electron. In my thesis book, I tried to convince my colleagues to adopt "lonely electron carrier" to cover all of these substances, but have met with no success, and have resigned myself (for now!) to the common, albeit somewhat sloppy, convention.
2. This curious fact is accepted to be an evolutionary leftover from the earliest days of evolution, when the ancestors of mitochondria were organisms in their own right, which formed a mutually beneficial relationship with cells that were the ancestors of all but the most primitive of organisms alive today. As we'll see, this turns out to be a rather important little item of trivia, with consequences much more serious than other evolutionary carryovers like the fact that humans still carry an appendix.
3. Harman D. The biologic clock: the mitochondria? *J Am Geriatr Soc* 1972;20(4):145–147.
4. Barja G. Rate of generation of oxidative stress-related damage and animal longevity. *Free Radic Biol Med* 2002;33(9):1167–1172.

5. Schriner SE, Linford NJ, Martin GM, Treuting P, Ogburn CE, Emond M, Coskun PE, Ladiges W, Wolf N, Van Remmen H, Wallace DC, Rabinovitch PS. Extension of murine life span by overexpression of catalase targeted to mitochondria. *Science* 2005;308(5730):1909–1911.
6. Schriner SE, Linford NJ. Extension of mouse lifespan by overexpression of catalase. *AGE* 2006;28(2):209–218.
7. Bandy B, Davison AJ. Mitochondrial mutations may increase oxidative stress: implications for carcinogenesis and aging? *Free Radic Biol Med* 1990;8(6):523–39.
8. Shigenaga MK, Hagen TM, Ames BN. Oxidative damage and mitochondrial decay in aging. *Proc Natl Acad Sci USA* 1994;91(23):10771–10778.
9. de Grey ADNJ. A proposed refinement of the mitochondrial free radical theory of aging. *BioEssays* 1997;19(2):161–166.
10. de Grey ADNJ. A mechanism proposed to explain the rise in oxidative stress during aging. *J Anti-Aging Med* 1998;1(1):53–66.
11. van Zutphen H, Cornwell DG. Some studies on lipid peroxidation in monomolecular and bimolecular lipid films. *J Membr Biol* 1973;13:79–88.
12. Kissova I, Deffieu M, Manon S, Camougrand N. Uth1p is involved in the autophagic degradation of mitochondria. *J Biol Chem* 2004;279(37):39068–39074.
13. Elmore SP, Qian T, Grissom SF, Lemasters JJ. The mitochondrial permeability transition initiates autophagy in rat hepatocytes. *FASEB J* 2001;15(12):2286–2287.
14. Meagher EA, Barry OP, Lawson JA, Rokach J, FitzGerald GA. Effects of vitamin E on lipid peroxidation in healthy persons. *JAMA* 2001;285(9):1178–1182.
15. Thomas SR, Stocker R. Molecular action of vitamin E in lipoprotein oxidation: implications for atherosclerosis. *Free Radic Biol Med* 2000;28(12):1795–1805.
16. de Grey ADNJ. The mitochondrial free radical theory of aging. 1999; Austin, TX: Landes Bioscience.
17. See, for example, Jacobs HT. The mitochondrial theory of aging: dead or alive? *Aging Cell* 2003;2(1):11–17.

6. Getting Off the Grid

1. Barja G. Rate of generation of oxidative stress-related damage and animal longevity. *Free Radic Biol Med* 2002;33(9):1167–1172.
2. Schriner SE, Linford NJ, Martin GM, Treuting P, Ogburn CE, Emond M, Coskun PE, Ladiges W, Wolf N, Van Remmen H, Wallace DC, Rabinovitch PS. Extension of murine life span by overexpression of catalase targeted to mitochondria. *Science* 2005;308(5730):1909–1911.
3. Bluher M, Kahn BB, Kahn CR. Extended longevity in mice lacking the insulin receptor in adipose tissue. *Science* 2003;299(5606):572–574.
4. Mitsui A, Hamuro J, Nakamura H, Kondo N, Hirabayashi Y, Ishizaki-Koizumi S, Hirakawa T, Inoue T, Yodoi J. Overexpression of human thioredoxin in transgenic mice controls oxidative stress and life span. *Antioxid Redox Signal* 2002;4(4):693–696.
5. Migliaccio E, Giorgio M, Mele S, Pelicci G, Reboldi P, Pandolfi PP, Lanfrancone L, Pelicci PG. The $p66^{shc}$ adaptor protein controls oxidative stress response and life span in mammals. *Nature* 1999;402(6759):309–313.
6. Huang TT, Carlson EJ, Gillespie AM, Shi Y, Epstein CJ. Ubiquitous overexpression of CuZn superoxide dismutase does not extend life span in mice. *J Gerontol A Biol Sci Med Sci* 2000;55(1):B5–B9.

7. Rhee SG. Redox signaling: hydrogen peroxide as intracellular messenger. *Exp Mol Med* 1999;31(2):53–59.
8. This could potentially cause a problem of too many instruction sets leading to too many proteins being produced—like putting out too many work orders to a factory with lots of capacity. While I don't think this is very likely for somewhat complex technical reasons, we should be able to find a way around any such problem even if it does occur by changing their regulation a bit so that each such copy produces a little bit less of its encoded protein than it normally would, keeping the *total* protein synthesis rate normal.
9. de Grey ADNJ. Forces maintaining organellar genomes: is any as strong as genetic code disparity or hydrophobicity? *BioEssays* 2005;27(4):436–446.
10. Indeed, the cases that we'll see in which allotopic expression has been performed have shown that it's almost trivial to express mitochondrial genes allotopically when the central problem is code disparity rather than hydrophobicity.
11. de Grey ADNJ. Mitochondrial gene therapy: an arena for the biomedical use of inteins. *Trends Biotechnol* 2000;18(9):394–399.
12. Ojaimi J, Pan J, Santra S, Snell WJ, Schon EA. An algal nucleus-encoded subunit of mitochondrial ATP synthase rescues a defect in the analogous human mitochondrial-encoded subunit. *Mol Biol Cell* 2002;13(11):3836–3844.
13. Daley DO, Clifton R, Whelan J. Intracellular gene transfer: reduced hydrophobicity facilitates gene transfer for subunit 2 of cytochrome c oxidase. *Proc Natl Acad Sci USA* 2002;99(16):10510–10515.
14. Manfredi G, Fu J, Ojaimi J, Sadlock JE, Kwong JQ, Guy J, Schon EA. Rescue of a deficiency in ATP synthesis by transfer of MTATP6, a mitochondrial DNA-encoded gene, to the nucleus. *Nat Genet* 2002;30(4):394–399.
15. Zullo SJ, Parks WT, Chloupkova M, Wei B, Weiner H, Fenton WA, Eisenstadt JM, Merril CR. Stable transformation of CHO cells and human NARP cybrids confers oligomycin resistance (oli^r) following transfer of a mitochondrial DNA-encoded oli^r ATPase6 gene to the nuclear genome: a model system for mtDNA gene therapy. *Rejuvenation Res* 2005;8(1):18–28.
16. Guy J, Qi X, Pallotti F, Schon EA, Manfredi G, Carelli V, Martinuzzi A, Hauswirth WW, Lewin AS. Rescue of a mitochondrial deficiency causing Leber Hereditary Optic Neuropathy. *Ann Neurol* 2002;52(5):534–542.
17. Ozawa T, Sako Y, Sato M, Kitamura T, Umezawa Y. A genetic approach to identifying mitochondrial proteins. *Nat Biotechnol* 2003;21(3):287–293.
18. Khan SM, Bennett JP. Development of mitochondrial gene replacement therapy. *J Bioenerg Biomembr* 2004;36(4):387–393.
19. Seo BB, Wang J, Flotte TR, Yagi T, Matsuno-Yagi A. Use of the NADH-quinone oxidoreductase (NDI1) gene of *Saccharomyces cerevisiae* as a possible cure for complex I defects in human cells. *J Biol Chem* 2000;275(48):37774–37778.

7. Upgrading the Biological Incinerators

1. Or sometimes LIP-oh-fuss-kin, or even lip-oh-FUSS-kin. I counsel against these pronunciations, however, as they may lead your interlocutor to think that you're talking about cosmetic fat-reduction surgery, which will not garner you respect amongst biogerontologists or the general public.

2. There is evidence to suggest that the single greatest contributor to the lipofuscin load is improperly degraded mitochondria: incorporated into the lysosome when they become defective (as I discussed in Chapter 5), mitochondrial parts that are not immediately degraded appear to undergo extensive internal chemical modifications, driven by their own generation of free radicals and accelerated by the permissive presence in the lysosome of "transition metals" like iron and copper that can catalyze more extensive free radical damage. These chemical reactions make the mitochondrial remnants resistant to degradation because of the chemical "cross-linking" of its molecules by these reactions—similar to the stiffening of old windscreen wipers.
3. The same, by the way, goes for experiments allegedly showing that various compounds—most famously, centrophenoxine (Lucidril)—reduce the accumulation of lipofuscin or even remove it from cells. It appears that these compounds simply increase the transport of the more transitory materials around in the cell. See: Porta EA. Pigments in aging: an overview. *Ann N Y Acad Sci.* 2002 Apr;959:57–65; also Katz, *Arch Gerontol Geriatr* 34(3):311–317; Dowson, *Exp Gerontol* 20(6):333–340; Andrews et al, *Neurobiol Aging* 7(2):107–113; Katz et al, *J Gerontol* 38(5):525-531; and Kano et al, *Neurosci Res* 33(3):207–213.
4. Brunk UT, Terman A. The mitochondrial-lysosomal axis theory of aging: accumulation of damaged mitochondria as a result of imperfect autophagocytosis. *Eur J Biochem* 2002;269(8):1996–2002.
5. de Grey ADNJ. The mitochondrial free radical theory of aging. 1999; Austin, TX: Landes Bioscience. (ISBN 1-57059-564-X).
6. Lusis AJ. Atherosclerosis. *Nature* 2000;407(6801):233–241.
7. de Grey ADNJ. Bioremediation meets biomedicine: therapeutic translation of microbial catabolism to the lysosome. *Trends Biotechnol* 2002;20(11):452–455.
8. de Grey ADNJ, Alvarez PJJ, Brady RO, Cuervo AM, Jerome WG, McCarty PL, Nixon RA, Rittmann BE, Sparrow JR. Medical bioremediation: prospects for the application of microbial catabolic diversity to aging and several major age-related diseases. *Ageing Res Rev* 2005;4(3):315–338.
9. de Grey ADNJ. Appropriating microbial catabolism: a proposal to treat and prevent neurodegeneration. *Neurobiol Aging* 2006;27(4):589–595.
10. Note that NFTs are not the same as the more famous amyloid plaque that enshrouds the outside of Alzheimer's patients—although this too is a problem that a more functional lysosome would probably fix, as we shall see in Chapter 8.
11. Terman A, Sandberg S. Proteasome inhibition enhances lipofuscin formation. *Ann N Y Acad Sci* 2002;973:309–312.
12. Nixon RA, Wegiel J, Kumar A, Yu WH, Peterhoff C, Cataldo A, Cuervo AM. Extensive involvement of autophagy in Alzheimer disease: an immuno-electron microscopy study. *J Neuropathol Exp Neurol* 2005;64(2):113–122.
13. As we'll see in other chapters, treatments based on attacking other contributors to neurodegenerative disease—notably, amyloid plaques in Alzheimer's disease—have been moving their way up through the drug development pipeline and have even entered clinical trials. But aside from the fact that this won't help us with Niemann-Pick disease or Parkinson's, it seems unlikely that clearance of amyloid alone will be enough to truly restore Alzheimer's patients to health. This is discussed later in the present chapter.
14. The need for vitamin A for this cycle is why deficiency in the vitamin can lead to blindness.

15. Butler D, Brown QB, Chin DJ, Batey L, Karim S, Mutneja MS, Karanian DA, Bahr BA. Degradative pathways responding to age-related protein accumulation involve autophagy and lysosomal enzyme activation. *Rejuvenation Res* 2005;8(4):227–237.
16. Du H, Schiavi S, Wan N, Levine M, Witte DP, Grabowski GA. Reduction of atherosclerotic plaques by lysosomal acid lipase supplementation. *Arterioscler Thromb Vasc Biol* 2004;24(1):147–154.
17. Nice as it might be, we can't simply use this technique to fill the body with healthy cells bearing the relevant enzymes, because the existing cells will still be stuffed with gunk that, by definition, will not just fade away on its own. And in the case of some cell types—neurons, most obviously—we would probably not want to remove the lysosomally-lame cells to replace them with new ones even if we could.

8. Cutting Free of the Cellular Spider Webs

1. Dobson CM. Getting out of shape. *Nature* 2002;418(6899):729–730.
2. Gamma-secretase is actually not one protein, but a multi-protein complex composed of *presenilin* (PS), *nicastrin* (NCT), *APH-1,* and *PEN-2*.
3. Also referred to as *memapsin, ASP-2,* or *BACE*.
4. Indeed, the most common *genetic* risk factor for Alzheimer's—carrying a less favorable version of the *apolipoprotein E* (APOE) gene—is not, itself, the kind of death warrant that (for example) the Huntington's disease gene is, because it is not actually essential for amyloid-beta plaque formation but instead makes the amyloid fibril protein more likely to be deposited in tissue.
5. Hashizume Y, Wang Y, Yoshida M. Neuropathological study in the central nervous system of centenarians. In: Tauchi H, Sato T, Watanabe T (eds). Japanese Centenarians: Medical Research for the Final Stages of Human Aging. 1999; Institute for Medical Science of Aging, Aichi, Japan:137–154.
6. Also unfortunately, other centenarian autopsy studies have largely been dissections that did not use the kinds of tools needed to identify underlying molecular problems like amyloidoses. Such data as are available confirm the *general* frailty suffered by these long-lived humans, caused by having an *entire body* whose structure is weakened by ubiquitous age-related molecular damage: half of all centenarians autopsied from 1958 to 1987, and a third of those examined from 1958–1995, had age-related "senescence and organ atrophy" as a major pathological finding—a result in accord with similar studies in Sweden and the United States. (Tauchi H, Sato T. Autopsy findings: outline and generational differences. In Tauchi et al, op cit:132–136).
7. http://www.supercentenarian-research-foundation.org/
8. Personal communication, Stanley Primmer, Supercentenarian Research Foundation, 2006-04–02.
9. Courtney C, Farrell D, Gray R, Hills R, Lynch L, Sellwood E, Edwards S, Hardyman W, Raftery J, Crome P, Lendon C, Shaw H, Bentham P; AD2000 Collaborative Group. Long-term donepezil treatment in 565 patients with Alzheimer's disease (AD2000): randomized double-blind trial. *Lancet* 2004;363(9427):2105–2115.
10. Reisberg B, Doody R, Stoffler A, Schmitt F, Ferris S, Mobius HJ. A 24-week open-label extension study of memantine in moderate to severe Alzheimer disease. *Arch Neurol* 2006;63(1):49–54.

11. D'Andrea MR, Nagele RG, Wang HY, Peterson PA, Lee DH. Evidence that neurones accumulating amyloid can undergo lysis to form amyloid plaques in Alzheimer's disease. *Histopathology* 2001;38(2):120–134.
12. Paresce DM, Chung H, Maxfield FR. Slow degradation of aggregates of the Alzheimer's disease amyloid beta-protein by microglial cells. *J Biol Chem* 1997;272(46):29390–29397.
13. There would still be the matter of the *neurofibrillary tangles* (NFTs) that we discussed in the last chapter—but of course, that same chapter also offered a solution to the problem. As we shall see, however, beta-amyloid immunization *itself* helps the brain to better clear out these aggregates, as if clearing the hurricane debris from your front door had allowed you to finally take out the trash.
14. Billings LM, Oddo S, Green KN, McGaugh JL, LaFerla FM. Intraneuronal Abeta causes the onset of early Alzheimer's disease-related cognitive deficits in transgenic mice. *Neuron* 2005;45(5):675–688.
15. There are other hypotheses for what is going on, but this remains the leading one.
16. Schenk D, Barbour R, Dunn W, Gordon G, Grajeda H, Guido T, Hu K, Huang J, Johnson-Wood K, Khan K, Kholodenko D, Lee M, Liao Z, Lieberburg I, Motter R, Mutter L, Soriano F, Shopp G, Vasquez N, Vandevert C, Walker S, Wogulis M, Yednock T, Games D, Seubert P. Immunization with amyloid-beta attenuates Alzheimer-disease-like pathology in the PDAPP mouse. *Nature* 1999;400(6740): 173–177.
17. Morgan D, Diamond DM, Gottschall PE, Ugen KE, Dickey C, Hardy J, Duff K, Jantzen P, DiCarlo G, Wilcock D, Connor K, Hatcher J, Hope C, Gordon M, Arendash GW. Abeta peptide vaccination prevents memory loss in an animal model of Alzheimer's disease. *Nature* 2000;408(6815):982–985.
18. Janus C, Pearson J, McLaurin J, Mathews PM, Jiang Y, Schmidt SD, Chishti MA, Horne P, Heslin D, French J, Mount HT, Nixon RA, Mercken M, Bergeron C, Fraser PE, St George-Hyslop P, Westaway D. Abeta peptide immunization reduces behavioural impairment and plaques in a model of Alzheimer's disease. *Nature* 2000;408(6815):979–982.
19. Orgogozo JM, Gilman S, Dartigues JF, Laurent B, Puel M, Kirby LC, Jouanny P, Dubois B, Eisner L, Flitman S, Michel BF, Boada M, Frank A, Hock C. Subacute meningoencephalitis in a subset of patients with AD after Abeta42 immunization. *Neurology* 2003;61(1):46–54.
20. Nicoll JA, Wilkinson D, Holmes C, Steart P, Markham H, Weller RO. Neuropathology of human Alzheimer disease after immunization with amyloid-beta peptide: a case report. *Nat Med* 2003;9(4):448–452.
21. Ferrer I, Boada Rovira M, Sanchez Guerra ML, et al. Neuropathology and pathogenesis of encephalitis following amyloid-beta immunization in Alzheimer's disease. *Brain Pathol* 2004;14(1):11–20.
22. Masliah E, Hansen L, Adame A, Crews L, Bard F, Lee C, Seubert P, Games D, Kirby L, Schenk D. Abeta vaccination effects on plaque pathology in the absence of encephalitis in Alzheimer disease. *Neurology* 2005;64(1):129–131.
23. Gilman S, Koller M, Black RS, Griffith SG, Fox NC, Eisner L, Kirby L, Rovira MB, Forette F, Orgogozo JM; AN1792(QS-21)-201 Study Team. Clinical effects of Abeta immunization (AN1792) in patients with AD in an interrupted trial. *Neurology* 2005;64(9):1553–1562.
24. Lemere CA, Beierschmitt A, Iglesias M, Spooner ET, Bloom JK, Leverone JF, Zheng JB, Seabrook TJ, Louard D, Li D, Selkoe DJ, Palmour RM, Ervin FR.

Alzheimer's disease abeta vaccine reduces central nervous system abeta levels in a nonhuman primate, the Caribbean vervet. *Am J Pathol* 2004;165(1):283–297.

25. The most prominent antibody currently under investigation may not actually work this way: it doesn't even appear to *get into* the brain, let alone actually activate microglia or bind to plaque. Instead, this antibody appears to work mostly by capturing soluble beta-amyloid in the fluids bathing the central nervous system, drawing it out of the brain and into the spinal fluid and eventually the periphery of the body by a sort of osmosis, ultimately lowering the burden of brain plaque and of beta-amyloid oligomers within neurons. This "peripheral sink" hypothesis is actually one interpretation of how the original vaccine, AN-1792, worked (see note 15).

26. Qu B, Boyer PJ, Johnston SA, Hynan LS, Rosenberg RN. Abeta42 gene vaccination reduces brain amyloid plaque burden in transgenic mice. *J Neurol Sci,* 2006;244(1–2):151–158.
27. Hrncic R, Wall J, Wolfenbarger DA, et al. Antibody-mediated resolution of light chain-associated amyloid deposits. *Am J Pathol* 2000;157(4):1239–1246.
28. Solomon A, Weiss DT, Wall JS. Immunotherapy in systemic primary (AL) amyloidosis using amyloid-reactive monoclonal antibodies. *Cancer Biother Radiopharm* 2003;18(6):853–860.

29. This is the same *recombinant DNA* biotechnology that is currently used to produce everything from insulin for diabetics to many artificial flavorings.
30. AA is the result of overexposure to SAA over a lifetime, which is usually the result of chronic inflammatory conditions. AA is much more common in the developing world than it is in many developed countries because people living without good sanitation infrastructure are much more subject to ongoing inflammatory stimuli—notably, repeated or chronic infections, including viral hepatitis. Thus, this amyloidosis is much less common in the United States, where it is usually linked to specific inflammatory diseases like rheumatoid arthritis.

31. Solomon A, Weiss DT, Wall JS. Therapeutic potential of chimeric amyloid-reactive monoclonal antibody 11-1F4. *Clin Cancer Res* 2003;9(10 Pt 2):3831S–3838S.
32. Wall J, Schell M, Hrncic R, Macy S, Wooliver C, Wolfenbarger D, Murphy C, Donnell R, Weiss D T, Solomon A. Treatment of amyloidosis using an anti-fibril monoclonal antibody: Preclinical efficacy in a murine model of AA-amyloidosis. In: Bély M, Apáthy A (eds). *Amyloid and Amyloidosis. The Proceedings of the IX International Symposium on Amyloidosis.* 2001; Budapest, Hungary: David Apáthy, pp. 158–160.
33. Lorenzo A, Razzaboni B, Weir GC, Yankner BA. Pancreatic islet cell toxicity of amylin associated with type-2 diabetes mellitus. *Nature* 1994;368(6473):756–760.
34. Badman MK, Pryce RA, Charge SB, Morris JF, Clark A. Fibrillar islet amyloid polypeptide (amylin) is internalized by macrophages but resists proteolytic degradation. *Cell Tissue Res* 1998;291(2):285–294.

9. Breaking the Shackles of AGE

1. Sell DR, Kleinman NR, Monnier VM. Longitudinal determination of skin collagen glycation and glycoxidation rates predicts early death in C57BL/6NNIA mice. *FASEB J* 2000;14(1):145–156.
2. Port SC, Goodarzi MO, Boyle NG, Jennrich RI. Blood glucose: a strong risk factor for mortality in nondiabetic patients with cardiovascular disease. *Am Heart J* 2005;150(2):209–214.

3. Khaw KT, Wareham N, Luben R, Bingham S, Oakes S, Welch A, Day N. Glycated haemoglobin, diabetes, and mortality in men in Norfolk cohort of european prospective investigation of cancer and nutrition (EPIC-Norfolk). *BMJ* 2001;322(7277):15–18.
4. Sell DR, Lane MA, Johnson WA, Masoro EJ, Mock OB, Reiser KM, Fogarty JF, Cutler RG, Ingram DK, Roth GS, Monnier VM. Longevity and the genetic determination of collagen glycoxidation kinetics in mammalian senescence. *Proc Natl Acad Sci USA* 1996;93(1):485–490.
5. UK Prospective Diabetes Study Group. Quality of life in type 2 diabetic patients is affected by complications but not by intensive policies to improve blood glucose or blood pressure control (UKPDS 37). *Diabetes Care* 1999; 22(7):1125–1136.
6. The Diabetes Control and Complications Trial Research Group. The effect of intensive treatment of diabetes on the development and progression of long-term complications in insulin-dependent diabetes mellitus. *N Engl J Med* 1993;329(14):977–986.
7. Beisswenger BG, Delucia EM, Lapoint N, Sanford RJ, Beisswenger PJ. Ketosis leads to increased methylglyoxal production on the Atkins diet. *Ann N Y Acad Sci* 2005;1043:201–210.
8. Lonn E, Yusuf S, Hoogwerf B, Yi Q, Zinman B, Bosch J, Dagenais G, Mann JF, Gerstein HC; HOPE Study; MICRO-HOPE Study. Effects of vitamin E on cardiovascular and microvascular outcomes in high-risk patients with diabetes: results of the HOPE study and MICRO-HOPE substudy. *Diabetes Care* 2002;25(11):1919–1927.
9. Boshtam M, Rafiei M, Golshadi ID, Ani M, Shirani Z, Rostamshirazi M. Long term effects of oral vitamin E supplement in type II diabetic patients. *Int J Vitam Nutr Res* 2005;75(5):341–346.
10. Manuel y Keenoy B, Vertommen J, De Leeuw I. The effect of flavonoid treatment on the glycation and antioxidant status in Type 1 diabetic patients. *Diabetes Nutr Metab* 1999;12(4):256–263.
11. Mustata GT, Rosca M, Biemel KM, Reihl O, Smith MA, Viswanathan A, Strauch C, Du Y, Tang J, Kern TS, Lederer MO, Brownlee M, Weiss MF, Monnier VM. Paradoxical effects of green tea (*Camellia sinensis*) and antioxidant vitamins in diabetic rats: improved retinopathy and renal mitochondrial defects but deterioration of collagen matrix glycoxidation and cross-linking. *Diabetes* 2005;54(2):517–526.
12. Anderson MM, Requena JR, Crowley JR, Thorpe SR, Heinecke JW. The myeloperoxidase system of human phagocytes generates Nepsilon-(carboxymethyl)lysine on proteins: a mechanism for producing advanced glycation end products at sites of inflammation. *J Clin Invest* 1999;104(1):103–113.
13. Brennan ML, Anderson MM, Shih DM, Qu XD, Wang X, Mehta AC, Lim LL, Shi W, Hazen SL, Jacob JS, Crowley JR, Heinecke JW, Lusis AJ. Increased atherosclerosis in myeloperoxidase-deficient mice. *J Clin Invest* 2001;107(4):419–430.
14. Cerami C, Founds H, Nicholl I, Mitsuhashi T, Giordano D, Vanpatten S, Lee A, Al-Abed Y, Vlassara H, Bucala R, Cerami A. Tobacco smoke is a source of toxic reactive glycation products. *Proc Natl Acad Sci USA* 1997;94(25):13915–13920.
15. Bolton WK, Cattran DC, Williams ME, Adler SG, Appel GB, Cartwright K, Foiles PG, Freedman BI, Raskin P, Ratner RE, Spinowitz BS, Whittier FC, Wuerth JP; ACTION I Investigator Group. Randomized trial of an inhibitor of formation of advanced glycation end products in diabetic nephropathy. *Am J Nephrol* 2004;24(1):32–40.
16. Alternative, though less likely, suspects (or possibly co-conspirators) in aminoguanidine's side effects include some of its antioxidant mechanisms, which

include quenching the Janus-faced *nitric oxide* molecule (an essential signaling molecule that is also a free radical and a major target of aminoguanidine) and the binding-up of transition metals.

17. Sell DR, Nelson JF, Monnier VM. Effect of chronic aminoguanidine treatment on age-related glycation, glycoxidation, and collagen cross-linking in the Fischer 344 rat. *J Gerontol A Biol Sci Med Sci* 2001;56(9):B405–B411.
18. Asif M, Egan J, Vasan S, Masurekar MR, Lopez S, Williams C, Torres RL, Wagle D, Ulrich P, Cerami A, Brines M, Regan TJ. An advanced glycation endproduct cross-link breaker can reverse age-related increases in myocardial stiffness. *Proc Natl Acad Sci USA* 2000;97(6):2809–2813. Researchers later confirmed that alagebrium could do the same thing in diabetic dogs; this result reinforced the drug's reputation as an AGE breaker, but was much less important from the perspective of using it to treat "normal" aging.
19. Vaitkevicius PV, Lane M, Spurgeon H, Ingram DK, Roth GS, Egan JJ, Vasan S, Wagle DR, Ulrich P, Brines M, Wuerth JP, Cerami A, Lakatta EG. A cross-link breaker has sustained effects on arterial and ventricular properties in older rhesus monkeys. *Proc Natl Acad Sci USA* 2001;98(3):1171–1175.
20. Kass DA, Shapiro EP, Kawaguchi M, Capriotti AR, Scuteri A, deGroof RC, Lakatta EG. Improved arterial compliance by a novel advanced glycation end-product crosslink breaker. *Circulation* 2001;104(13):1464–1470.
21. Little WC, Zile MR, Kitzman DW, Hundley WG, O'Brien TX, Degroof RC. The effect of alagebrium chloride (ALT-711), a novel glucose cross-link breaker, in the treatment of elderly patients with diastolic heart failure. *J Card Fail* 2005;11(3): 191–195.
22. Bakris GL, Bank AJ, Kass DA, Neutel JM, Preston RA, Oparil S. Advanced glycation end-product cross-link breakers. A novel approach to cardiovascular pathologies related to the aging process. *Am J Hypertens* 2004;17(12 Pt 2):23S–30S.
23. Melenovsky V, Clattenburg L, Corretti M, Fitzgerald P, Capriotti A, Gerstenblith G, Kass D, Zieman S. Improved flow-mediated arterial vasodilation by advanced glycation crosslink breaker, alagebrium chloride (ALT-711), in older adults with isolated systolic hypertension. Presented at the American Heart Association Annual Scientific Session, November 13–16, 2005. Presentation Number 2875.
24. Thohan V, Koerner MM, Pratt CM, Torre GA. Improvements in diastolic function among patients with advanced systolic heart failure utilizing alagebrium, an oral advanced glycation end-product crosslink breaker. Presented at the American Heart Association Annual Scientific Session, November 13–16, 2005. Presentation Number 2647.
25. Biemel KM, Friedl DA, Lederer MO. Identification and quantification of major maillard cross-links in human serum albumin and lens protein. Evidence for glucosepane as the dominant compound. *J Biol Chem* 2002;277(28):24907–24915.
26. Cheng R, Feng Q, Argirov OK, Ortwerth BJ. Structure elucidation of a novel yellow chromophore from human lens protein. *J Biol Chem* 2004;279(44):45441–45449.

10. Putting the Zombies to Rest

1. Bower CE, Olson JC. Pneumonia and influenza. In: Hofmann M, Hooper MA, eds. *Connecticut Women's Health*. 2001; Hartford, CT: Connecticut Department of Public Health, 157–160.

2. Centers for Disease Control and Prevention. Disaster Center. Death rates for twelve age groups from pneumonia and influenza (480–487). Online at http:// www.disastercenter.com/cdc/1pneumo.html Accessed 2006-09-13.
3. Delarosa O, Pawelec G, Peralbo E, Wikby A, Mariani E, Mocchegiani E, Tarazona R, Solana R. Immunological biomarkers of ageing in man: changes in both innate and adaptive immunity are associated with health and longevity. *Biogerontology* 2006;7(5-6):471–481. As is so often the case when discussing what changes in the body are and are not "really" due to aging, the distinction being made is a slippery one. Zinc absorption, and vitamin D production in response to sunlight, are both compromised by biological aging. The point is, however, that if actual aging *damage* were undone, then the *effects* of aging damage (such as these effects on zinc and vitamin D) would be reversed—and so would their downstream effects, such as those on the innate immune system. These deficiencies could also be addressed in large part through supplementation.
4. Lynch MD. How does cellular senescence prevent cancer? *DNA Cell Biol* 2006;25(2):69–78.
5. Khan N, Shariff N, Cobbold M, Bruton R, Ainsworth JA, Sinclair AJ, Nayak L, Moss PA. Cytomegalovirus seropositivity drives the CD8 T cell repertoire toward greater clonality in healthy elderly individuals. *J Immunol* 2002;169(4):1984–1992.
6. McVoy MA, Adler SP. Immunologic evidence for frequent age-related cytomegalovirus reactivation in seropositive immunocompetent individuals. *J Infect Dis* 1989;160(1):1–10.
7. Pawelec G, Akbar A, Caruso C, Solana R, Grubeck-Loebenstein B, Wikby A. Human immunosenescence: is it infectious? *Immunol Rev* 2005;205:257–268.
8. Karrer U, Sierro S, Wagner M, Hengel H, Koszinowski UH, Phillips RE, Klenerman P. Memory inflation: continuous accumulation of antiviral CD8+ T cells over time. *J Immunol* 2003;170(4):2022–2029.
9. Ouyang Q, Wagner WM, Voehringer D, Wikby A, Klatt T, Walter S, Muller CA, Pircher H, Pawelec G. Age-associated accumulation of CMV-specific CD8+ T cells expressing the inhibitory killer cell lectin-like receptor G1 (KLRG1). *Exp Gerontol* 2003;38(8):911–920.
10. Koch S, Solana R, Dela Rosa O, Pawelec G. Human cytomegalovirus infection and T cell immunosenescence: a minireview. *Mech Ageing Dev* 2006;127(6): 538–543.
11. Fletcher JM, Vukmanovic-Stejic M, Dunne PJ, Birch KE, Cook JE, Jackson SE, Salmon M, Rustin MH, Akbar AN. Cytomegalovirus-specific CD4+ T cells in healthy carriers are continuously driven to replicative exhaustion. *J Immunol* 2005;175(12):8218–8225.
12. Munks MW, Cho KS, Pinto AK, Sierro S, Klenerman P, Hill AB. Four distinct patterns of memory CD8 T cell responses to chronic murine cytomegalovirus infection. *J Immunol* 2006;177(1):450–458.
13. Ouyang Q, Wagner WM, Zheng W, Wikby A, Remarque EJ, Pawelec G. Dysfunctional CMV-specific $CD8^+$ T cells accumulate in the elderly. *Exp Gerontol* 2004;39(4):607–613.
14. Hadrup SR, Strindhall J, Kollgaard T, Seremet T, Johansson B, Pawelec G, thor Straten P, Wikby A. Longitudinal studies of clonally expanded CD8 T cells reveal a repertoire shrinkage predicting mortality and an increased number of dysfunctional cytomegalovirus-specific T cells in the very elderly. *J Immunol* 2006;176(4):2645–2653.
15. Eaton SM, Burns EM, Kusser K, Randall TD, Haynes L. Age-related defects in

CD4 T cell cognate helper function lead to reductions in humoral responses. *J Exp Med* 2004;200(12):1613–1622.
16. Yang X, Stedra J, Cerny J. Relative contribution of T and B cells to hypermutation and selection of the antibody repertoire in germinal centers of aged mice. *J Exp Med* 1996;183(3):959–970.
17. The converse was also true: even though animals getting old B cells would mostly *initially* produce a more effective version of the antibody gene when they received young T cells along with them, still the rate of hypermutation was lower in old than in young B cells. So there does appear to be some additional defect in the B cells taken from older animals; it's unclear whether this is something intrinsic to those cells, or a result of other, primary aging processes.
18. Messaoudi I, Lemaoult J, Guevara-Patino JA, Metzner BM, Nikolich-Zugich J. Age-related CD8 T cell clonal expansions constrict CD8 T cell repertoire and have the potential to impair immune defense. *J Exp Med* 2004;200(10):1347–1358.
19. Trzonkowski P, Mysliwska J, Szmit E, Wieckiewicz J, Lukaszuk K, Brydak LB, Machala M, Mysliwski A. Association between cytomegalovirus infection, enhanced proinflammatory response and low level of anti-hemagglutinins during the anti-influenza vaccination—an impact of immunosenescence. *Vaccine* 2003;21(25-26):3826–3836.
20. Khan N, Hislop A, Gudgeon N, Cobbold M, Khanna R, Nayak L, Rickinson AB, Moss PA. Herpesvirus-specific CD8 T cell immunity in old age: cytomegalovirus impairs the response to a coresident EBV infection. *J Immunol* 2004;173(12):7481–7489.
21. McElhaney JE. The unmet need in the elderly: designing new influenza vaccines for older adults. *Vaccine* 2005;23(Suppl 1):S10–S25.
22. Effros RB, Dagarag M, Spaulding C, Man J. The role of CD8+ T-cell replicative senescence in human aging. *Immunol Rev* 2005;205:147–157.
23. Jonasson L, Tompa A, Wikby A. Expansion of peripheral CD8+ T cells in patients with coronary artery disease: relation to cytomegalovirus infection. *J Intern Med.* 2003;254(5):472–478.
24. Olsson J, Wikby A, Johansson B, Lofgren S, Nilsson BO, Ferguson FG. Age-related change in peripheral blood T-lymphocyte subpopulations and cytomegalovirus infection in the very old: the Swedish longitudinal OCTO immune study. *Mech Ageing Dev* 2000;121(1–3):187–201.
25. Wikby A, Nilsson BO, Forsey R, Thompson J, Strindhall J, Lofgren S, Ernerudh J, Pawelec G, Ferguson F, Johansson B. The immune risk phenotype is associated with IL–6 in the terminal decline stage: findings from the Swedish NONA immune longitudinal study of very late life functioning. *Mech Ageing Dev* 2006;127(8):695–704.
26. Hadrup SR, Strindhall J, Kollgaard T, Seremet T, Johansson B, Pawelec G, thor Straten P, Wikby A. Longitudinal studies of clonally expanded CD8 T cells reveal a repertoire shrinkage predicting mortality and an increased number of dysfunctional cytomegalovirus-specific T cells in the very elderly. *J Immunol* 2006;176(4):2645–2653.
27. Kerkela R, Grazette L, Yacobi R, Iliescu C, Patten R, Beahm C, Walters B, Shevtsov S, Pesant S, Clubb FJ, Rosenzweig A, Salomon RN, Van Etten RA, Alroy J, Durand JB, Force T. Cardiotoxicity of the cancer therapeutic agent imatinib mesylate. *Nat Med* 2006;12(8):908–916.
28. In the real world, of course, we have learned that actual "smart bombs" are not

nearly accurate enough to merit use as a metaphor for the kind of precise, focused killing required here. But I digress.

29. Kukowska-Latallo JF, Candido KA, Cao Z, Cao Z, Nigavekar SS, Majoros IJ, Thomas TP, Balogh LP, Khan MK, Baker JR. Nanoparticle targeting of anticancer drug improves therapeutic response in animal model of human epithelial cancer. *Cancer Res* 2005;65(12):5317–5324.
30. ibid.
31. Yang W, Barth RF, Wu G, Sferra TJ, Bandyopadhyaya AK, Tjarks W, Ferketich AK, Moeschberger ML, Binns PJ, Riley KJ, Coderre JA, Ciesielski MJ, Fenstermaker RA, Wikstrand CJ. Molecular targeting and treatment of EGFRvIII-positive gliomas using boronated monoclonal antibody L8A4. *Clin Cancer Res* 2006;12(12):3792–3802.
32. Byrnes AP. Challenges and future prospects in gene therapy. *IDrugs.* 2005;8(12): 993–996.
33. Gu J, Andreeff M, Roth JA, Fang B. hTERT promoter induces tumor-specific Bax gene expression and cell killing in syngenic mouse tumor model and prevents systemic toxicity. *Gene Ther* 2002;9(1):30–37.
34. Takeda T, Inaba H, Yamazaki M, Kyo S, Miyamoto T, Suzuki S, Ehara T, Kakizawa T, Hara M, DeGroot LJ, Hashizume K. Tumor-specific gene therapy for undifferentiated thyroid carcinoma utilizing the telomerase reverse transcriptase promoter. *J Clin Endocrinol Metab* 2003;88(8):3531–3538.
35. Markert ML, Alexieff MJ, Li J, Ozaki DA, Devlin BH, Sedlak DA, Sempowski GD, Hale LP, Rice HE, Mahaffey SM, Skinner MA. Postnatal thymus transplantation with immunosuppression as treatment for DiGeorge syndrome. *Blood* 2004;104(8):2574–2581.
36. Basu R, Breda E, Oberg AL, Powell CC, Dalla Man C, Basu A, Vittone JL, Klee GG, Arora P, Jensen MD, Toffolo G, Cobelli C, Rizza RA. Mechanisms of the age-associated deterioration in glucose tolerance: contribution of alterations in insulin secretion, action, and clearance. *Diabetes* 2003;52(7).1738–1748.
37. Cefalu WT, Wang ZQ, Werbel S, Bell-Farrow A, Crouse JR 3rd, Hinson WH, Terry JG, Anderson R. Contribution of visceral fat mass to the insulin resistance of aging. *Metabolism* 1995;44(7):954–959.
38. Barzilai N, Banerjee S, Hawkins M, Chen W, Rossetti L. Caloric restriction reverses hepatic insulin resistance in aging rats by decreasing visceral fat. *J Clin Invest* 1998;101(7):1353–1361.
39. Klein S, Fontana L, Young VL, Coggan AR, Kilo C, Patterson BW, Mohammed BS. Absence of an effect of liposuction on insulin action and risk factors for coronary heart disease. *N Engl J Med.* 2004;350(25):2549–2557.
40. Wang SS, Brownell KD. Public policy and obesity: the need to marry science with advocacy. *Psychiatr Clin North Am* 2005;28(1):235–252. For a popular treatment of this approach, see Brownell KD, Horgen KB. *Food Fight: The Inside Story of the Food Industry, America's Obesity Crisis, and What We Can Do About It.* 2004; New York, New York: Contemporary Books.
41. Chen G, Koyama K, Yuan X, Lee Y, Zhou YT, O'Doherty R, Newgard CB, Unger RH. Disappearance of body fat in normal rats induced by adenovirus-mediated leptin gene therapy. *Proc Natl Acad Sci USA* 1996;93(25):14795–14799.
42. Orci L, Cook WS, Ravazzola M, Wang MY, Park BH, Montesano R, Unger RH. Rapid transformation of white adipocytes into fat-oxidizing machines. *Proc Natl Acad Sci USA* 2004;101(7):2058–2063.

43. Westerterp-Plantenga MS, Saris WH, Hukshorn CJ, Campfield LA. Effects of weekly administration of pegylated recombinant human OB protein on appetite profile and energy metabolism in obese men. *Am J Clin Nutr* 2001;74(4): 426–434.
44. Wang MY, Orci L, Ravazzola M, Unger RH. Fat storage in adipocytes requires inactivation of leptin's paracrine activity: implications for treatment of human obesity. *Proc Natl Acad Sci USA* 2005;102(50):18011–18016.
45. Abate, Cook, and Livingston, personal communication; cited in *Proc Natl Acad Sci USA* 2005;102(50):18011–18016.
46. Though she is quite rarely cited in this chapter, my understanding of the biology and role in carcinogenesis of senescent cells, and of the potential remedies for them, has been heavily influenced by the work of Dr. Judy Campisi of the Lawrence Berkeley National Laboratory and the Buck Institute for Age Research, a coauthor with me of the original SENS scientific review and manifesto. Key publications of Judy's on which I have relied in the preparation of this chapter include: *Mech Ageing Dev* 126(10):1040–1045; *Cell* 120(4):513–522; and *Int J Biochem Cell Biol* 34(11):1401–1414.
47. However, the frequency is higher in the cushioning cartilage of the joints, and some researchers argue that this low frequency may be simply the result of relying too exclusively on a single marker of senescence. A recent study in our close relatives, the baboons, found that almost one cell in four was already senescent in young adulthood, and that one in twenty was living in the twilight zone by the last years of life.
48. Coppe JP, Kauser K, Campisi J, Beausejour CM. Secretion of vascular endothelial growth factor by primary human fibroblasts at senescence. *J Biol Chem.* 2006;281(40):29568–29574.
49. Orimo A, Gupta PB, Sgroi DC, Arenzana-Seisdedos F, Delaunay T, Naeem R, Carey VJ, Richardson AL, Weinberg RA. Stromal fibroblasts present in invasive human breast carcinomas promote tumor growth and angiogenesis through elevated SDF–1/CXCL12 secretion. *Cell* 2005;121(3):335–348.
50. Lee BY, Han JA, Im JS, Morrone A, Johung K, Goodwin EC, Kleijer WJ, DiMaio D, Hwang ES. Senescence-associated beta-galactosidase is lysosomal beta-galactosidase. *Aging Cell* 2006;5(2):187–195.
51. Yang NC, Hu ML. The limitations and validities of senescence associated-beta-galactosidase activity as an aging marker for human foreskin fibroblast Hs68 cells. *Exp Gerontol* 2005;40(10):813–819.
52. Krishna DR, Sperker B, Fritz P, Klotz U. Does pH 6 beta-galactosidase activity indicate cell senescence? *Mech Ageing Dev* 1999;109(2):113–123.
53. Severino J, Allen RG, Balin S, Balin A, Cristofalo VJ. Is beta-galactosidase staining a marker of senescence in vitro and in vivo? *Exp Cell Res* 2000;257(1):162–171.
54. Litaker JR, Pan J, Cheung Y, Zhang DK, Liu Y, Wong SC, Wan TS, Tsao SW. Expression profile of senescence-associated beta-galactosidase and activation of telomerase in human ovarian surface epithelial cells undergoing immortalization. *Int J Oncol* 1998;13(5):951–956.
55. Wei W, Sedivy JM. Differentiation between senescence (M1) and crisis (M2) in human fibroblast cultures. *Exp Cell Res* 1999;253(2):519–522.
56. Herbig U, Ferreira M, Condel L, Carey D, Sedivy JM. Cellular senescence in aging primates. *Science* 2006;311(5765):1257.

57. An additional feature of senescent baboon cells is that a remarkable 95 percent of

them also have abnormal chromatin "scaffolding" on their DNA, which in principle might also be used as part of a molecular profile for identifying senescence (although a mechanism whereby it would be turned from a marker into a targeting molecule is unclear at this time).

11. New Cells for Old

1. Stem cells derived from umbilical cords are intermediate in plasticity between embryonic and adult stem cells.
2. Lindvall O, Kokaia Z, Martinez-Serrano A. Stem cell therapy for human neurodegenerative disorders—how to make it work. *Nat Med* 2004;10(Suppl):S42–S50.
3. Deten A, Volz HC, Clamors S, Leiblein S, Briest W, Marx G, Zimmer HG. Hematopoietic stem cells do not repair the infarcted mouse heart. *Cardiovasc Res* 2005;65(1):52–63.
4. Murry CE, Soonpaa MH, Reinecke H, Nakajima H, Nakajima HO, Rubart M, Pasumarthi KB, Virag JI, Bartelmez SH, Poppa V, Bradford G, Dowell JD, Williams DA, Field LJ. Haematopoietic stem cells do not transdifferentiate into cardiac myocytes in myocardial infarcts. *Nature* 2004;428(6983):664–668.
5. Balsam LB, Wagers AJ, Christensen JL, Kofidis T, Weissman IL, Robbins RC. Haematopoietic stem cells adopt mature haematopoietic fates in ischaemic myocardium. Nature 2004;428(6983):668–673.
6. Nygren JM, Jovinge S, Breitbach M, Sawen P, Roll W, Hescheler J, Taneera J, Fleischmann BK, Jacobsen SE. Bone marrow-derived hematopoietic cells generate cardiomyocytes at a low frequency through cell fusion, but not transdifferentiation. *Nat Med 2004*;10(5):494–501.
7. Wang X, Willenbring H, Akkari Y, Torimaru Y, Foster M, Al-Dhalimy M, Lagasse E, Finegold M, Olson S, Grompe M. Cell fusion is the principal source of bone-marrow-derived hepatocytes. *Nature* 2003;422(6934):897–901.
8. Alvarez-Dolado M, Pardal R, Garcia-Verdugo JM,Fike JR, Lee HO, Pfeffer K, Lois C, Morrison SJ, Alvarez-Buylla A. Fusion of bone-marrow-derived cells with Purkinje neurons, cardiomyocytes and hepatocytes. *Nature* 2003;425(6961): 968–973.
9. Vassilopoulos G, Wang PR, Russell DW. Transplanted bone marrow regenerates liver by cell fusion. *Nature* 2003;422(6934):901–904.
10. Wagers AJ, Sherwood RI, Christensen JL, Weissman IL. Little evidence for developmental plasticity of adult hematopoietic stem cells. *Science* 2002;297(5590): 2256–2259.
11. Mangi AA, Noiseux N, Kong D, Mangi AA, Noiseux N, Kong D. Mesenchymal stem cells modified with Akt prevent remodeling and restore performance of infarcted hearts. *Nat Med* 2003;9(9):1195–1201.
12. Lunde K, Solheim S, Aakhus S, Arnesen H, Abdelnoor M, Egeland T, Endresen K, Ilebekk A, Mangschau A, Fjeld JG, Smith HJ, Taraldsrud E, Grogaard HK, Bjornerheim R, Brekke M, Muller C, Hopp E, Ragnarsson A, Brinchmann JE, Forfang K. Intracoronary injection of mononuclear bone marrow cells in acute myocardial infarction. *N Engl J Med* 2006;355(12):1199–1209.
13. Schachinger V, Erbs S, Elsasser A, Haberbosch W, Hambrecht R, Holschermann H, Yu J, Corti R, Mathey DG, Hamm CW, Suselbeck T, Assmus B, Tonn T, Dimmeler S, Zeiher AM; REPAIR-AMI Investigators. Intracoronary bone marrow-

derived progenitor cells in acute myocardial infarction. *N Engl J Med* 2006;355(12):1210–1221.

14. Assmus B, Honold J, Schachinger V, Britten MB, Fischer-Rasokat U, Lehmann R, Teupe C, Pistorius K, Martin H, Abolmaali ND, Tonn T, Dimmeler S, Zeiher AM. Transcoronary transplantation of progenitor cells after myocardial infarction. *N Engl J Med* 2006;355(12):1222–1232.
15. Schwartz RS. The politics and promise of stem-cell research. *N Engl J Med* 2006;355(12):1189–1191.
16. Menard C, Hagege AA, Agbulut O, Menard C, Hagege AA, Agbulut O. Transplantation of cardiac-committed mouse embryonic stem cells to infarcted sheep myocardium: a preclinical study. *Lancet* 2005;366(9490):1005–1012.
17. Kolossov E, Bostani T, Roell W, Breitbach M, Pillekamp F, Nygren JM, Sasse P, Rubenchik O, Fries JW, Wenzel D, Geisen C, Xia Y, Lu Z, Duan Y, Kettenhofen R, Jovinge S, Bloch W, Bohlen H, Welz A, Hescheler J, Jacobsen SE, Fleischmann BK. Engraftment of engineered ES cell-derived cardiomyocytes but not BM cells restores contractile function to the infarcted myocardium. *J Exp Med* 2006;203(10):2315–2327.
18. Hori Y, Rulifson IC, Tsai BC, Heit JJ, Cahoy JD, Kim SK. Growth inhibitors promote differentiation of insulin-producing tissue from embryonic stem cells. *Proc Natl Acad Sci USA* 2002;99(25):16105–16110.
19. McDonald JW, Liu XZ, Qu Y, Liu S, Mickey SK, Turetsky D, Gottlieb DI, Choi DW. Transplanted embryonic stem cells survive, differentiate and promote recovery in injured rat spinal cord. *Nat Med* 1999;5(12):1410–1412.
20. Keirstead HS, Nistor G, Bernal G, Totoiu M, Cloutier F, Sharp K, Steward O. Human embryonic stem cell-derived oligodendrocyte progenitor cell transplants remyelinate and restore locomotion after spinal cord injury. *J Neurosci* 2005;25(19):4694–4705.
21. Liu S, Qu Y, Stewart TJ, Howard MJ, Chakrabortty S, Holekamp TF, McDonald JW. Embryonic stem cells differentiate into oligodendrocytes and myelinate in culture and after spinal cord transplantation. *Proc Natl Acad Sci USA* 2000;97(11):6126–6131.
22. Srivastava AS, Shenouda S, Mishra R, Carrier E. Transplanted embryonic stem cells successfully survive, proliferate, and migrate to damaged regions of the mouse brain. *Stem Cells* 2006;24(7):1689–1694.
23. Park KI, Teng YD, Snyder EY. The injured brain interacts reciprocally with neural stem cells supported by scaffolds to reconstitute lost tissue. *Nat Biotechnol* 2002;20(11):1111–1117.
24. Ikeda R, Kurokawa MS, Chiba S, Yoshikawa H, Ide M, Tadokoro M, Nito S, Nakatsuji N, Kondoh Y, Nagata K, Hashimoto T, Suzuki N. Transplantation of neural cells derived from retinoic acid-treated cynomolgus monkey embryonic stem cells successfully improved motor function of hemiplegic mice with experimental brain injury. *Neurobiol Dis* 2005;20(1):38–48.
25. Ben-Hur T, Idelson M, Khaner H, Pera M, Reinhartz E, Itzik A, Reubinoff BE. Transplantation of human embryonic stem cell-derived neural progenitors improves behavioral deficit in Parkinsonian rats. *Stem Cells* 2004;22(7):1246–1255.
26. Deshpande DM, Kim YS, Martinez T, Carmen J, Dike S, Shats I, Rubin LL, Drummond J, Krishnan C, Hoke A, Maragakis N, Shefner J, Rothstein JD, Kerr DA. Recovery from paralysis in adult rats using embryonic stem cells. *Ann Neurol* 2006;60(1):32–44.
27. Lund RD, Wang S, Klimanskaya I, Holmes T, Ramos-Kelsey R, Lu B, Girman S,

Bischoff N, Sauvé Y, Lanza R. Human embryonic stem cell-derived cells rescue visual function in dystrophic RCS rats. *Cloning Stem Cells* 2006;8(3):189–199.

28. Takagi Y, Takahashi J, Saiki H, Morizane A, Hayashi T, Kishi Y, Fukuda H, Okamoto Y, Koyanagi M, Ideguchi M, Hayashi H, Imazato T, Kawasaki H, Suemori H, Omachi S, Iida H, Itoh N, Nakatsuji N, Sasai Y, Hashimoto N. Dopaminergic neurons generated from monkey embryonic stem cells function in a Parkinson primate model. *J Clin Invest* 2005;115(1):102–109.
29. Hoffman DI, Zellman GL, Fair CC, Mayer JF, Zeitz JG, Gibbons WE, Turner TG Jr; Society for Assisted Reproduction Technology (SART) and RAND. Cryopreserved embryos in the United States and their availability for research. *Fertil Steril* 2003;79(5):1063–1069.
30. Hodgson DM, Behfar A, Zingman LV, Kane GC, Perez-Terzic C, Alekseev AE, Puceat M, Terzic A. Stable benefit of embryonic stem cell therapy in myocardial infarction. *Am J Physiol Heart Circ Physiol* 2004;287(2):H471–H479.
31. Min JY, Yang Y, Sullivan MF, Ke Q, Converso KL, Chen Y, Morgan JP, Xiao YF. Long-term improvement of cardiac function in rats after infarction by transplantation of embryonic stem cells. *J Thorac Cardiovasc Surg* 2003;125(2):361–369.
32. Bonde S, Zavazava N. Immunogenicity and engraftment of mouse embryonic stem cells in allogeneic recipients. *Stem Cells* 2006;24(10):2192–2201.
33. Taylor CJ, Bolton EM, Pocock S, Sharples LD, Pedersen RA, Bradley JA. Banking on human embryonic stem cells: estimating the number of donor cell lines needed for HLA matching. *Lancet* 2005;366(9502):2019–2025.
34. Lanza RP, Chung HY, Yoo JJ, Wettstein PJ, Blackwell C, Borson N, Hofmeister E, Schuch G, Soker S, Moraes CT, West MD, Atala A. Generation of histocompatible tissues using nuclear transplantation. *Nat Biotechnol* 2002;20(7):689–696.
35. Barberi T, Klivenyi P, Calingasan NY, Lee H, Kawamata H, Loonam K, Perrier AL, Bruses J, Rubio ME, Topf N, Tabar V, Harrison NL, Beal MF, Moore MA, Studer L. Neural subtype specification of fertilization and nuclear transfer embryonic stem cells and application in parkinsonian mice. *Nat Biotechnol* 2003;21(10): 1200–1207.
36. Lanza R, Moore MA, Wakayama T, Perry AC, Shieh JH, Hendrikx J, Leri A, Chimenti S, Monsen A, Nurzynska D, West MD, Kajstura J, Anversa P. Regeneration of the infarcted heart with stem cells derived by nuclear transplantation. *Circ Res* 2004;94(6):820–827.
37. Rideout WM, Hochedlinger K, Kyba M, Daley GQ, Jaenisch R. Correction of a genetic defect by nuclear transplantation and combined cell and gene therapy. *Cell* 2002;109(1):17–27.
38. Brambrink T, Hochedlinger K, Bell G, Jaenisch R. ES cells derived from cloned and fertilized blastocysts are transcriptionally and functionally indistinguishable. *Proc Natl Acad Sci USA* 2006;103(4):933–938.
39. de Grey ADNJ. Inter-species therapeutic cloning: the looming problem of mitochondrial DNA and two possible solutions. *Rejuvenation Res* 2004;7(2):95–98.
40. Dyce PW, Wen L, Li J. In vitro germline potential of stem cells derived from fetal porcine skin. *Nat Cell Biol* 2006;8(4):384–390.
41. Biever C. US stem cells tainted by mouse material. *New Scientist.* 2004 Nov 1. Online at http://www.newscientist.com/article.ns?id=dn6604. Accessed 2006–09–05; Weiss R. Approved Stem Cells' Potential Questioned. *Washington Post.* 2004 Oct 29; A03. Online at http://www.washingtonpost.com/wp-dyn/articles/A7420–2004Oct28.html. Accessed 2006–09–05; Carol Ware, personal correspondence.

42. Martin MJ, Muotri A, Gage F, Varki A. Human embryonic stem cells express an immunogenic nonhuman sialic acid. *Nat Med* 2005;11(2):228–232.
43. Klimanskaya I, Chung Y, Meisner L, Johnson J, West MD, Lanza R. Human embryonic stem cells derived without feeder cells. *Lancet* 2005;365(9471):1636–1641.
44. Nisbet MC. Public opinion about stem cell research and human cloning. *Public Opinion Quarterly* 2004;68(1):131–154.
45. Cibelli JB, Kiessling A, Cunniff K, Richards C, Lanza RP, West MD. Somatic cell nuclear transfer in humans: pronuclear and early embryonic development. e-biomed: *J Regenerative Med* 2001;2:25–31.
46. Stojkovic M, Stojkovic P, Leary C, Hall VJ, Armstrong L, Herbert M, Nesbitt M, Lako M, Murdoch A. Derivation of a human blastocyst after heterologous nuclear transfer to donated oocytes. *Reprod Biomed Online* 2005;11(2):226–231.
47. Klimanskaya I, Chung Y, Becker S, Lu SJ, Lanza R. Human embryonic stem cell lines derived from single blastomeres. *Nature* 2006, in press.
48. Takahashi K, Yamanaka S. Induction of pluripotent stem cells from mouse embryonic and adult fibroblast cultures by defined factors. *Cell* 2006;126(4):663–676.
49. Atala A. Tissue engineering and regenerative medicine: concepts for clinical application. *Rejuvenation Res* 2004;7(1):15–31.

12. Nuclear Mutations and the Total Defeat of Cancer

1. Some DNA encodes RNA that has functions other than protein manufacture.
2. Freitas RA. The future of nanofabrication and molecular scale devices in nanomedicine. *Stud Health Technol Inform*. 2002;80:45–59.
3. de Grey ADNJ. Falsifying falsifications: the most critical task of theoreticians in biology. *Med Hypotheses*. 2004;62(6):1012–1020. This paper explains the use and misuse of different tests to prove or disprove the involvement of some process in aging.
4. Klungland A, Rosewell I, Hollenbach S, Larsen E, Daly G, Epe B, Seeberg E, Lindahl T, Barnes DE. Accumulation of premutagenic DNA lesions in mice defective in removal of oxidative base damage. *Proc Natl Acad Sci USA* 1999;96(23): 13300–13305.
5. Dollé ME, Busuttil RA, Garcia AM, Wijnhoven S, van Drunen E, Niedernhofer LJ, van der Horst G, Hoeijmakers JH, van Steeg H, Vijg J. Increased genomic instability is not a prerequisite for shortened lifespan in DNA repair deficient mice. *Mutat Res* 2006;596(1–2):22–35.
6. Schriner SE, Linford NJ, Martin GM, Treuting P, Ogburn CE, Emond M, Coskun PE, Ladiges W, Wolf N, Van Remmen H, Wallace DC, Rabinovitch PS. Extension of murine lifespan by overexpression of catalase targeted to mitochondria. *Science* 2005;308(5730):1909–1911.
7. Schriner SE, Linford NJ. Extension of mouse lifespan by overexpression of catalase. *AGE* 2006;28(2): 209–218.
8. Vijg J, Dollé ME. Large genome rearrangements as a primary cause of aging. *Mech Ageing Dev* 2002;123(8):907–915.
9. Dollé ME, Giese H, Hopkins CL, Martus HJ, Hausdorff JM, Vijg J. Rapid accumulation of genome rearrangements in liver but not in brain of old mice. *Nat Genet* 1997;17(4):431–434.
10. Dollé ME, Vijg J. Genome dynamics in aging mice. *Genome Res* 2002;12(11): 1732–1738.

11. Bennett-Baker PE, Wilkowski J, Burke DT. Age-associated activation of epigenetically repressed genes in the mouse. *Genetics* 2003;165(4):2055–2062.
12. Fraga MF, Ballestar E, Paz MF, Ropero S, Setien F, Ballestar ML, Heine-Suner D, Cigudosa JC, Urioste M, Benitez J, Boix-Chornet M, Sanchez-Aguilera A, Ling C, Carlsson E, Poulsen P, Vaag A, Stephan Z, Spector TD, Wu YZ, Plass C, Esteller M. Epigenetic differences arise during the lifetime of monozygotic twins. *Proc Natl Acad Sci* USA 2005;102(30):10604–10609.
13. Lu T, Pan Y, Kao SY, Li C, Kohane I, Chan J, Yankner BA. Gene regulation and DNA damage in the ageing human brain. *Nature* 2004;429(6994):883–891.
14. Weindruch R, Kayo T, Lee CK, Prolla TA. Gene expression profiling of aging using DNA microarrays. *Mech Ageing Dev* 2002;123(2–3):177–193.
15. Dhahbi JM, Tsuchiya T, Kim HJ, Mote PL, Spindler SR. Gene expression and physiologic responses of the heart to the initiation and withdrawal of caloric restriction. *J Gerontol A Biol Sci Med Sci* 2006;61(3):218–231.
16. Spindler SR. Rapid and reversible induction of the longevity, anticancer and genomic effects of caloric restriction. *Mech Ageing Dev* 2005;126(9):960–966.
17. Bahar R, Hartmann CH, Rodriguez KA, Denny AD, Busuttil RA, Dolle ME, Calder RB, Chisholm GB, Pollock BH, Klein CA, Vijg J. Increased cell-to-cell variation in gene expression in ageing mouse heart. *Nature* 2006;441(7096):1011–1014.
18. Herndon LA, Schmeissner PJ, Dudaronek JM, Brown PA, Listner KM, Sakano Y, Paupard MC, Hall DH, Driscoll M. Stochastic and genetic factors influence tissue-specific decline in ageing *C. elegans*. *Nature* 2002;419(6909):808–814.
19. Somel M, Khaitovich P, Bahn S, Paabo S, Lachmann M. Gene expression becomes heterogeneous with age. *Curr Biol* 2006;16(10):R359–R360.
20. Rubin H. What keeps cells in tissues behaving normally in the face of myriad mutations? *BioEssays* 2006;28(5):515–524.
21. Trosko JE, Ruch RJ. Cell-cell communication in carcinogenesis. *Front Biosci* 1998;3:d208–d236.
22. de Grey AD. Protagonistic pleiotropy: why cancer may be the only pathogenic effect of accumulating nuclear mutations and epimutations in aging. *Mech Ageing Dev* 2007 (in press).
23. Beyond evolutionary considerations, there's one likely reason why the burden of cells with nuclear mutations doesn't rise much with age: the fact that some mutations will kill the cell instead of crippling it, either because the mutation is so severe that the cell just can't survive it, or because the mutation sets off alarms inside the cell, leading to apoptosis (the cell's in-built anticancer "death program") or to the activation of the senescence response (in which case, at the very least, their mutations will not be passed on to any of their progeny, because no further progeny will ensue). The number of cells with mutations obviously doesn't go up when the cells that contain them are lost altogether. Obviously, cell loss is a bad outcome, and to the extent that it happens it must eventually be fixed. But it changes nothing from our practical perspective as anti-aging engineers in respect of mutations, because in these cases there are no mutations to repair or to obviate, because there is no cell to rescue by fixing them any more. Fortunately, as you'll recall, we have a foreseeable fix for cell loss of whatever origin: stem cells.
24. See its dedicated Web site at http://www.cancer.gov/aboutnci/2015
25. Miller M. 2015: a target date for eliminating suffering and death due to cancer. *BenchMarks*. 2003 May 16;3(2). Accessed online 2006–10–27 at http://www.cancer.gov/newscenter/archive/benchmarks-vol3-issue2/page1

26. This, incidentally, is the part of the answer to many of the standard "intelligent design" challenges to evolutionary theory, to the effect that such complex structures as the eye could not have evolved when there was no previous purpose for each of its component parts.
27. de Grey ADNJ, Ames BN, Andersen JK, Bartke A, Campisi J, Heward CB, McCarter RJM, Stock G. Time to talk SENS: critiquing the immutability of human aging. *Ann N Y Acad Sci* 2002;959:452–462.
28. Bielas JH, Loeb LA. Mutator phenotype in cancer: timing and perspectives. *Environ Mol Mutagen* 2005;45(2–3):206–213.
29. de Grey ADNJ, Campbell FC, Dokal I, Fairbairn LJ, Graham GJ, Jahoda CAB, Porter ACG. Total deletion of in vivo telomere elongation capacity: an ambitious but possibly ultimate cure for all age-related human cancers. *Ann N Y Acad Sci* 2004;1019:147–170.
30. A similar case is the venerable Leonard Hayflick, who was part of the discussion at the second SENS workshop, and who also refused to sign at least in part because of opposition to success. He also claims not to think it'll work, but then again, he is making the extreme claim that the slowing, arrest, or reversal of biological aging by *any* means is contrary to the laws of physics: Hayflick L. "Anti-aging" is an oxymoron. *J Gerontol A Biol Sci Med Sci* 2004;59(6):B573–B578.
31. Some updates on WILT appeared in: de Grey ADNJ. Whole-body interdiction of lengthening of telomeres: a proposal for cancer prevention. *Front Biosci* 2005;10: 2420–2429.
32. Rudolph KL, Millard M, Bosenberg MW, DePinho RA. Telomere dysfunction and evolution of intestinal carcinoma in mice and humans. *Nat Genet* 2001;28(2): 155–159.
33. Gonzalez-Suarez E, Samper E, Flores JM, Blasco MA. Telomerase-deficient mice with short telomeres are resistant to skin tumorigenesis. *Nat Genet* 2000;26(1): 114–117.
34. One variety of DKC also increases the risk of cancer, but these patients don't actually have mutations in telomerase itself, but in a helper-protein called dyskerin, which is also needed for other purposes not involving telomerase, so we can't say for sure what's going on here; also, there have been so few cases that there has been no chance to test them for activation of ALT.
35. Vulliamy TJ, Marrone A, Knight SW, Walne A, Mason PJ, Dokal I. Mutations in dyskeratosis congenita: their impact on telomere length and the diversity of clinical presentation. *Blood* 2006;107(7):2680–2685.
36. Vulliamy T, Marrone A, Szydlo R, Walne A, Mason PJ, Dokal I. Disease anticipation is associated with progressive telomere shortening in families with dyskeratosis congenita due to mutations in TERC. *Nat Genet* 2004;36(5):447–449.
37. Stelzner M, Hoagland VD, Woolman JD. Identification of optimal harvest sites of ileal stem cells for treatment of bile acid malabsorption in a dog model. *J Gastrointest Surg* 2003;7(4):516–522.

13. Getting from Here to There: The War on Aging

1. The saga of the acceptance of the chemiosmotic theory of mitochondrial function was so colorful that whole books have been written about it. A concise summary is: Prebble J. Peter Mitchell and the ox phos wars. *Trends Biochem Sci* 2002;27(4):209–212.

2. Warner H, Anderson J, Austad S, Bergamini E, Bredesen D, Butler R, Carnes BA, Clark BF, Cristofalo V, Faulkner J, Guarente L, Harrison DE, Kirkwood T, Lithgow G, Martin G, Masoro E, Melov S, Miller RA, Olshansky SJ, Partridge L, Pereira-Smith O, Perls T, Richardson A, Smith J, von Zglinicki T, Wang E, Wei JY, Williams TF. Science fact and the SENS agenda. What can we reasonably expect from ageing research? *EMBO Rep*. 2005;6(11):1006–1008.
3. Warner H. Scientific and ethical concerns regarding engineering human longevity. *Rejuvenation Res* 2006;9(4):440–442.
4. de Grey ADNJ. SENS is hard, yes, but not too hard to try: a reply to Warner. *Rejuvenation Res* 2006;9(4):443–445.
5. The Alliance presented a letter entitled "Pursuing the Longevity Dividend" to a group of U.S. Congressmen at a special symposium held on Capitol Hill on September 12, 2006. I was one of nearly one hundred scientists who signed it. Its full text can be found at http://www.agingresearch.org/longevitydividend/overview.pdf.
6. The war on cancer officially began with President Nixon's State of the Union address in 1971 and led to that year's National Cancer Act. There remains a wide spread of opinions as to whether the trebling (or so) of public funding for cancer research that occurred at that time was well spent. A recent account is Faguet GB. *The War on Cancer: An Anatomy of Failure, a Blueprint for the Future.* 2006; New York, NY: Springer.
7. Somia N, Verma IM. Gene therapy: trials and tribulations. *Nat Rev Genet.* 2000;1(2):91–99.
8. Ruwart M. October 11, 2005. The law most likely to kill you. Available online at http://www.lewrockwell.com/orig3/ruwart2.html

14. Bootstrapping Our Way to an Ageless Future

1. I first used the phrase "escape velocity" in print in the paper arising from the second SENS workshop—de Grey ADNJ, Baynes JW, Berd D, Heward CB, Pawelec G, Stock G. Is human aging still mysterious enough to be left only to scientists? *BioEssays* 2002;24(7):667–676. My first thorough description of the concept, however, didn't appear until two years later: de Grey ADNJ. Escape velocity: why the prospect of extreme human life extension matters now. *PLoS Biology* 2004;2(6):e187.
2. Gates disburses these funds through the Bill and Melinda Gates Foundation, http://www.gatesfoundation.org/
3. Buffett's decision to donate most of his wealth to the Gates Foundation was announced in June 2006 and is the largest act of charitable giving in United States history.

Afterword

1. Dufour E, Terzioglu M, Hansson FS, Sörensen L, Galter D, Olson L, Wilbertz J, Larsson NG. Age-associated mosaic respiratory chain deficiency causes transneuronal degeneration. *Hum Mol Genet* 2008, in press.
2. Vermulst M, Wanagat J, Kujoth GC, Bielas JH, Rabinovitch PS, Prolla TA, Loeb LA. DNA deletions and clonal mutations drive premature aging in mitochondrial mutator mice. *Nat Genet* 2008;40(4):392–394.

3. Kaltimbacher V, Bonnet C, Lecoeuvre G, Forster V, Sahel JA, Corral-Debrinski M. mRNA localization to the mitochondrial surface allows the efficient translocation inside the organelle of a nuclear recoded ATP6 protein. RNA 2006;12(7):1408–1417.
4. Bonnet C, Kaltimbacher V, Ellouze S, Augustin S, Bénit P, Forster V, Rustin P, Sahel JA, Corral-Debrinski M. Allotopic mRNA localization to the mitochondrial surface rescues respiratory chain defects in fibroblasts harboring mitochondrial DNA mutations affecting complex I or V subunits. *Rejuvenation Res* 2007;10(2):127–144.
5. Ellouze S, Bonnet C, Augustin S, Kaltimbacher V, Forster V, Simonutti M, Sahel JA, Corral-Debrinski M. Allotopic mRNA localization to the mitochondrial surface: a tool for rescuing respiration deficiencies. *Rejuvenation Res* 2007;10(Suppl 1):S24.
6. Qi X, Sun L, Lewin AS, Hauswirth WW, Guy J. The mutant human ND4 subunit of complex I induces optic neuropathy in the mouse. *Invest Ophthalmol Vis Sci* 2007;48(1):1–10.
7. Mahata B, Mukherjee S, Mishra S, Bandyopadhyay A, Adhya S. Functional delivery of a cytosolic tRNA into mutant mitochondria of human cells. *Science* 2006;314(5798): 471–474.
8. Mukherjee S, Mahata B, Mahato B, Adhya S. Use of a parasite-derived protein complex to modulate the function of mitochondria in human cells. *Rejuvenation Res* 2007;10(Suppl 1):S19.
9. de Grey ADNJ. Mitochondrial gene therapy: an arena for the biomedical use of inteins. *Trends Biotechnol* 2000;18(9):394–399.
10. Katrangi E, D'Souza G, Boddapati SV, Kulawiec M, Singh KK, Bigger B, Weissig V. Xenogenic transfer of isolated murine mitochondria into human rho^0 cells can improve respiratory function. *Rejuvenation Res* 2007;10(4):561–570.
11. Spees JL, Olson SD, Whitney MJ, Prockop DJ. Mitochondrial transfer between cells can rescue aerobic respiration. *Proc Natl Acad Sci USA* 2006;103(5):1283–1288.
12. Rittmann BE, Schloendorn J. Engineering away lysosomal junk: medical bioremediation. *Rejuvenation Res* 2007;10(3):359–365.
13. Mathieu J, Schloendorn J, Rittmann BE, Alvarez PJJ. Microbial degradation of 7-ketocholesterol. *Biodegradation* 2008, in press.
14. In order to avoid prejudicing scientific publication of recent studies, in adherence to nondisclosure agreements with investigators and institutions with possible and actual intellectual property issues, and with an eye to the possibility of future licensing agreements, I've agreed to keep the identity of all these enzymes under wraps for the moment. Identities of enzymes, and due credit to some scientists left anonymous in this account, will be disclosed when the project is ripe.
15. de Grey ADNJ, Alvarez PJJ, Brady RO, Cuervo AM, Jerome WG, McCarty PL, Nixon RA, Rittmann BE, Sparrow JR. Medical bioremediation: prospects for the application of microbial catabolic diversity to aging and several major age-related diseases. *Ageing Res Rev* 2005;4(3):315–338.
16. Scaffidi P, Misteli T. Lamin A-dependent nuclear defects in human aging. *Science* 2006;312(5776):1059–1063.
17. Bombois S, Maurage CA, Gompel M, Deramecourt V, Mackowiak-Cordoliani MA, Black RS, Lavielle R, Delacourte A, Pasquier F. Absence of beta-amyloid deposits after immunization in Alzheimer disease with Lewy body dementia. *Arch Neurol* 2007;64(4):583–587.
18. Streffer JR, Treyer V, Schmidt ME, Blagoev M, Hintermann S, Auberson Y, Nitsch RM, Ametamey SM, Buck A, Hock C. Imaging Alzheimer's disease amyloid pathology. *Alzheimers Dement* 2006; 2(3 Suppl 1):S358.
19. Thal L, Black R, Vellas B, Fox N, Daniels M, McLennan G, Tompkins C, Leibman

C, Pomfret M, Grundman M. Long Term Follow Up of Patients Immunized with AN1972(QS-21): Reduced functional decline in antibody responders. *Alzheimers Dement* 2007;3(3 Supp 1):S196.

20. Pride M, Seubert P, Grundman M, Hagen M, Eldridge J, Black RS. Progress in the active immunotherapeutic approach to Alzheimer's disease: clinical investigations into AN1792-associated meningoencephalitis. *Neurodegener Dis* 2008;5(3-4): 194–196.
21. Elan and Wyeth to Initiate Phase 3 Clinical Trial of Bapineuzumab (AAB-001) in Alzheimer's Disease. *http://www.wyeth.com/news?nav=display&navTo=/wyeth_html/home/news/pressreleases/2007/1180013248454.html* Accessed online 2008-04-20.
22. Braam B, Verhaar MC, Blankestijn P, Boer WH, Joles JA. Technology insight: Innovative options for end-stage renal disease—from kidney refurbishment to artificial kidney. *Nat Clin Pract Nephrol* 2007;3(10):564–572.
23. Mendelson K, Aikawa E, Mettler BA, Sales V, Martin D, Mayer JE, Schoen FJ. Healing and remodeling of bioengineered pulmonary artery patches implanted in sheep. *Cardiovasc Pathol* 2007;16(5):277–282.
24. O'Rourke MF, Hashimoto J. Mechanical factors in arterial aging: a clinical perspective. *J Am Coll Cardiol* 2007;50(1):1–13.
25. O'Rourke MF. Arterial aging: pathophysiological principles. *Vasc Med* 2007;12(4): 329–341.
26. Fain JN. Release of interleukins and other inflammatory cytokines by human adipose tissue is enhanced in obesity and primarily due to the nonfat cells. *Vitam Horm* 2006;74:443–477.
27. Cancello R, Clement K. Is obesity an inflammatory illness? Role of low-grade inflammation and macrophage infiltration in human white adipose tissue. *BJOG* 2006;113(10):1141–1147.
28. Lumeng CN, Bodzin JL, Saltiel AR. Obesity induces a phenotypic switch in adipose tissue macrophage polarization. *J Clin Invest* 2007;117(1):175–184.
29. Nguyen MT, Favelyukis S, Nguyen AK, Reichart D, Scott PA, Jenn A, Liu-Bryan R, Glass CK, Neels JG, Olefsky JM. A subpopulation of macrophages infiltrates hypertrophic adipose tissue and is activated by free fatty acids via Toll-like receptors 2 and 4 and JNK-dependent pathways. *J Biol Chem* 2007;282(48):35279–35292.
30. Byrne JA, Pedersen DA, Clepper LL, Nelson M, Sanger WG, Gokhale S, Wolf DP, Mitalipov SM. Producing primate embryonic stem cells by somatic cell nuclear transfer. *Nature* 2007;450(7169):497–502.
31. Takahashi K, Yamanaka S. Induction of pluripotent stem cells from mouse embryonic and adult fibroblast cultures by defined factors. *Cell* 2006;126(4):663–676.
32. Okita K, Ichisaka T, Yamanaka S. Generation of germline-competent induced pluripotent stem cells. *Nature* 2007;448(7151):313–317.
33. Wernig M, Meissner A, Foreman R, Brambrink T, Ku M, Hochedlinger K, Bernstein BE, Jaenisch R. In vitro reprogramming of fibroblasts into a pluripotent ES-cell-like state. *Nature* 2007;448(7151):318–324.
34. Maherali N, Sridharan R, Xie W, Utikal J, Eminli S, Arnold K, Stadtfeld M, Yachechko R, Tchieu J, Jaenisch R, Plath K, Hochedlinger K. Directly reprogrammed fibroblasts show global epigenetic remodeling and widespread tissue contribution. *Cell Stem Cell* 2007;1(1):55–70.
35. Takahashi K, Tanabe K, Ohnuki M, Narita M, Ichisaka T, Tomoda K, Yamanaka S. Induction of pluripotent stem cells from adult human fibroblasts by defined factors. *Cell* 2007;131(5):861–872.
36. Yu J, Vodyanik MA, Smuga-Otto K, Antosiewicz-Bourget J, Frane JL, Tian S, Nie J,

Jonsdottir GA, Ruotti V, Stewart R, Slukvin II, Thomson JA. Induced pluripotent stem cell lines derived from human somatic cells. *Science* 2007;318(5858):1917–1920.
37. Hanna J, Wernig M, Markoulaki S, Sun CW, Meissner A, Cassady JP, Beard C, Brambrink T, Wu LC, Townes TM, Jaenisch R. Treatment of sickle cell anemia mouse model with iPS cells generated from autologous skin. *Science* 2007;318(5858): 1920–1923.
38. Tabar V, Tomishima M, Panagiotakos G, Wakayama S, Menon J, Chan B, Mizutani E, Al-Shamy G, Ohta H, Wakayama T, Studer L. Therapeutic cloning in individual Parkinsonian mice. *Nat Med* 2008;14(4):379–381.
39. Wernig M, Zhao JP, Pruszak J, Hedlund E, Fu D, Soldner F, Broccoli V, Constantine-Paton M, Isacson O, Jaenisch R. Neurons derived from reprogrammed fibroblasts functionally integrate into the fetal brain and improve symptoms of rats with Parkinson's disease. *Proc Natl Acad Sci USA* 2008, in press.
40. Masaki H, Ishikawa T, Takahashi S, Okumura M, Sakai N, Haga M, Kominami K, Migita H, McDonald F, Shimada F, Sakurada K. Heterogeneity of pluripotent marker gene expression in colonies generated in human iPS cell induction culture. *Stem Cell Res* 2008, in press.
41. As we get closer and closer to cures, and even begin to see cures accomplished with cell therapies, the potential for saving additional lives and alleviating more misery from tissues made with such cells may well become a powerful political force before we are able to generate isolated tissues of sufficient sophistication to meet the clear potential of these cells. Tissue engineering is a science still in its infancy, and it will probably be the case for many years that "the body is the best bioreactor"—the complex physical, chemical, and regulatory environment of a functioning body helps to guide cells into their proper developmental role and physical structure in a way that will continue to challenge our efforts to replicate for a long time. If so, this will inevitably create pressure to reexamine our limits on the use of nascent tissues (as opposed to cells only) developing from iPS and other sources, rendering them more directly and broadly therapeutic.
42. Ott HC, Matthiesen TS, Goh SK, Black LD, Kren SM, Netoff TI, Taylor DA. Perfusion-decellularized matrix: using nature's platform to engineer a bioartificial heart. *Nat Med* 2008;14(2):213–221.
43. Johnson J, Canning J, Kaneko T, Pru JK, Tilly JL. Germline stem cells and follicular renewal in the postnatal mammalian ovary. *Nature* 2004;428(6979):145–150.
44. Lee HJ, Selesniemi K, Niikura Y, Niikura T, Klein R, Dombkowski DM, Tilly JL. Bone marrow transplantation generates immature oocytes and rescues long-term fertility in a preclinical mouse model of chemotherapy-induced premature ovarian failure. *J Clin Oncol* 2007;25(22):3198–3204.
45. Conboy IM, Conboy MJ, Wagers AJ, Girma ER, Weissman IL, Rando TA. Rejuvenation of aged progenitor cells by exposure to a young systemic environment. *Nature* 2005;433(7027):760–764.
46. Carlson ME, Conboy IM. Loss of stem cell regenerative capacity within aged niches. *Aging Cell* 2007;6(3):371–382.
47. Adler AS, Sinha S, Kawahara TL, Zhang JY, Segal E, Chang HY. Motif module map reveals enforcement of aging by continual NF-kappaB activity. *Genes Dev* 2007; 21(24):3244–3257.
48. Miller JC, Holmes MC, Wang J, Guschin DY, Lee YL, Rupniewski I, Beausejour CM, Waite AJ, Wang NS, Kim KA, Gregory PD, Pabo CO, Rebar EJ. An improved zinc-finger nuclease architecture for highly specific genome editing. *Nat Biotechnol* 2007;25(7):778–785.

49. Schaetzlein S, Kodandaramireddy NR, Ju Z, Lechel A, Stepczynska A, Lilli DR, Clark AB, Rudolph C, Kuhnel F, Wei K, Schlegelberger B, Schirmacher P, Kunkel TA, Greenberg RA, Edelmann W, Rudolph KL. Exonuclease-1 deletion impairs DNA damage signaling and prolongs lifespan of telomere-dysfunctional mice. *Cell* 2007;130(5):863–877.
50. Bonnet S, Archer SL, Allalunis-Turner J, Haromy A, Beaulieu C, Thompson R, Lee CT, Lopaschuk GD, Puttagunta L, Bonnet S, Harry G, Hashimoto K, Porter CJ, Andrade MA, Thebaud B, Michelakis ED. A mitochondria-K+ channel axis is suppressed in cancer and its normalization promotes apoptosis and inhibits cancer growth. *Cancer Cell* 2007;11(1):37–51.
51. Cui Z, Molnar I, Willingham IC, Pomper GJ, Stehle JR, Blanks M. From a newly discovered innate anticancer immune response in mice to a new treatment for human cancers. *Rejuvenation Res* 2007;10(Suppl 1):S25.
52. Phoenix CR, de Grey ADNJ. A model of aging as accumulated damage matches observed mortality patterns and predicts the life-extending effects of prospective interventions. *AGE* 2007;29(4):133–189.

Glossary

2-deoxyglucose (2-DG): a compound that closely resembles blood sugar, but unlike sugar cannot be metabolized into energy in the *mitochondria*. For a while, it looked as if 2-DG were a *calorie restriction mimetic.*

Active vaccination: giving a person the *antigen* against which doctors want to provide immune protection, to trigger the immune system to produce *antibodies* that will attack it.

Adaptive immune system: the branch of the immune system that learns about and targets specific foreign and "hijacked" cells though their *antigens.* The existence of the adaptive immune system provides the basis of vaccines.

Adenosine triphosphate: see *ATP*.

Adult stem cell: stem cells found in mature bodies, that can give rise to only a relatively narrow range of specialized, mature cells. Compare *embryonic stem cell.*

Advanced glycation endproducts (AGE): highly chemically stable compound that results as the "end product" of a series of chemical reactions in *glycation* chemistry.

Advanced lipoxidation endproducts (ALEs): stable chemical damage to proteins caused by their interactions with fats; similar to *advanced glycation endproducts*.

Aging damage: from the practical viewpoint of the *engineer's school* of anti-aging medicine, changes at the molecular level to the structure of the organism that distinguish an organism that has been around for a long time from one that has not. On an abstract, theoretical level, we would only want to include those changes that actually contribute to *biological aging*—i.e., those that contribute to the increasing risk of disease, dysfunction, and death as we go on living. From the point of view of *anti-aging medicine,* however, this distinction becomes problematic, because it would require us to *first* determine which changes genuinely contribute to *biological aging* and which do not—which would require many more decades of research. The *engineer's school* of anti-aging medicine sidesteps this ignorance by classifying *all* aging changes that might contribute to biological aging as "damage," and then intervening to repair or obviate it.

Aging: the word is used in many senses, but the one that's important for our purpose is *biological aging.*

Alagebrium: a drug that appears to break *advanced glycation end-product cross-links.* Chemically, 4,5-dimethyl-3-(2-oxo-2-phenylethyl)-thiazolium chloride. Originally code-named ALT-711.

Allotopic expression (AE): Expression of proteins from a different (Greek *allo-*) place (*topos*). The creation of "backup copies" in the *nucleus* of the protein-coding genes now housed in mitochondrial DNA.

Alpha-diketone: an *advanced glycation end-product* cross-link. Hypothetically, the AGE broken by *alagebrium.*

Alpha-secretase: an enzyme required for the normal processing of *Amyloid Precursor Protein.* APP does not form *beta-amyloid protein* when it is processed by this enzyme.

ALT: see *Alternative Lengthening of Telomeres* pathway.

ALT-711: the original code-name of *alagebrium.*

Alternative Lengthening of Telomeres (ALT): a poorly understood mechanism whereby some cancer cells relengthen their *telomeres* without using *telomerase.*

Amadori product: a somewhat stable intermediate compound of *glycation* chemistry, linking *Schiff bases* to *advanced glycation endproducts (AGE). Glycated hemoglobin* or *HbA1c* is an Amadori product.

Aminoguanidine: drug that inhibits the formation of *advanced glycation end products*, first inhibiting the initial *glycation* reaction, and then reducing *glycoxidation* through *antioxidant* and metal-*chelating* effects, and above all through its ability to mop up *oxoaldehydes.* Trade name *Pimagidine.*

Amyloid beta: see *beta-amyloid protein.*

Amyloid precursor protein (APP): a protein produced in the brain that, when damaged, can form *beta-amyloid protein.* Has *some* essential function in our bodies, possibly including being necessary to allow *neurons* to rewire themselves in response to new learning and to grow out *neurites.*

Amyloid protein A amyloidosis (AA): an *amyloidosis* caused by the excessive production of *amyloid A protein*, a protein involved in the inflammatory response, usually resulting from chronic inflammatory conditions. The most common amyloid disorder outside the United States. Also termed "inflammatory" or "secondary" amyloidosis.

Amyloid: any one of a range of cell-snaring chains of molecules that are created by damage to healthy proteins naturally present in the body, causing them to become twisted out of their proper configuration in ways that cause them to undergo toxic interactions with each other, or with other constituents of the cell. Typically form chemically "sticky" "webs" around cell structures that inhibit their functions. See for example *beta-amyloid protein*.

Amyloidoses: Diseases caused by the accumulation of *amyloids*.

Anergic T cells: T cells that can no longer carry out their duties.

Angiogenesis: the growth of new blood vessels.

Anti-aging medicine: Biomedical attempts to intervene in the pathological effects of *biological aging*. See *geriatrician's school, gerontologists' school,* and *engineer's school* of anti-aging medicine. Most of what is called "anti-aging medicine" in the marketplace today is either sheer hokum, or is a mixture of (a) crude and unproven attempts at the *gerontologist's school* of anti-aging medicine (hormone shots, antioxidant vitamins), combined with basic (though unfortunately rarely practiced) preventive medicine (improved diet and exercise) and the *geriatrician's school* of anti-aging medicine.

Antibodies: Proteins that specifically recognize and bind to *antigens* of the organisms they're intended to fight, either by flagging the invader for attack by other components of the immune system, or by blocking receptors and other proteins that are needed for the pathogen's survival.

Antigen: a protein recognized by the immune system.

Antigen-presenting cells (APCs): the immune system's reconnaissance teams, which identify enemy combatants' antigens through direct encounters with them or by digging through the rubble of old battlegrounds (the remains of cells ravaged by them)and then alert T cells specialized in waging war against the specific invaders that carry them.

Antisense mRNA: genetic material that binds to a pre-defined transcribed gene as it emerges, preventing it from being used to make the protein that it encodes.

Apoptosis: often referred to as "programmed cell death" or "cellular suicide." A carefully orchestrated programmed process of self-destruction carried out in cells that have been hijacked by "enemy forces" (viruses or cancer, for example) to prevent them from threatening the organism as a whole. Also used extensively during development to cull unnecessary cells. Apoptosis is carefully sequenced to prevent damage to surrounding cells (see *necrosis*).

ASP-2: see *beta-secretase*.

ATP: adenosine triphosphate. The cellular "energy currency." Just as you can use many different fuels (enriched uranium, coal, solar energy) and turn it into useable "universal energy" (electricity) to fuel a wide range of devices (DVD players, food processors, washing machines), so the body uses the chemical energy stored in various fuels (glucose, amino acids, fatty acids, etc) to make ATP to drive many of its metabolic processes.

ATPase, vacuolar: an energy- (that is, ATP-) consuming pump located on the membrane of the *lysosome* that drives extra protons into the lysosome from the main body of the cell, increasing its acidity.

B cells: cells of the immune system that recognise specific markers (*antigens*) on the surfaces of invaders that mark them as foreign, and churn out *antibodies* to them. Mostly responsible for defending us against pathogens like bacteria and parasites that are purely foreign to the body, and that can therefore be targeted *directly* for destruction.

BACE: see *beta-secretase.*

Basement membrane: the biological filter material of the kidney.

Beta-amyloid protein (also called "amyloid beta"): a peptide formed from *amyloid precursor protein* that accumulates as the waxy "senile plaques" that cluster around the brain cells of people with Alzheimer's disease, choking off the neurons' nourishment and preventing their normal functioning.

Beta-secretase: an enzyme with an uncertain function that sometimes mistakenly processes *Amyloid Precursor Protein,* contributing to the formation of *beta-amyloid protein.*

Biogerontology: The study of biological aging aimed at understanding it better.

Biological aging: the universal, progressive, and deleterious process of escalating loss of molecular fidelity with age, resulting stochastically from the intrinsic metabolic processes of the organism, that degrades its ability to maintain homeostasis in the face of environmental stressors, leading to increased intrinsic vulnerability to pathology and mortality.

Biomedical gerontology: The study of biological aging aimed at combating it in humans.

Blastocyst: the very primitive ball of cells that is formed within just a few days after sperm meets egg. The embryo only remains in this stage of development very briefly; it has developed much further by the time the embryo is implanted in the womb.

Blood-brain barrier: the protective layer of cells surrounding the blood vessels that feed the brain, that denies many molecules in the circulation free access to the brain.

C. elegans: see *Caenorhabditis elegans.*

Caenorhabditis elegans (C. elegans): a nematode (roundworm) that is now widely used to study genetic pathways involved in the rate of aging.

Calorie restriction (CR): Reducing the amount of food energy (calories) in the diet while maintaining adequate levels of essential vitamins, minerals, fats, and protein. In laboratory rodents and many other species, calorie restriction slows down aging—extending life beyond its 'natural' limits while preserving youthful functionality and protecting against nearly all age-related diseases and degenerative processes—in direct proportion to the level of restriction: fewer calories lead to more healthy, youthful life span. The first and best-studied way to slow down aging in mammals. Also known as "dietary restriction," "energy restriction," or "food restriction."

Calorie restriction mimetic (CR mimetic): A substance that would induce metabolic changes that would reproduce the essential anti-aging features of *calorie restriction*.

Carbonyl: an organic molecule with a carbon atom double-bonded to an oxygen atom. This structure makes many carbonyls highly biologically active, and many virulent precursors of *advanced glycation end products* are carbonyls such as *oxoaldehydes*.

Carboxymethyllysine (CML): a common *advanced glycation endproduct* derived from *glycoxidation*. Can also be an *advanced lipoxidation endproduct*.

Catalase: an enzyme produced by the body that detoxifies *hydrogen peroxide* by breaking it down into water and oxygen.

CD28: a *receptor* on the surface of *T cells* that allows *antigen-presenting cells* to identify the *CD8* cells that target the *antigen* found by the APCs.

CD4 cells: cells of the immune systems that help other immune cells to ramp up their counteroffensive when pathogens first invade. Also known as "T-helper" cells.

CD8 cells: see *cytotoxic T cells*.

Cerebral amyloid angiopathy (CAA): a disease caused when beta-amyloid binds up the interior surfaces of the brain's blood vessels, crusting them up, weakening them, and reducing their ability to flex in response to the surging flow of the pulse. This leaves them vulnerable to bursting open in a bleeding stroke.

Cerebrospinal fluid (CSF): the fluids bathing the brain and spinal cord.

Ceroid: substances that share many of *lipofuscin*'s properties (and are therefore often confused for it) but are much easier for the cell to break down and do not accumulate in "normal" *biological aging*.

Chelate: to tie up a metal in unreactive form.

Clonal expansion: the process whereby a single cell (such as a *memory cytotoxic T cell*) reproduces itself in large numbers, creating a "clone" of identical cells.

CML: see *carboxymethyllysine*.

CMV: see *cytomegalovirus*.

Code disparity: differences in the "languages" made from the DNA "letters" of genes in the mitochondria and cell nucleus. Code disparity makes some of the genes in the mitochondria unreadable by the *expression* machinery of the *nucleus* (and vice versa).

Complex V: the mitochondrial "turbine" that uses the flow of *protons* to fuel the storage of energy as *ATP*.

CR: see *Calorie restriction*.

Creatinine: a waste product of protein breakdown that healthy kidneys efficiently remove, and is therefore used as a blood test of kidney function.

Crescentic glomerulonephritis: a form of highly inflammatory kidney disease named for the crescent-shaped abnormalities that are seen in biopsies of victims' kidneys. Once it develops, crescentic glomerulonephritis leads to very rapid loss of kidney function.

CR mimetic: see *Calorie restriction mimetic*.

Cross-link: a molecular "handcuff" between adjacent proteins.

Crystallin: the clear, flexible proteins that make up the structure of the lens of the eye.

Cytochrome c oxidase: one of the pumping complexes of the *electron transport chain*.

Cytokine: an inflammatory signalling molecule of the immune system.

Cytomegalovirus (CMV): a persistent virus in the herpes family.

Cytotoxic T cells: *T cells* responsible for rooting out cells that are native to the body but that have now been turned against it, such as cancer cells or cells hijacked by viruses. Also called *CD8 cells* because of the characteristic receptor they bear.

Dauer pathway: a hibernation-like state in roundworms like *C. elegans*. In the dauer pathway, a larva suspends its development for a period than can last much longer than the entire lifetime of a nematode that follows the normal, nondauer trajectory.

Deletion: *mutation* that involves the total removal of large stretches of DNA, annihilating many genes at once even though it is, strictly speaking, only a single mutation.

Dendrimer: a tiny *nanotechnology* structure with exquisitely complex branching structures that extend outward like bushes, forming a spherical shape. Dendrimers' branches are engineered in a way that allows us to bind a wide range of molecules to them.

Diastolic heart failure (DHF): heart failure when the pumping chamber of the heart can't expand sufficiently well to take in the required volume of blood.

Diastolic pressure: the second number that you get from a blood pressure cuff, like the "80" in "110 over 80." The baseline pressure in the arteries at rest.

DNA polymerase: the enzyme responsible for making a new copy of a cell's DNA.

EGFR and EGFRvIII: see *epidermal growth factor receptor*.

Electron transport chain (ETC): the series of "pumps" that use food energy in the form of *electrons* provided by *NADH* to fill up a "reservoir" of *protons* held back by the *mitochondrial inner membrane* (the mitochondrial "dam"), to drive the production of energy in the form of *ATP* by *complex V* via *oxidative phosphorylation*.

Electron: a charged particle in an atom. The flow of electrons is the basis for electricity and of *oxidative phosphorylation*.

Embryonic stem cell: the primordial "master cells" from which our mature cells spring. *Stem cells* apparently found only in embryos, that have the ability to become *any* of the many different mature cells of the body. Compare *adult stem cell*.

Engineer's school of anti-aging medicine: direct intervention in the molecular *damage* of aging. Leaving metabolism alone to wreak damage on our molecular structures, but preventing the ensuing damage from leading to pathology, either by cleaning up the damage (keeping it (or restoring it to) below the threshold at which it becomes pathological), or by devising ways of rendering the damage itself harmless.

Enzyme: a biological catalyst that facilitates a chemical reaction in the body.

Epidermal growth factor receptor (EGFR): a *receptor* that stimulates cell growth. Excessive production of EGFR, or production of a mutated form called *EGFRvIII*, is implicated in many cancers.

Epigenetic structures: the "scaffolding" that is anchored to the DNA in our chromosomes. Epigenetic structures help determine which genes are turned on in a cell and which are turned off, allowing the same overall DNA to be used to create cells as diverse as liver, heart, and kidney cells.

Epimutation: a permanent, unprogrammed structural alteration in the "superstructure" or "scaffolding" that controls the expression of genes. Because changes in this "scaffolding" induce changes in the regulation of gene expression (turning genes on or off), epimutations are *functionally* the same as mutations and are treated as such by *SENS*.

Expression: the process by which the "blueprints" present in genes are executed in the creation of proteins.

Fenton reaction: chemical reaction in which *transition metals* make preexisting, but relatively harmless, free radicals become more virulent.

F_o/F_1 ATP synthase: see *complex V*.

Free radicals: Usually defined as an electrically neutral atom or molecule containing an electron that is missing its twinned "pair." This deficiency makes free radicals unstable and highly chemically reactive. Many free radicals "steal" electrons from other molecules to stabilize themselves, in the process damaging the molecule from which the electrons are "stolen" and often turning that molecule into a free radical itself. An excess of free radicals in the cell can cause *oxidative stress,* leading to dysfunctional metabolic imbalances and direct damage to key structural components such as proteins, DNA, and fatty membranes. Many substances behave like "free radicals" and are often *called* free radicals but don't, strictly, meet this definition (e.g., *reduced iron ions* (Fe^{2+}); equally, some molecules that strictly are free radicals are not harmful at all.

Gamma-secretase: an enzyme required for the normal processing of *Amyloid Precursor Protein* but that can also play a role in its abnormal processing into *beta-amyloid protein,* particularly if the enzyme is mutated and thus produces an abnormal form of the enzyme.

Geriatrician's school of anti-aging medicine: attempting to interfere with the pathological *consequences* of aging as they come up, by treating them medically on a one-by-one basis.

Gerontologists' school of anti-aging medicine: attempting to interfere with the *mechanisms* of aging, by cleaning up the biochemically "messy" processes of *metabolism* or by neutralizing the reactive by-products of metabolism before they cause damage to our molecular structures.

Gerontology: the study of "aging," in any of its aspects or meanings: the psychology of the old in our society; the prejudices against old people; the social structures that support or restrict the access of the frail elderly to society; and the biology of aging (*biogerontology*).

Glial cell: part of the caretaking support staff of *neurons.*

Glioblastoma multiforme: a rare, extremely aggressive, and hard-to-treat brain cancer.

Glucosepane: a complex *advanced glycation end-product* that is the single most important contributor to the body's AGE burden known to date, tying up as much as one out of every five molecules of the key structural protein *collagen* in old, nondiabetic humans.

Glycated hemoglobin (HbA1c): an *Amadori product* that forms on red blood cells. Because HbA1c persists for 2-3 months, it reflects average blood sugar levels for that time period, and so is used as a lab test to measure overall blood sugar control.

Glycation: the spontaneous chemical bonding of sugar molecules to proteins.

Glycoxidation: the accelerated conversion of some precursors *of advanced glycation end-products* into those end-products by *free radicals.*

HbA1c: see *glycated hemoglobin.*

Herceptin: a *monoclonal antibody* used as a *targeted cancer therapy*. Herceptin targets a *receptor* called HER-2 that stimulates cell growth. By tying up HER-2, herceptin prevents the excessive growth of cancer cells dependent on overstimulation from HER-2 to keep up their very rapid rate of reproduction.

Homeostasis: the ability of the cell or organism to maintain a defined metabolic equilibrium.

Hydrogen peroxide: a molecule that acts like a free radical. Produced from the breakdown of *superoxide radicals*.

Hydrolase: an *enzyme* that breaks down a compound by combining it with water molecules. Most of the *enzymes* of the *lysosome* are hydrolases.

Innate immune system: the branch of the immune system that doesn't have to "learn" to identify a specific enemy. Its job is similar to that of regular soldiers on patrol in a demilitarized zone, trying to maintain order but unsure of who might be the enemy, ready to confront anything suspicious-looking that it happens upon.

Insulin: the hormone whose job it is to move carbohydrates and amino acids into fat and muscle cells.

Intein: sequences that are inserted temporarily into some proteins when they are first synthesised, possibly to help the protein mature into its final form properly, and are then snipped out once they've served their purpose.

Isolated systolic hypertension (ISH): the kind of high blood pressure in which a person's *systolic* reading (the first of the two numbers that you get from a blood pressure cuff, like the "110" in "110 over 80") is high, even though their *diastolic* pressure (the second number) is fine

K2P: a major *advanced glycation endproduct* in the lenses of our eyes and possibly other tissues.

KLRG1: a *receptor* that prevents *cytotoxic T cells* from proliferating when no infection is present. Healthy cells bearing KLRG1 are able to reproduce when a threat is actually present; *anergic* CD8 cells cannot, because of another receptor called CD57.

LEV: see *Longevity Escape Velocity.*

Lipofuscin (lip-oh-FEW-sin): a catch-all term for the mixture of stubborn waste products that builds up in the *lysosomes* of long-lived cells like those of the heart and the brain as we age. A chemical hodgepodge of fatty and proteinaceous materials derived from membranes, reactive metals like iron and copper, and a variety of other organic molecules that the normal complement of *enzymes* in the *lysosome* doesn't know how to deal with and so refuse to be broken down after being sent there. Glows red when exposed to light of a particular wavelength. Called popularly, "age pigment."

Longevity escape velocity (LEV): a threshold rate of biomedical progress that will allow us to stave off aging indefinitely. The point at which each successive round of refinements to the SENS age-reversing toolkit is buying us more time than we need to develop the next round of refinements, until we can eventually escape age-related decline indefinitely, however old we become in purely chronological terms.

Lysosomal storage diseases (LSDs): a range of genetic disorders caused by failure, by one mechanism or another, of the *lysosomes*. Many sufferers completely lack the gene for a lysosomal *enzyme,* or bear a mutated copy of it, resulting in a misshapen and ineffective version of the protein. In other cases, the problem is that one of the specialized transport proteins on the surface of the lysosomal membrane is missing or defective, so that the lysosome can't bring the junk *into* itself to break it down. LSDs all result in deadly degenerative diseases, with specific symptoms varying based on which organs a given mutation affects, and how severely.

Lysosome: An acidic *organelle* that uses *enzymes* to break damaged cellular components down at the molecular level into more basic constituents that can be used as raw materials for the biosynthesis of new cellular membranes, enzymes, and other important components of the cellular machinery. The biological "garbage incinerator" or "recycling center." The extra protons that create the lysosome's acidity are actively pumped out of the main chamber of the cell and into the lysosome by an energy- (that is, ATP-) consuming pump located on its membrane (the *vacuolar ATPase*).

Maillard reaction: a major form of *glycation* chemistry, in which a molecule of sugar opens its structure and glues onto a protein molecule, forming a *Schiff base.* This structure is relatively unstable, so the Schiff base will often spontaneously fall apart. Sometimes, however, it will collapse into a more stable *Amadori product.*

Matrix metalloproteinases (MMPs): protein-digesting enzymes that act as the "demolition teams" of tissue remodeling, clearing away the old, damaged "scaffolding" in which cells are embedded in a tissue, making space for new growth.

Maximum life span: see *species maximum life span.*

Memapsin: see *beta-secretase.*

Memory cytotoxic T cells: specialized to attack hijacked cells bearing antigens encountered in previous immunological battles.

Meningoencephalitis: life-threatening swelling of the brain, apparently as a result of an overreaction of the immune system inside the brain.

Metabolism: the sum of the physical and chemical processes that occur in the living body, including the breakdown and buildup of body structures and proteins, and the uptake, distribution within the body, chemical and physical transformation, and ultimate elimination of food, air, and other compounds taken in from the environment.

Metastasis: the process whereby cancerous cells escape from the restraints of the tissue in which they were originally embedded, and begin a new colony in tissues far removed from the original cancer site.

Methylglyoxal: a major *oxoaldehyde* precursor of *advanced glycation end products,* up to 40,000 times more reactive than blood sugar.

Microglia: the immune cells of the brain.

Mitochondria: the cellular "power plants" that take the body's raw fuels (glucose, amino acids, fatty acids, etc) and turn them into useable cellular energy (*adenosine triphosphate*—ATP).

Mitochondrial inner membrane: mitochondrial "dam" that holds back the "reservoir" of *protons* used to drive the production of ATP by *Complex V* (the "turbine" of the mitochondrial "hydroelectrical dam").

Mitochondriopathies: a class of diseases caused by defects in the *inherited* mitochondrial DNA (or, more rarely, by mutations acquired through causes independent of the aging process). These mutations lead to a failure of energy production that causes a spectrum of dysfunctions in various organs, depending on the exact mutation involved.

MMPs: see *matrix metalloproteinases.*

Monoclonal antibody: an *antibody* produced on a mass scale in the laboratory to target a specific *antigen.*

Monomer, beta-amyloid: an individual fragment of *beta-amyloid protein* that initially floats free in the brain.

mRNA: the cell's transcripts of the DNA instructions.

Mutation: a permanent, unprogrammed structural alteration in the DNA. As used in Chapter 12, includes *epimutations.*

Myelin sheath: insulating material that surrounds nerves.

Myeloperoxidase: an enzyme used by macrophages to kill bacteria by generating toxic hypochlorous acid.

NAD+/NADH: A biological "carrier molecule" that shuttles *electrons* from food into the *mitochondria*.

Naïve cytotoxic T cells: a reserve of as-yet-unspecialized *cytotoxic T cells* that are ready to identify new threats, "learn" about their key antigens, and then mount an attack.

Nanotechnology: engineering performed at the molecular level.

Necrosis: traumatic, uncontrolled death of a cell. Necrosis usually causes the cell to swell and break open, harmfully releasing its contents and damaging neighboring cells.

Neurite: the branching "fiber optic cables" that allow *neurons* to talk with one another.

Neuron: a main type of cell of the brain and nervous system, specialized to receive and transmit information from the body and from other neurons.

Neuropathy, diabetic: the debilitating damage to the *nerves* that is suffered by so many diabetics, resulting from the degeneration of the insulating *myelin sheaths*, complicated by the slow shrinking away of the "electric cables" (dendrites and axons) through which nerves communicate with one another.

Notch receptor 1 (NOTCH1), a protein required for the activation of *stem cells,* the growth of new blood vessels, and the maturation of some kinds of immune cells.

N-phenacylthiazolium bromide (PTB): a *thiazolium* breaker of *advanced glycation end products*. Chemical cousin and predecessor of *alagebrium*.

Nucleus: the part of the cell that houses its central DNA genetic instructions.

Oligomer, beta-amyloid: a short chain of monomers of *beta-amyloid protein* that can still float free in the brain.

Organelle: a self-contained cellular "factory" that exists outside the *nucleus* and carries out specific metabolic functions for the cell as a whole. Examples include *mitochondria* and *lysosomes*.

Oxidant: substances that chemically "need" *electrons*. Includes many *free radicals*, but also normal biochemical intermediates that are sometimes required to be in this state.

Oxidating agent: see *oxidant*.

Oxidative phosphorylation (OXPHOS): The "charging up" of the "battery" of *ATP* by *complex V,* achieved by adding phosphorus to its precursor molecule. Consumes oxygen and produces carbon dioxide and water, and thus called *cellular respiration*.

Oxidative stress: the imbalance of those substances in the body that tend to chemically "need" electrons (*oxidants* including *free radicals,* but also normal biochemical intermediates that are sometimes required to be in this state) relative to substances that chemically "want" to donate them (*reductants* or *reducing agents*). Oxidative stress increases the risk that *free radicals* will damage key cellular components instead of being detoxified, and can also cause dysfunctional imbalances in metabolic processes by shifting the electrical homeostasis of the cell.

Oxoaldehyde: a highly reactive *carbonyl* intermediate product of *glycation* chemistry.

OXPHOS: see *oxidative phosphorylation.*

Passive vaccination: directly providing antibodies against an *antigen* that we want to target for immune attack, bringing out the immune response that the same antibodies elicit when they are produced by the body.

Pentosidine: one of the more easily measured *advanced glycation end products*.

Pimagidine: see *aminoguanidine*.

Point mutations: *mutations* that change only one, or a few, of the "letters" in one "word" in the "sentence" of instructions comprising an individual gene.

Prodrug: a substances that is inactive until metabolized in some way, whereupon it is chemically transformed into a pharmacologically active product.

Proton: an electrically charged particle present in the atom. The flow of protons across the *mitochondrial inner membrane* through *complex V* provides the power to drive the storage of cellular energy as *ATP*.

Receptor: a molecular "lock" on the surface of the cell that responds to the correct molecular "key" by performing functions like opening the cell up to a needed nutrient or inducing a signaling cascade.

Reductant: a substance that characteristically "wants" to give electrons to other compounds.

Retinopathy, diabetic: vision loss in diabetics linked to damage to the fine blood vessels feeding the light-absorbing tissues at the back of the eyeball.

RMR: see *robust mouse rejuvenation.*

Robust mouse rejuvenation (RMR): the irrefutable *reversal* of aging in mice. RMR will be achieved when we can take at least twenty mice of the species *Mus musculus* (the common house or laboratory mouse), of a healthy strain (one with a normal average life span of at least three years), and administer anti-aging treatments starting only when they are at least two years old that lead them to be able to live in good health to an average of five years.

SA-beta-gal: see *Senescence-associated beta-galactosidase.*

Schiff base: a compound that can result from various *glycation* reactions that contains a double bond between carbon and nitrogen atom, with specific classes of compounds connected to it in turn through the nitrogen atom. The Schiff bases that result from glycation are unstable and tend to go on to form *Amadori products*.

SCNT: see *somatic cell nuclear transfer.*

Senescence-associated beta-galactosidase (SA-beta-gal): an enzyme whose activity identifies senescent cells.

Senescent cells: cells that have lost the ability to divide as a result of aging damage (and, usually, the body's *response* to that damage).

SENS: Strategies for Engineered Negligible Senescence. The scientific platform for *anti-aging medicine* based on the heuristic of the *engineer's school* of *anti-aging medicine*.

Serotonin: a chemical messenger in *neurons* (and elsewhere) involved in mood, appetite, thought, and sensory perception. Serotonin is the substance whose metabolism is modulated by drugs like Prozac.

Somatic cell nuclear transfer (SCNT): The process of making new, perfectly-matched *embryonic stem cells* out of a specialized mature cell (a "somatic cell") from a patient by fusing it with an egg cell, provided by a prospective stem cell recipient, whose cell nucleus is removed to make way for the one from the patient's cell. When the fused cell begins dividing, it creates ESCs with the donor's genetic code, and so with absolutely no risk of rejection.

Species maximum life span: how long the oldest old of the species (and not just of a particular strain of them, or of the cohort of animals in which the intervention was tested) can live under the best possible conditions.

Stem cell: an early, unspecialized cell that can renew itself indefinitely and develop into one or more of the highly specialized, mature cells of each tissue in the body. Includes *embryonic stem cells* and *adult stem cells*.

Subcutaneous fat: the fat that lies under your skin all over the body, producing a "pear shape" when present in excess. Compare *visceral fat.*

Superoxide radical: a free radical produced by the "fumbling" of *electrons* by the *electron transport chain* in the *mitochondria.*

Systemic AL amyloidosis: the most common form of *amyloidosis* in the United States and some other industrialised countries, caused by overproduction by a kind of blood cell of a component of a class of *antibodies* called "immunoglobulin light chain" (L—thus "AL," for "Amyloidosis Light-chain").

Systolic blood pressure: the first of the two numbers that you get from a blood pressure cuff, like the "110" in "110 over 80." A measure of how much pressure is applied to the artery wall by the surge of blood into the vessel as the heart contracts.

T cells: immune cells that mature in the thymus. Includes *cytotoxic T cells* and *helper T cells.*

Targeted cancer therapy: a selective cancer treatment that "targets" cancer cells selectively. Usually refers specifically to drugs that interfere with specific *receptors* or signaling processes upon which a particular cancer relies strongly.

Targeting sequence: a special string of amino acids that when appended to the "nose" of a protein produced in the main body of the cell, directs it into a particular location such as the mitochondria.

Telomerase: the enzyme that relengthens shortened *telomeres.*

Telomeres: long stretches of DNA present at the ends of all our chromosomes that contain no protein blueprints. Telomeres are worn down with each round of DNA replication.

T-helper cells: see *CD4 cells.*

Therapeutic cloning: see *somatic cell nuclear transfer.*

Thiazolium: member of a class of compounds with a chemical structure related to thiamin (vitamin B1).

Thymidine kinase (TK): an enzyme that is required for the synthesis of DNA.

TIM/TOM complex: short for "Translocase of the Inner Mitochondrial membrane" (TIM) and the "Translocase of the Outer Mitochondrial membrane" (TOM). The elaborate machinery that specifically moves ("translocates") proteins through the mitochondrial membranes.

TK: see *thymidine kinase.*

Transition metals: elements like iron and copper that can intensify oxidative stress through their role in the *Fenton reaction.*

Translocase of the Inner Mitochondrial membrane (TIM): see *TIM/TOM Complex.*

Translocase of the Outer Mitochondrial membrane (TOM): see *TIM/TOM Complex.*

Triglycerides: fats; especially fats circulating in the blood.

Vacuolar ATPase: see *ATPase, vacuolar.*

Vascular endothelial growth factor (VEGF): a chemical messenger that stimulates new blood vessel growth.

VEGF: see *vascular endothelial growth factor.*

Visceral fat: fat tissue that surrounds your internal organs, as opposed to the *subcutaneous* fat that lies under your skin all over the body. Too much visceral fat is responsible for an "apple-shaped" or "beer-bellied" overweight appearance. Visceral fat appears to be implicated in much of the metabolic derangement of diabetes and of aging.

Index